Time Series Analysis

Texts in Statistical Science

Time Series Analysis

Henrik Madsen
Technical University of Denmark

Chapman & Hall/CRC
Taylor & Francis Group
Boca Raton London New York

Chapman & Hall/CRC is an imprint of the
Taylor & Francis Group, an informa business

Chapman & Hall/CRC
Taylor & Francis Group
6000 Broken Sound Parkway NW, Suite 300
Boca Raton, FL 33487-2742

© 2008 by Taylor & Francis Group, LLC
Chapman & Hall/CRC is an imprint of Taylor & Francis Group, an Informa business

No claim to original U.S. Government works
Printed in the United States of America on acid-free paper
10 9 8 7 6 5 4 3 2 1

International Standard Book Number-13: 978-1-4200-5967-0 (Hardcover)

This book contains information obtained from authentic and highly regarded sources. Reprinted material is quoted with permission, and sources are indicated. A wide variety of references are listed. Reasonable efforts have been made to publish reliable data and information, but the author and the publisher cannot assume responsibility for the validity of all materials or for the consequences of their use.

Except as permitted under U.S. Copyright Law, no part of this book may be reprinted, reproduced, transmitted, or utilized in any form by any electronic, mechanical, or other means, now known or hereafter invented, including photocopying, microfilming, and recording, or in any information storage or retrieval system, without written permission from the publishers.

For permission to photocopy or use material electronically from this work, please access www.copyright.com (http://www.copyright.com/) or contact the Copyright Clearance Center, Inc. (CCC) 222 Rosewood Drive, Danvers, MA 01923, 978-750-8400. CCC is a not-for-profit organization that provides licenses and registration for a variety of users. For organizations that have been granted a photocopy license by the CCC, a separate system of payment has been arranged.

Trademark Notice: Product or corporate names may be trademarks or registered trademarks, and are used only for identification and explanation without intent to infringe.

Library of Congress Cataloging-in-Publication Data

Madsen, Henrik, 1955-
Time series analysis / Henrik Madsen.
p. cm. -- (Chapman & Hall/CRC texts in statistical science series ; v. 72)
Includes bibliographical references and index.
ISBN 978-1-4200-5967-0 (hardback : alk. paper)
1. Time-series analysis. I. Title. II. Series.

QA280.M32 2007
519.5'5--dc22 2007036211

Visit the Taylor & Francis Web site at
http://www.taylorandfrancis.com

and the CRC Press Web site at
http://www.crcpress.com

Contents

Preface

The aim of this book is to give an introduction to time series analysis. The emphasis is on methods for modeling of linear stochastic systems. Both time domain and frequency domain descriptions will be given; however, emphasis is on the time domain description. Due to the highly different mathematical approaches needed for linear and non-linear systems, it is instructive to deal with them in seperate textbooks, which is why non-linear time series analysis is not a topic in this book—instead the reader is referred to Madsen, Holst, and Lindström (2007).

Theorems are used to emphasize the most important results. Proofs are given only when they clarify the results. Small problems are included at the end of most chapters, and a separate chapter with real-life problems is included as the final chapter of the book. This also serves as a demonstration of the many possible applications of time series analysis in areas such as physics, engineering, and econometrics.

During the sequence of chapters, more advanced stochastic models are gradually introduced; with this approach, the family of linear time series models and methods is put into a clear relationship. Following an initial chapter covering static models and methods such as the use of the general linear model for time series data, the rest of the book is devoted to stochastic dynamic models which are mostly formulated as difference equations, as in the famous ARMA or vector ARMA processes. It will be obvious to the reader of this book that even knowing how to solve difference equations becomes important for understanding the behavior of important aspects such as the autocovariance functions and the nature of the optimal predictions.

The important concept of time-varying systems is dealt with using a state space approach and the Kalman filter. However, the strength of also using adaptive estimation methods for on-line forecasting and control is often not adequately recognized. For instance, in finance the classical methods for forecasting are often not very useful, but, by using adaptive techniques, interesting results are often obtained.

The last chapter of this book is devoted to problems inspired by real life. Solutions to the problems are found at `http://www.imm.dtu.dk/~hm/time.series.analysis`. This home page also contains additional exercises, called assignments, intended for being solved using a computer with dedicated software for time series analysis.

I am grateful to all who have contributed with useful comments and suggestions for improvement. Especially, I would like to thank my colleagues Jan Holst, Henrik Spliid, Leif Mejlbro, Niels Kjølstad Poulsen, and Henrik Aalborg Nielsen for their valuable comments and suggestions. Furthermore, I would like to thank former students Morten Høier Olsen, Rasmus Tamstorf, and Jan Nygaard Nielsen for their great effort in proofreading and improving the first manuscript in Danish. For this 2007 edition in English, I would like to thank Devon Yates, Stig Mortensen, and Fannar Örn Thordarson for proofreading and their very useful suggestions. In particular, I am grateful to Anna Helga Jónsdóttir for her assistance with figures and examples. Finally, I would like to thank Morten Høgholm for both proofreading and for proposing and creating a new layout in LaTeX.

Lyngby, Denmark *Henrik Madsen*

Notation

All vectors are column vectors. Vectors and matrices are emphasized using a bold font. Lowercase letters are used for vectors and uppercase letters are used for matrices. Transposing is denoted with the upper index ^{T}.

Random variables are always written using uppercase letters. Thus, it is not possible to distinguish between a multivariate random variable (random vector) and a matrix. However, random variables are assigned to letters from the last part of the alphabet (X, Y, Z, U, V, ...), while deterministic terms are assigned to letters from the first part of the alphabet (a, b, c, d, ...). Thus, it should be possible to distinguish between a matrix and a random vector.

CHAPTER 1

Introduction

Time series analysis deals with statistical methods for analyzing and modeling an ordered sequence of observations. This modeling results in a stochastic process model for the system which generated the data. The ordering of observations is most often, but not always, through time, particularly in terms of equally spaced time intervals. In some applied literature, time series are often called signals. In more theoretical literature a time series is just an observed or measured realization of a stochastic process.

This book on time series analysis focuses on modeling using linear models. During the sequence of chapters more and more advanced models for dynamic systems are introduced; by this approach the family of linear time series models and methods are placed in a structured relationship. In a subsequent book, non-linear time series models will be considered.

At the same time the book intends to provide the reader with an understanding of the mathematical and statistical background for time series analysis and modeling. In general the theory in this book is kept in a second order theory framework, focussing on the second order characteristics of the persistence in time as measured by the autocovariance and autocorrelation functions.

The separation of linear and non-linear time series analysis into two books facilitates a clear demonstration of the highly different mathematical approaches that are needed in each of these two cases. In linear time series analysis some of the most important approaches are linked to the fact that superposition is valid, and that classical frequency domain approaches are directly usable. For non-linear time series superposition is not valid and frequency domain approaches are in general not very useful.

The book can be seen as a text for graduates in engineering or science departments, but also for statisticians who want to understand the link between models and methods for linear dynamical systems and linear stochastic processes. The intention of the approach taken in this book is to bridge the gap between scientists or engineers, who often have a good understanding of methods for describing dynamical systems, and statisticians, who have a good understanding of statistical theory such as likelihood-based approaches.

In classical statistical analysis the correlation of data in time is often disregarded. For instance in regression analysis the assumption about serial uncorrelated residuals is often violated in practice. In this book it will be demonstrated that it is crucial to take this autocorrelation into account in the modeling procedure. Also for applications such as simulations and forecasting, we will most often be able to provide much more reasonable and realistic results by taking the autocorrelation into account.

On the other hand adequate methods and models for time series analysis can often be seen as a simple extension of linear regression analysis where previous observations of the dependent variable are included as explanatory variables in a simple linear regression type of model. This facilitates a rather easy approach for understanding many methods for time series analysis, as demonstrated in various chapters of this book.

There are a number of reasons for studying time series. These include a characterization of time series (or signals), understanding and modeling the data generating system, forecasting of future values, and optimal control of a system.

In the rest of this chapter we will first consider some typical time series and briefly mention the reasons for studying them and the methods to use in each case. Then some of the important methodologies and models are introduced with the help of an example where we wish to predict the monthly wheat prices. Finally the contents of the book is outlined while focusing on the model structures and their basic relations.

1.1 Examples of time series

In this section we will show examples of time series, and at the same time indicate possible applications of time series analysis. The examples contain both typical examples from economic studies and more technical applications.

1.1.1 Dollar to Euro exchange rate

The first example is the daily US dollar to Euro interbank exchange rate shown in Figure 1.1. This is a typical economic time series where time series analysis could be used to formulate a model for forecasting future values of the exchange rate. The analysis of such a problem relates to the models and methods described in Chapters 3, 5, and 6.

1.1.2 Number of monthly airline passengers

Next we consider the number of monthly airline passengers in the US shown in Figure 1.2. For this series a clear annual variation is seen. Again it might be useful to construct a model for making forecasts of the future number of airline passengers. Models and methods for analyzing time series with seasonal variation are described in Chapters 3, 5, and 6.

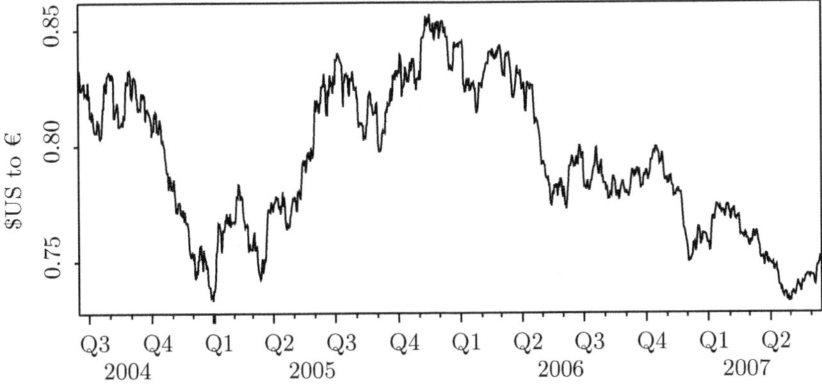

Figure 1.1: *Daily US dollar to Euro interbank exchange rate.*

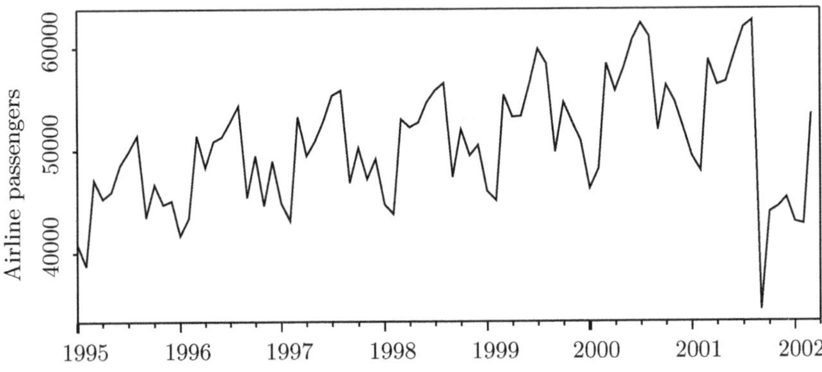

Figure 1.2: *Number of monthly airline passengers in the US. A clear annual variation can be seen in the series.*

1.1.3 Heat dynamics of a building

Now let us consider a more technical example. Figure 1.3 on the following page shows measurements from an unoccupied test building. The data on the lower plot show the indoor air temperature, while on the upper plot the ambient air temperature, the heat supply, and the solar radiation are shown.

For this example it might be interesting to characterize the thermal behavior of the building. As a part of that the so-called resistance against heat flux from inside to outside can be estimated. The resistance characterizes the insulation of the building. It might also be useful to establish a dynamic model for the building and to estimate the time constants. Knowledge of the time constants can be used for designing optimal controllers for the heat supply.

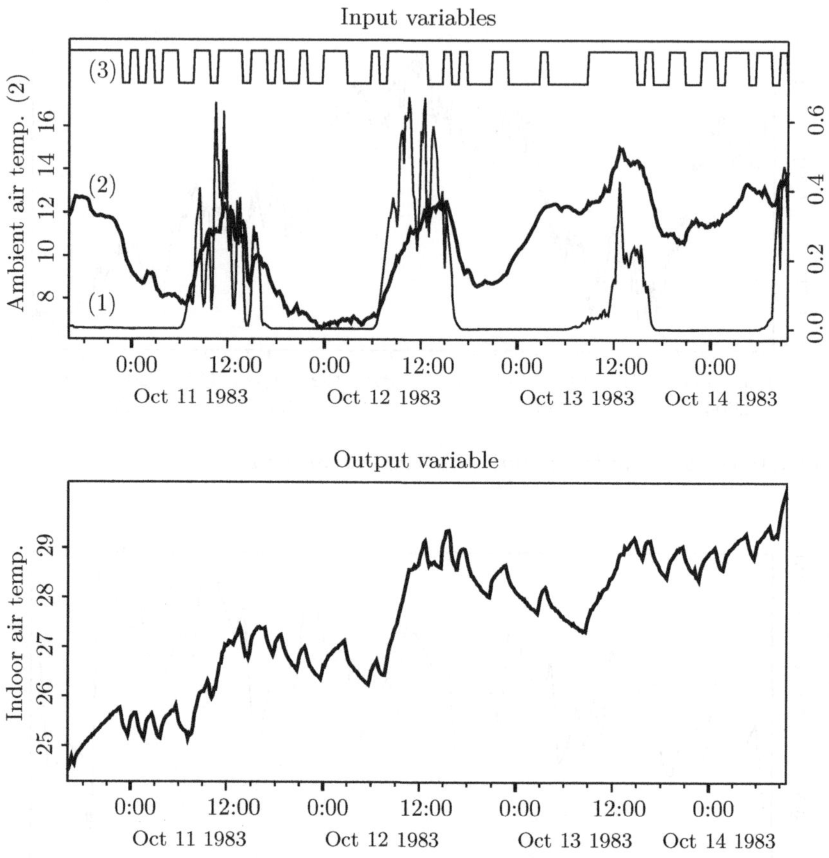

Figure 1.3: *Measurements from an unoccupied test building. The input variables are (1) solar radiation, (2) ambient air temperature, and (3) heat input. The output variable is the indoor air temperature.*

For this case methods for transfer function modeling as described in Chapter 8 can be used, where the input (explanatory) variables are the solar radiation, heat input, and outdoor air temperature, while the output (dependent) variable is the indoor air temperature. For the methods in Chapter 8 it is crucial that all the signals can be classified as either input or output series related to the system considered.

1.1.4 Predator-prey relationship

This example illustrates a typical multivariate time series, since it is not possible to classify one of the series as input and the other series as output. Figure 1.4 shows a widely studied predator-prey case, namely the series of annually traded skins of muskrat and mink by the Hudson's Bay Company

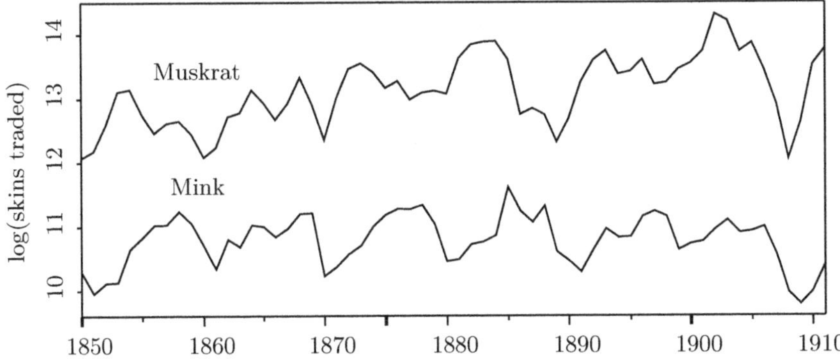

Figure 1.4: *Annually traded skins of muskrat and mink by the Hudson's Bay Company after logarithmic transformation. It is not possible to classify one of the series as input and the other series as output.*

during the 62 year period 1850–1911. In fact the population of muskrats depends on the population of mink, and the population of mink depends on the number of muskrats. In such cases both series must be included in a multivariate time series. This series has been considered in many texts on time series analysis, and the purpose is to describe in general the relation between populations of muskrat and mink. Methods for analyzing such multivariate series are considered in Chapter 9.

1.2 A first crash course

Let us introduce some of the most important concepts of time series analysis by considering an example where we look for simple models for predicting the monthly prices of wheat.

In the following, let P_t denote the price of wheat at time (month) t. The first naive guess would be to say that the price next month is the same as in this month. Hence, the *predictor* is

$$\widehat{P}_{t+1|t} = P_t \,. \tag{1.1}$$

This predictor is called the *naive predictor* or the *persistent predictor*. The syntax used is short for a prediction (or estimate) of P_{t+1} given the observations $P_t, P_{t-1}, \ldots$.

Next month, i.e., at time $t+1$, the actual price is P_{t+1}. This means that the *prediction error* or *innovation* may be computed as

$$\varepsilon_{t+1} = P_{t+1} - \widehat{P}_{t+1|t} \,. \tag{1.2}$$

By combining Equations (1.1) and (1.2) we obtain the *stochastic model* for the wheat price

$$P_t = P_{t-1} + \varepsilon_t \tag{1.3}$$

If $\{\varepsilon_t\}$ is a sequence of uncorrelated zero mean random variables (*white noise*), the process (1.3) is called a *random walk*. The random walk model is very often seen in finance and econometrics. For this model the optimal predictor is the naive predictor (1.1).

The random walk can be rewritten as

$$P_t = \varepsilon_t + \varepsilon_{t-1} + \cdots \tag{1.4}$$

which shows that the random walk is an integration of the noise, and that the variance of P_t is unbounded; therefore, no stationary distribution exists. This is an example of a *non-stationary process*.

However, it is obvious to try to consider the more general model

$$P_t = \varphi P_{t-1} + \varepsilon_t \tag{1.5}$$

called the $AR(1)$ *model* (the autoregressive first order model). For this process a stationary distribution exists for $|\varphi| < 1$. Notice that the random walk is obtained for $\varphi = 1$.

Another candidate for a model for wheat prices is

$$P_t = \psi P_{t-12} + \varepsilon_t \tag{1.6}$$

which assumes that the price this month is explained by the price in the same month last year. This seems to be a reasonable guess for a simple model, since it is well known that wheat price exhibits a *seasonal variation*. (The noise processes in (1.5) and (1.6) are, despite the notation used, of course, not the same).

For wheat prices it is obvious that both the actual price and the price in the same month in the previous year might be used in a description of the expected price next month. Such a model is obtained if we assume that the innovation ε_t in model (1.5) shows an annual variation, i.e., the combined model is

$$(P_t - \varphi P_{t-1}) - \psi(P_{t-12} - \varphi P_{t-13}) = \varepsilon_t\,. \tag{1.7}$$

Models such as (1.6) and (1.7) are called *seasonal models*, and they are used very often in econometrics.

Notice, that for $\psi = 0$ we obtain the AR(1) model (1.5), while for $\varphi = 0$ the most simple seasonal model in (1.6) is obtained.

By introducing the *backward shift operator* B by

$$\mathrm{B}^k P_t = P_{t-k} \tag{1.8}$$

the models can be written in a more compact form. The AR(1) model can be written as $(1 - \varphi \mathrm{B})P_t = \varepsilon_t$, and the seasonal model in (1.7) as

$$(1 - \varphi \mathrm{B})(1 - \psi \mathrm{B}^{12})P_t = \varepsilon_t \tag{1.9}$$

If we furthermore introduce the *difference operator*

$$\nabla = (1 - \mathrm{B}) \tag{1.10}$$

then the random walk can be written $\nabla P_t = \varepsilon_t$ using a very compact notation. In this book these kinds of notations will be widely used in order to obtain compact equations.

Given a *time series* of observed monthly wheat prices, $P_1, P_2, \ldots, P_N$, the *model structure* can be identified, and, for a given model, the time series can be used for *parameter estimation.*

The *model identification* is most often based on the estimated autocorrelation function, since, as it will be shown in Chapter 6, the autocorrelation function fulfils the same difference equation as the model. The autocorrelation function shows how the price is correlated to previous prices; more specifically the autocorrelation in lag k, called $\rho(k)$, is simply the correlation between P_t and P_{t-k} for stationary processes. For the monthly values of the wheat price we might expect a dominant annual variation and, hence, that the autocorrelation in lag 12, i.e., $\rho(12)$ is high.

The models above will, of course, be generalized in the book. It is important to notice that these processes all belong to the more general class of linear processes, which again is strongly related to the theory of linear systems as demonstrated in the book.

1.3 Contents and scope of the book

As mentioned previously, this book will concentrate on analyzing and modeling dynamical systems using statistical methods. The approach taken will focus on the formulation of appropriate models, their theoretical characteristics, and on links between the members of the class of stochastic dynamic models considered. In general, the models considered are all linear and formulated in discrete time. However, some results related to continuous time models are provided.

This section describes the contents of the subsequent chapters. In order to illustrate the relation between various models, some fundamental examples of the considered models are outlined in the following section. However, for more rigorous descriptions of the details related to the models we refer to the following chapters.

In Chapter 2 the concept of multivariate random variables is introduced. This chapter also introduces necessary fundamental concepts such as the conditional mean and the linear projection. In general, the chapter provides the formulas and methods for adapting a second order approach for characterising random variables. The second order approach limits the attention to first and second order central moments of the density related to the random variable. This approach links closely to the very important second order characterisation of stochastic processes by the autocovariance function in subsequent chapters.

Although time series are realizations of dynamical phenomena, non-dynamical methods are often used. Chapter 3 is devoted to describing *static models* applied for time series analysis. However, in the rest of the book dynamical models will be considered. The methods introduced in Chapter 3 are all linked to the class of regression models, of which the general linear model is the most important member. A brief description of the general linear model follows here.

In the following, let Y_t denote the dependent variable and $\boldsymbol{x}_t = (x_{1t}, x_{2t}, \ldots, x_{pt})^T$ a known vector of p explanatory (or independent) variables indexed by the time t. The *general linear model (GLM)* is a linear relation between the variables which can be written

$$Y_t = \sum_{k=1}^{p} x_{kt}\theta_k + \varepsilon_t \tag{1.11}$$

where ε_t is a zero mean random variable, and $\boldsymbol{\theta} = (\theta_1, \theta_2, \ldots, \theta_p)^T$ is a vector of the p parameters of the model. Notice that the model (1.11) is a static model since all the variables refer to the same point in time.

On-line and recursive methods are very important for time series analysis. These methods provide us with the possibility of always using the most recent data, e.g., for on-line predictions. Furthermore, changes in time of the considered phenomena calls for adaptive models, where the parameters typically are allowed to vary slowly in time. For on-line predictions and control, adaptive estimation of parameters in relatively simple models is often to be preferred, since the alternative is a rather complicated model with explicit time-varying parameters. Adaptive methods for estimating parameters in the general linear model are considered in Chapter 3. This approach introduces *exponential smoothing*, the *Holt-Winter procedure*, and *trend models* as important special cases.

The remaining chapters of the book consider linear systems and appropriate related *dynamical models*. A linear system converts an input series to an output series as illustrated in Figure 1.5.

In Chapter 4 we introduce linear dynamic deterministic systems. In this chapter, one should note that for random variables capital letters are used whereas for deterministic variables we use lower case letters.

As a background for Chapter 4, one should be aware that for linear and time-invariant systems the fundamental relation between the deterministic

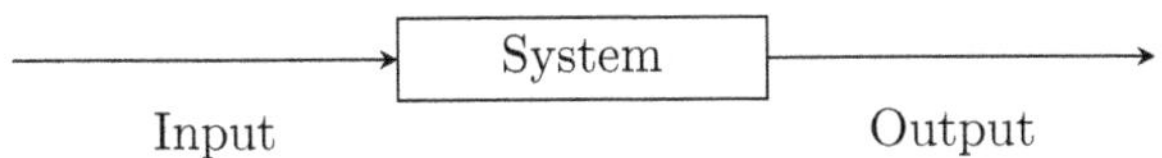

Figure 1.5: *Schematic representation of a linear system.*

input x_t and the corresponding output y_t is the *convolution*

$$y_t = \sum_{k=-\infty}^{\infty} h_k x_{t-k} \tag{1.12}$$

The sequence $\{h_k\}$ is called the *impulse response function* for the linear dynamic system. For physical systems where the output does not depend on future values of the input, the sum in (1.12) is from $k = 0$. Based on the impulse response function we will obtain the *frequency response function* by a Fourier transformation, and the *transfer function* by using the z transformation.

A very important model belonging to the model class described by (1.12) is the *linear difference equation*

$$y_t + \varphi_1 y_{t-1} + \cdots + \varphi_p y_{t-p} = \omega_0 x_t + \omega_1 x_{t-1} + \cdots + \omega_q x_{t-q} \tag{1.13}$$

Chapter 5 considers stochastic processes, and the focus is on the linear stochastic process $\{Y_t\}$ which is defined by the convolution

$$Y_t = \sum_{k=0}^{\infty} \psi_k \varepsilon_{t-k} \tag{1.14}$$

where $\{\varepsilon_t\}$ is the so-called white noise process, i.e., a sequence of mutually uncorrelated identically distributed zero mean random variables. Equation (1.14) defines a zero mean process, however, if the mean is not zero the mean μ_Y is just added on the right hand side of (1.14).

Notice the similarity between (1.14) and (1.12). This implies that a transfer function can be defined for the linear process as for the deterministic linear systems. Stochastic processes with a rational transfer function are the ARMA(p, q) process, the ARIMA(p, d, q) process, and the multiplicative seasonal processes. These important processes are considered in detail. As an example the process $\{Y_t\}$ given by

$$Y_t + \phi_1 Y_{t-1} + \cdots + \phi_p Y_{t-p} = \varepsilon_t + \theta_1 \varepsilon_{t-1} + \cdots + \theta_q \varepsilon_{t-q}, \tag{1.15}$$

where $\{\varepsilon_t\}$ is white noise, is known as the *ARMA(p, q) process*. Notice the similarity between (1.14) and (1.15). Such a process is useful for describing the data related to the dollar to Euro exchange rate in Section 1.1.1, and the seasonal process models are useful for modeling the monthly number of airline passengers in Section 1.1.2.

Given a time series of observations $Y_1, Y_2, \ldots, Y_N$, Chapter 6 deals with identification, estimation, and model checking for finding an appropriate model for the underlying stochastic process. This chapter focuses on time domain methods where the autocorrelation function is the key to an identification. Frequency domain methods are typically linked to the spectral analysis which is the subject of Chapter 7.

The so-called *transfer function models* are considered in Chapter 8. This class of models describes the relation between a stochastic input process $\{X_t\}$ and the output process $\{Y_t\}$. Basically the models can be written

$$Y_t = \sum_{k=0}^{\infty} h_k X_{t-k} + N_t \tag{1.16}$$

where $\{N_t\}$ is a correlated noise process, e.g., an ARMA(p, q) process. This gives rise to the so-called *Box-Jenkins transfer function model*, which can be seen as a combination of (1.12) and (1.14) on the previous page. It is relatively straightforward to include a number of input processes by adding the corresponding number of extra convolutions on the right hand side of (1.16).

An important assumption related to the Box-Jenkins transfer function models is that the output process does not influence the input process. Hence for the heat dynamics of a building example in Section 1.1.3, a transfer function model for the relation between the outdoor air temperature and the indoor air temperature can be formulated. This model can be extended to also include the solar radiation and the heat supply (provided that no feedback exists from the indoor air temperature to the heat supply).

In the case of multiple processes with no obvious split in input and output processes, the multivariate approach must be considered. In Chapter 9 the *multivariate linear process* is introduced as an m-dimensional stochastic process $\{\boldsymbol{Y}_t\}$ defined by the multivariate convolution

$$\boldsymbol{Y}_t = \sum_{k=0}^{\infty} \boldsymbol{\psi}_k \boldsymbol{\varepsilon}_{t-k} \tag{1.17}$$

where $\boldsymbol{\psi}$ is a coefficient matrix, and $\{\boldsymbol{\varepsilon}_t\}$ the multivariate white noise process. This formulation is used in Chapter 9 as the background for formulating the *multivariate ARMA(p, q) process* (also called the Vector-ARMA(p, q) process) and other related models.

As mentioned previously, the muskrat-mink case represents a problem which must be formulated as a multivariate process, simply because the population of minks influences the population of muskrats and vice versa.

Until now all the models can be considered as input-output models. The purpose of the modeling procedure is simply to find an appropriate model which relates the output to the input process, which in many cases is simply the white noise process. An important class of models which not only focuses on the input-output relations, but also on the internal state of the system, is the class of *state space models* introduced in Chapter 10.

A state space model in discrete time is formulated using a first order (multivariate) difference equation describing the dynamics of the *state vector*, which we shall denote $\boldsymbol{X}_t$, and a static relation between the state vector and the (multivariate) observation $\boldsymbol{Y}_t$. More specifically the *linear state space model* consists of the *system equation*

$$\boldsymbol{X}_t = \boldsymbol{A}\boldsymbol{X}_{t-1} + \boldsymbol{B}\boldsymbol{u}_{t-1} + \boldsymbol{e}_{1,t}, \tag{1.18}$$

and the *observation equation*

$$\boldsymbol{Y}_t = \boldsymbol{C}\boldsymbol{X}_t + \boldsymbol{e}_{2,t}, \tag{1.19}$$

where $\boldsymbol{X}_t$ is the m-dimensional, latent (not directly observable), random *state vector*. Furthermore $\boldsymbol{u}_t$ is a deterministic *input vector*, $\boldsymbol{Y}_t$ is a vector of observable (measurable) stochastic output, and $\boldsymbol{A}$, $\boldsymbol{B}$, and $\boldsymbol{C}$ are known matrices of suitable dimensions. Finally, $\{\boldsymbol{e}_{1,t}\}$ and $\{\boldsymbol{e}_{2,t}\}$ are vector white noise processes.

For linear state space models the *Kalman filter* is used to estimate the latent state vector and for providing predictions. The *Kalman smoother* can be used to estimate the values of the latent state vector, given all N values of the time series, for $\boldsymbol{Y}_t$.

To illustrate an example of application of the state space model, consider again the heat dynamics of the test building in Section 1.1.3. Madsen and Holst (1995) shows that a second order system is needed to describe the dynamics. Furthermore it is suggested to define the two elements of the state vector as the indoor air temperature and the temperature of the heat accumulating concrete floor. The input vector $\boldsymbol{u}_t$ consists of the ambient air temperature, the solar radiation, and the heat input. Only the indoor air temperature is observed, and hence, $\boldsymbol{Y}_t$ is the measured indoor air temperature. Using the state space approach gives us a possibility of estimating the temperature of heat accumulating in the concrete floor using the so-called Kalman filter technique.

In general, the parameters of the models are assumed constant in time. However, in practice, it is often observed that the dynamical characteristics change with time. In Chapter 11 *recursive and adaptive* methods are introduced. Basically the adaptive schemes introduce a time window related to the data, such that newer data obtains more influence than older data. This leads to methods for adaptive forecasting and control.

CHAPTER 2

Multivariate random variables

An *n-dimensional random variable (random vector)* is a vector of n scalar random variables. The random vector is written

$$\boldsymbol{X} = \begin{pmatrix} X_1 \\ X_2 \\ \vdots \\ X_n \end{pmatrix} \tag{2.1}$$

Random vectors will also be denoted *multivariate random variables*.

2.1 Joint and marginal densities

Every random variable has a distribution function. The n-dimensional random variable $\boldsymbol{X}$ has the *joint distribution function*

$$F(x_1, \ldots, x_n) = \mathrm{P}\{X_1 \le x_1, \ldots, X_n \le x_n\} \tag{2.2}$$

If $\boldsymbol{X}$ is defined on a continuous sample space, the *joint (probability) density function* is defined as

$$f(x_1, \ldots, x_n) = \frac{\partial^n F(x_1, \ldots, x_n)}{\partial x_1 \cdots \partial x_n} \tag{2.3}$$

and the relation between the distribution and density function is

$$F(x_1, \ldots, x_n) = \int_{-\infty}^{x_1} \cdots \int_{-\infty}^{x_n} f(t_1, \ldots, t_n)\, dt_1 \ldots dt_n \tag{2.4}$$

A random variable, $\boldsymbol{X}$, is called *discrete* if it takes values on a discrete (countable) sample space. In this case the joint density function (or *mass function*) is defined as

$$f(x_1, \ldots, x_n) = \mathrm{P}\{X_1 = x_1, \ldots, X_n = x_n\} \tag{2.5}$$

The joint distribution and mass functions are related by

$$F(x_1, \ldots, x_n) = \sum_{t_1 \le x_1} \cdots \sum_{t_n \le x_n} f(t_1, \ldots, t_n) \tag{2.6}$$

For a sub-vector $(X_1, \ldots, X_k)^T$ $(k < n)$ of the random vector $\boldsymbol{X}$ the *marginal density function* is

$$f_S(x_1, \ldots, x_k) = \int_{-\infty}^{\infty} \cdots \int_{-\infty}^{\infty} f(x_1, \ldots, x_n)\, dx_{k+1} \cdots dx_n \tag{2.7}$$

in the continuous case, and

$$f_S(x_1, \ldots, x_k) = \sum_{x_{k+1}} \cdots \sum_{x_n} f(x_1, \ldots, x_n) \tag{2.8}$$

if $\boldsymbol{X}$ is discrete. In both cases the *marginal distribution function* is

$$F_S(x_1, \ldots, x_k) = F(x_1, \ldots, x_k, \infty, \ldots, \infty) \tag{2.9}$$

Please note that we will write $f_X(x)$ and $F_X(x)$ instead of $f(x)$ and $F(x)$, respectively, whenever it is necessary to emphasize to which random variable the function belongs.

2.2 Conditional distributions

In time series analysis conditional distributions play an important role, especially in relation to prediction and filtering. For instance, in prediction, where we want to give a statement about future values of the time series given past observations, then the conditional distribution contains all available information about the future value.

Let A and B denote some *events*. If $\mathrm{P}(B) > 0$ then the *conditional probability* of A occurring given that B occurs is

$$\mathrm{P}(A \mid B) = \frac{\mathrm{P}(A \cap B)}{\mathrm{P}(B)} \tag{2.10}$$

We interpret $\mathrm{P}(A \mid B)$ as "the probability of A given B."

Suppose that the continuous random variables X and Y have joint density, $f_{X,Y}$. If we wish to use (2.10) to determine the conditional distribution of Y given that X takes the value x, we have the problem that the probability $\mathrm{P}(Y \leq y \mid X = x)$ is undefined as we may only condition on events which have a strictly positive probability, and $\mathrm{P}(X = x) = 0$. However, $f_X(x)$ is positive for some x values. Therefore, for both discrete and continuous random variables we use the following definition.

DEFINITION 2.1 (CONDITIONAL DENSITY)
The *conditional density (function)* of Y given $X = x$ is

$$f_{Y|X=x}(y) = \frac{f_{X,Y}(x, y)}{f_X(x)}, \quad (f_X(x) > 0) \tag{2.11}$$

where $f_{X,Y}$ is the joint density function of X and Y. Both X and Y may be multivariate random variables.

The *conditional distribution function* is then found by integration or summation as previously described in (2.4) and (2.5) for the continuous and discrete cases, respectively.

It follows from (2.11) that

$$f_{X,Y}(x,y) = f_{Y|X=x}(y) f_X(x) \tag{2.12}$$

and by interchanging X and Y on the right hand side of (2.12) we get *Bayes' rule*:

$$f_{Y|X=x}(y) = \frac{f_{X|Y=y}(x) f_Y(y)}{f_X(x)} \tag{2.13}$$

We now define independence:

DEFINITION 2.2 (INDEPENDENCE)
$\boldsymbol{X}$ and $\boldsymbol{Y}$ are *independent* if

$$f_{X,Y}(x,y) = f_X(x) f_Y(y) \tag{2.14}$$

which corresponds to

$$F_{X,Y}(x,y) = F_X(x) F_Y(y) \tag{2.15}$$

If X and Y are independent, it is clearly seen that

$$f_{Y|X=x}(y) = f_Y(y) \tag{2.16}$$

Bear in mind that *if two random variables are independent, then they are also uncorrelated*, while uncorrelated variables are not necessarily independent.

2.3 Expectations and moments

For a discrete variable X, the expectation is

$$\mathrm{E}[X] = \sum_x x P\{X = x\},$$

i.e., an average of the possible values of X, each value being weighted by its probability.

For continuous variables, expectations are defined as integrals.

DEFINITION 2.3 (EXPECTATION)
The *expectation* (or *mean value*) of a continuous variable X with density function f_X is

$$\mathrm{E}[X] = \int_{-\infty}^{\infty} x f_X(x)\, dx \tag{2.17}$$

whenever this integral exists, i.e., if

$$\mathrm{E}[|x|] < \infty \tag{2.18}$$

Note that usually we allow the existence of $\int g(x)\,dx$ only if $\int |g(x)|\,dx < \infty$.

▸ **Remark 2.1**
The *expectation operator* E can more generally be defined by a *Stieltjes integral*, i.e.,

$$\mathrm{E}[X] = \int_{-\infty}^{\infty} x\,dF_X(x), \tag{2.19}$$

where F_X is the distribution function for X. This definition covers both discrete and continuous variables. ◂

The expectation is called the *first moment*, because it is the first moment of f_X with respect to the line $x = 0$.

If X is a random variable and g is a function, then $Y = g(X)$ is also a random variable. To calculate the expectation of Y, we could find f_Y and use (2.17). However, the process of finding f_Y can be complicated; instead, we can exploit that

$$\mathrm{E}[Y] = \mathrm{E}[g(X)] = \int_{-\infty}^{\infty} g(x) f_X(x)\,dx. \tag{2.20}$$

This provides a method for calculating the "moments" of a distribution.

DEFINITION 2.4 (MOMENTS)
The *n'th moment* of X is

$$\mathrm{E}[X^n] = \int_{-\infty}^{\infty} x^n f_X(x)\,dx, \tag{2.21}$$

and the *n'th central moment* is

$$\mathrm{E}[(X - \mathrm{E}[X])^n] = \int_{-\infty}^{\infty} (x - \mathrm{E}[X])^n f_X(x)\,dx. \tag{2.22}$$

The second central moment is also called the *variance*, and the variance of X is given by

$$\mathrm{Var}[X] = \mathrm{E}[(X - \mathrm{E}[X])^2] = \mathrm{E}[X^2] - (\mathrm{E}[X])^2. \tag{2.23}$$

Now let X be an n-dimensional random variable and put $Y = g(X)$, where g is a function. As in (2.20) we have

$$\begin{aligned} \mathrm{E}[Y] &= \mathrm{E}[g(X)] \\ &= \int_{-\infty}^{\infty} \cdots \int_{-\infty}^{\infty} g(x_1, \ldots, x_n) f_X(x_1, \ldots, x_n)\, dx_1 \cdots dx_n \end{aligned} \tag{2.24}$$

In particular, setting $g(x_1, x_2) = ax_1 + bx_2$ implies that

$$\mathrm{E}[aX_1 + bX_2] = a\,\mathrm{E}[X_1] + b\,\mathrm{E}[X_2] \tag{2.25}$$

Thus, *the expectation operator is a linear operator.*

The elements X_i and X_j in the random vector $\boldsymbol{X}$ (2.24) can be used for defining *mixed moments*

$$\mathrm{E}\left[X_i^\alpha X_j^\beta\right], \tag{2.26}$$

where α and β are integers and similarly for the corresponding central moments. Of special interest is the *covariance* of X_i and X_j

$$\begin{aligned} \mathrm{Cov}[X_i, X_j] &= \mathrm{E}[(X_i - \mathrm{E}[X_i])(X_j - \mathrm{E}[X_j])] \\ &= \mathrm{E}[X_i X_j] - \mathrm{E}[X_i]\,\mathrm{E}[X_j]. \end{aligned} \tag{2.27}$$

The covariance gives information about the simultaneous variation of two variables and is useful for finding interdependencies.

By applying (2.27) and the linearity of the expectation operator, we obtain the following important *calculation rule for the covariance*:

$$\begin{aligned} \mathrm{Cov}[aX_1 + bX_2, cX_3 + dX_4] = {} & ac\,\mathrm{Cov}[X_1, X_3] + ad\,\mathrm{Cov}[X_1, X_4] \\ & + bc\,\mathrm{Cov}[X_2, X_3] + bd\,\mathrm{Cov}[X_2, X_4] \end{aligned} \tag{2.28}$$

where $X_1, \ldots, X_4$ are random variables and $a, \ldots, d$ are constants.

2.4 Moments of multivariate random variables

We now consider multivariate random variables. In time series analysis we often use only the first moment (the mean value) and the second central moment (the variance). Therefore, it is helpful to have the following definitions:

DEFINITION 2.5 (EXPECTATION OF THE RANDOM VECTOR)
The *expectation* (or the *mean value*) of the random vector $\boldsymbol{X}$ is

$$\boldsymbol{\mu} = \mathrm{E}[\boldsymbol{X}] = \begin{pmatrix} \mathrm{E}[X_1] \\ \mathrm{E}[X_2] \\ \vdots \\ \mathrm{E}[X_n] \end{pmatrix} \tag{2.29}$$

DEFINITION 2.6 (COVARIANCE)
The *covariance (matrix)* of $\boldsymbol{X}$ is

$$\begin{aligned}\boldsymbol{\Sigma}_{\boldsymbol{X}} &= \mathrm{Var}[\boldsymbol{X}] \\ &= \mathrm{E}[(\boldsymbol{X}-\boldsymbol{\mu})(\boldsymbol{X}-\boldsymbol{\mu})^T] \\ &= \begin{pmatrix} \mathrm{Var}[X_1] & \mathrm{Cov}[X_1,X_2] & \dots & \mathrm{Cov}[X_1,X_n] \\ \mathrm{Cov}[X_2,X_1] & \mathrm{Var}[X_2] & \dots & \mathrm{Cov}[X_2,X_n] \\ \vdots & & & \vdots \\ \mathrm{Cov}[X_n,X_1] & \mathrm{Cov}[X_n,X_2] & \dots & \mathrm{Var}[X_n] \end{pmatrix}\end{aligned} \tag{2.30}$$

$\boldsymbol{\Sigma}_{\boldsymbol{X}}$ is called the covariance matrix of $\boldsymbol{X}$.

Sometimes we shall use the notation

$$\boldsymbol{\Sigma} = \begin{pmatrix} \sigma_1^2 & \sigma_{12} & \dots & \sigma_{1n} \\ \sigma_{21} & \sigma_2^2 & \dots & \sigma_{2n} \\ \vdots & \vdots & & \vdots \\ \sigma_{n1} & \sigma_{n2} & \dots & \sigma_n^2 \end{pmatrix} \tag{2.31}$$

For the variance, σ_i^2, we sometimes use the notation σ_{ii}.

The correlation of two random variables, X_i and X_j, is a normalization of the covariance and can be written

$$\rho_{ij} = \frac{\mathrm{Cov}[X_i,X_j]}{\sqrt{\mathrm{Var}[X_i]\,\mathrm{Var}[X_j]}} = \frac{\sigma_{ij}}{\sigma_i\sigma_j} \tag{2.32}$$

Equations (2.30) and (2.32) lead to the definition

DEFINITION 2.7 (CORRELATION MATRIX)
The *correlation matrix* for $\boldsymbol{X}$ is

$$\boldsymbol{R} = \boldsymbol{\rho} = \begin{pmatrix} 1 & \rho_{12} & \dots & \rho_{1n} \\ \rho_{21} & 1 & \dots & \rho_{2n} \\ \vdots & \vdots & & \vdots \\ \rho_{n1} & \rho_{n2} & \dots & 1 \end{pmatrix} \tag{2.33}$$

THEOREM 2.1
The covariance matrix $\boldsymbol{\Sigma}$ and the correlation matrix $\boldsymbol{R}$ are (a) symmetric and (b) positive semi-definite.

Proof (a) The symmetry is obvious. (b) Use $\mathrm{Var}[z^T X] \geq 0$ for all values of z. ∎

▸ **Remark 2.2**
If, for instance, $\boldsymbol{\Sigma}$ is positive definite, we often write $\boldsymbol{\Sigma} > 0$. ◂

DEFINITION 2.8
The *covariance matrix* of a random vector $\boldsymbol{X}$ with dimension p and mean $\boldsymbol{\mu}$, and a random vector $\boldsymbol{Y}$ with dimension q and mean $\boldsymbol{\nu}$ is

$$\boldsymbol{\Sigma}_{\boldsymbol{XY}} = \mathrm{C}[\boldsymbol{X}, \boldsymbol{Y}] = \mathrm{E}\left[(\boldsymbol{X} - \boldsymbol{\mu})(\boldsymbol{Y} - \boldsymbol{\nu})^T\right] = \begin{pmatrix} \mathrm{Cov}[X_1, Y_1] & \dots & \mathrm{Cov}[X_1, Y_q] \\ \vdots & & \vdots \\ \mathrm{Cov}[X_p, Y_1] & \dots & \mathrm{Cov}[X_p, Y_q] \end{pmatrix} \tag{2.34}$$

It can be clearly seen that $\mathrm{C}[\boldsymbol{X}, \boldsymbol{X}] = \mathrm{Var}[\boldsymbol{X}]$.

THEOREM 2.2 (CALCULATION RULES FOR THE COVARIANCE)
Let $\boldsymbol{X}$ and $\boldsymbol{Y}$ be defined as in Definition 2.8, and let $\boldsymbol{A}$ and $\boldsymbol{B}$ be $n \times p$ and $m \times q$ real matrices. Let $\boldsymbol{U}$ and $\boldsymbol{V}$ be p- and q-dimensional random vectors. Then

$$\begin{aligned} \mathrm{C}[\boldsymbol{A}(\boldsymbol{X} + \boldsymbol{U}), \boldsymbol{B}(\boldsymbol{Y} + \boldsymbol{V})] = {} & \boldsymbol{A}\,\mathrm{C}[\boldsymbol{X}, \boldsymbol{Y}]\boldsymbol{B}^T + \boldsymbol{A}\,\mathrm{C}[\boldsymbol{X}, \boldsymbol{V}]\boldsymbol{B}^T \\ & + \boldsymbol{A}\,\mathrm{C}[\boldsymbol{U}, \boldsymbol{Y}]\boldsymbol{B}^T + \boldsymbol{A}\,\mathrm{C}[\boldsymbol{U}, \boldsymbol{V}]\boldsymbol{B}^T \end{aligned} \tag{2.35}$$

Important special cases are

$$\mathrm{Var}[\boldsymbol{A}\boldsymbol{X}] = \boldsymbol{A}\,\mathrm{Var}[\boldsymbol{X}]\boldsymbol{A}^T \tag{2.36}$$

$$\mathrm{C}[\boldsymbol{X} + \boldsymbol{U}, \boldsymbol{Y}] = \mathrm{C}[\boldsymbol{X}, \boldsymbol{Y}] + \mathrm{C}[\boldsymbol{U}, \boldsymbol{Y}] \tag{2.37}$$

Proof Follows directly from the definition of the covariance and the linearity of the expectation operator. ∎

Compare the rules above with the rules for scalar random variables given in (2.28).

Example 2.1 (Linear transformations of random variables)
Let $\boldsymbol{X}$ be an n-dimensional random variable with mean $\boldsymbol{\mu}_{\boldsymbol{X}}$ and covariance $\boldsymbol{\Sigma}_{\boldsymbol{X}}$. Now we introduce a new random variable $\boldsymbol{Y} = (Y_1, \dots, Y_k)^T$ by the linear transformation

$$\boldsymbol{Y} = \boldsymbol{a} + \boldsymbol{B}\boldsymbol{X} \tag{2.38}$$

where $\boldsymbol{a}$ is a $(k \times 1)$ vector and $\boldsymbol{B}$ is a $(k \times n)$ matrix. By using the fact that the expectation operator is linear we find

$$\boldsymbol{\mu}_{\boldsymbol{Y}} = \mathrm{E}[\boldsymbol{Y}] = \mathrm{E}[\boldsymbol{a} + \boldsymbol{B}\boldsymbol{X}] = \boldsymbol{a} + \boldsymbol{B}\,\mathrm{E}[\boldsymbol{X}] = \boldsymbol{a} + \boldsymbol{B}\boldsymbol{\mu}_{\boldsymbol{X}} \tag{2.39}$$

and by using the state rules (2.35) and (2.37) for calculating the covariance

$$\boldsymbol{\Sigma}_{\boldsymbol{Y}} = \mathrm{Var}[\boldsymbol{a} + \boldsymbol{B}\boldsymbol{X}] = \mathrm{Var}[\boldsymbol{B}\boldsymbol{X}] = \boldsymbol{B}\,\mathrm{Var}[\boldsymbol{X}]\boldsymbol{B}^T = \boldsymbol{B}\boldsymbol{\Sigma}_{\boldsymbol{X}}\boldsymbol{B}^{\boldsymbol{T}} \tag{2.40}$$

Please note that sometimes $\mathrm{Cov}[\cdot,\cdot]$ is used also in the multivariate case.

2.5 Conditional expectation

The conditional expectation is the expectation of a random variable, given values of another random variable. Later it will be shown that the optimal predictor (in terms of minimum variance) is the conditional mean. Conditional means are also used in filtering.

DEFINITION 2.9 (CONDITIONAL EXPECTATION)
The *conditional expectation* (or *conditional mean*) of the random variable Y given $X = x$ is

$$\mathrm{E}[Y|X = x] = \int_{-\infty}^{\infty} y f_{Y|X=x}(y)\,dy \tag{2.41}$$

▸ **Remark 2.3**
If we know the value $X = x_1$ and the conditional density function, then we are able to calculate the conditional mean. For another value $X = x_2$ we will get a different value for the conditional mean. Hence, the conditional expectation of Y given $X = x$ is written $\psi(x) = \mathrm{E}[Y|X = x]$. Because the conditional expectation depends on the value x taken by X, we can also think of the conditional expectation as a function $\psi(X)$ of X itself. ◂

THEOREM 2.3 (PROPERTIES OF CONDITIONAL MEANS)
Let X, Y and Z be random variables (with joint density $f_{X,Y,Z}$), a and b are real numbers, and g is a real function. Then

$$\mathrm{E}[Y|X] = \mathrm{E}[Y], \quad \textit{if } X \textit{ and } Y \textit{ are independent} \tag{2.42}$$
$$\mathrm{E}[Y] = \mathrm{E}[\mathrm{E}[Y|X]] \tag{2.43}$$
$$\mathrm{E}[g(X)Y|X] = g(X)\,\mathrm{E}[Y|X] \tag{2.44}$$
$$\mathrm{E}[g(X)Y] = \mathrm{E}[g(X)\,\mathrm{E}[Y|X]] \tag{2.45}$$
$$\mathrm{E}[a|X] = a \tag{2.46}$$
$$\mathrm{E}[g(X)|X] = g(X) \tag{2.47}$$
$$\mathrm{E}[cX + dZ|Y] = c\,\mathrm{E}[X|Y] + d\,\mathrm{E}[Z|Y] \tag{2.48}$$

Proof Omitted—but follows from (2.11) on page 14 and (2.41). ■

Equation (2.48) shows that the conditional expectation operator is linear.

DEFINITION 2.10 (CONDITIONAL VARIANCE)
The *conditional variance* of $\boldsymbol{Y}$ given $\boldsymbol{X}$ is

$$\mathrm{Var}[\boldsymbol{Y}|\boldsymbol{X}] = \mathrm{E}[(\boldsymbol{Y} - \mathrm{E}[\boldsymbol{Y}|\boldsymbol{X}])(\boldsymbol{Y} - \mathrm{E}[\boldsymbol{Y}|\boldsymbol{X}])^T|\boldsymbol{X}] \tag{2.49}$$

and the *conditional covariance* between $\boldsymbol{Y}$ and $\boldsymbol{Z}$ given $\boldsymbol{X}$ is

$$\mathrm{C}[\boldsymbol{Y}, \boldsymbol{Z}|\boldsymbol{X}] = \mathrm{E}[(\boldsymbol{Y} - \mathrm{E}[\boldsymbol{Y}|\boldsymbol{X}])(\boldsymbol{Z} - \mathrm{E}[\boldsymbol{Z}|\boldsymbol{X}])^T|\boldsymbol{X}] \tag{2.50}$$

From Remark 2.3 it is clearly seen that $\mathrm{Var}[\boldsymbol{Y}|\boldsymbol{X}]$ and $\mathrm{C}[\boldsymbol{Y}, \boldsymbol{Z}|\boldsymbol{X}]$ are matrices of random variables.

THEOREM 2.4 (THE VARIANCE SEPARATION THEOREM)
Let $\boldsymbol{X}$, $\boldsymbol{Y}$ and $\boldsymbol{Z}$ be random variables. Then

$$\mathrm{Var}[\boldsymbol{Y}] = \mathrm{E}[\mathrm{Var}[\boldsymbol{Y}|\boldsymbol{X}]] + \mathrm{Var}[\mathrm{E}[\boldsymbol{Y}|\boldsymbol{X}]] \tag{2.51}$$

$$\mathrm{C}[\boldsymbol{Y}, \boldsymbol{Z}] = \mathrm{E}[\mathrm{C}[\boldsymbol{Y}, \boldsymbol{Z}|\boldsymbol{X}]] + \mathrm{C}[\mathrm{E}[\boldsymbol{Y}|\boldsymbol{X}], \mathrm{E}[\boldsymbol{Z}|\boldsymbol{X}]] \tag{2.52}$$

Proof Omitted—see, for instance, Jazwinski (1970). ■

Example 2.2 (Linear model)
Assume Y is defined by the linear model

$$Y = X\theta + \varepsilon$$

where X and ε are mutually independent random variables with mean μ_X and $\mu_\varepsilon = 0$, and variance σ_X^2 and σ_ε^2, respectively. We assume that θ is known. By using the theorems above it follows that

$$\mathrm{E}[Y|X] = \mathrm{E}[X\theta + \varepsilon|X] = X\theta$$

$$\mathrm{Var}[Y|X] = \mathrm{Var}[X\theta + \varepsilon|X] = \sigma_\varepsilon^2$$

Hence, we see that for a given $X = x$, we have the "prediction" $\mathrm{E}[Y|X = x] = x\theta$, and the corresponding uncertainty is given by the variance $\mathrm{Var}[Y|X = x] = \sigma_\varepsilon^2$.

The marginal mean of Y is

$$\mathrm{E}[Y] = \mathrm{E}[\mathrm{E}[Y|X]] = \mu_X\theta,$$

and by using (2.51) we get the marginal variance

$$\begin{aligned}\mathrm{Var}[Y] &= \mathrm{E}[\mathrm{Var}[Y|X]] + \mathrm{Var}[\mathrm{E}[Y|X]] \\ &= \sigma_\varepsilon^2 + \theta^2\sigma_X^2\end{aligned}$$

Hence, the variance separation theorem yields that the marginal variance of Y is composed of the variance of ε plus a scaled contribution from the variance of X.

2.6 The multivariate normal distribution

In time series analysis the normal distribution and the distributions derived from the normal distribution are of major interest. For example, the multivariate normal distribution is the fundamental tool used for formulating the likelihood function in later chapters.

We assume that $X_1, X_2, \ldots, X_n$ are independent random variables with means $\mu_1, \mu_2, \ldots, \mu_n$, and variances $\sigma_1^2, \sigma_2^2, \ldots, \sigma_n^2$. We write $X_i \in \mathrm{N}(\mu_i, \sigma_i^2)$. Now, define the random vector $\boldsymbol{X} = (X_1, X_2, \ldots, X_n)^T$. Because the random variables are independent, it follows from (2.14) on page 15 that

$$\begin{aligned}
f_X(x_1, \ldots, x_n) &= f_{X_1}(x_1) \cdots f_{X_n}(x_n) \\
&= \prod_{i=1}^{n} \frac{1}{\sigma_i \sqrt{2\pi}} \exp\left[-\frac{(x_i - \mu_i)^2}{2\sigma_i^2}\right] \\
&= \frac{1}{\left(\prod_{i=1}^{n} \sigma_i\right)(2\pi)^{n/2}} \exp\left[-\frac{1}{2}\sum_{i=1}^{n}\left[\frac{x_i - \mu_i}{\sigma_i}\right]^2\right]
\end{aligned}$$

By introducing the mean $\boldsymbol{\mu} = (\mu_1, \ldots, \mu_n)^T$ and the covariance $\boldsymbol{\Sigma_X} = \mathrm{Var}[\boldsymbol{X}] = \mathrm{diag}(\sigma_1^2, \ldots, \sigma_n^2)$, this is written

$$f_X(x) = \frac{1}{(2\pi)^{n/2}\sqrt{\det \boldsymbol{\Sigma_X}}} \exp\left[-\tfrac{1}{2}(\boldsymbol{x} - \boldsymbol{\mu})^T \boldsymbol{\Sigma_X}^{-1} (\boldsymbol{x} - \boldsymbol{\mu})\right] \tag{2.53}$$

A generalization to the case where the covariance matrix is a full matrix leads to the following:

DEFINITION 2.11 (THE MULTIVARIATE NORMAL DISTRIBUTION)
The joint density function for the n-dimensional random variable $\boldsymbol{X}$ with mean $\boldsymbol{\mu}$ and covariance $\boldsymbol{\Sigma}$ is

$$f_X(x) = \frac{1}{(2\pi)^{n/2}\sqrt{\det \boldsymbol{\Sigma}}} \exp\left[-\tfrac{1}{2}(\boldsymbol{x} - \boldsymbol{\mu})^T \boldsymbol{\Sigma}^{-1} (\boldsymbol{x} - \boldsymbol{\mu})\right] \tag{2.54}$$

where $\boldsymbol{\Sigma} > 0$. We write $\boldsymbol{X} \in \mathrm{N}(\boldsymbol{\mu}, \boldsymbol{\Sigma})$. If $\boldsymbol{X} \in \mathrm{N}(\boldsymbol{0}, \boldsymbol{I})$ we say that $\boldsymbol{X}$ is standardized normally distributed.

THEOREM 2.5
Any n-dimensional normally distributed random variable with mean $\boldsymbol{\mu}$ and covariance $\boldsymbol{\Sigma}$ can be written as

$$\boldsymbol{X} = \boldsymbol{\mu} + \boldsymbol{T}\varepsilon \tag{2.55}$$

where $\varepsilon = (\varepsilon_1, \ldots, \varepsilon_n) \in \mathrm{N}(\boldsymbol{0}, \mathbf{I})$.

Proof Due to the symmetry of $\boldsymbol{\Sigma}$, there always exists a real matrix so that $\boldsymbol{\Sigma} = \boldsymbol{T}\boldsymbol{T}^T$. Then the result follows from (2.40) on page 20 and (2.39) on page 19. ■

2.7 Distributions derived from the normal distribution

Most of the test quantities used in time series analysis are based on the normal distribution or on one of the distributions derived from the normal distribution.

Any linear combination of normally distributed random variables is normal. If, for instance, $\boldsymbol{X} \in \mathrm{N}(\boldsymbol{\mu}, \boldsymbol{\Sigma})$, then the linear transformation $\boldsymbol{Y} = \boldsymbol{a} + \boldsymbol{B}\boldsymbol{X}$ defines a normally distributed random variable as

$$\boldsymbol{Y} \in \mathrm{N}(\boldsymbol{a} + \boldsymbol{B}\boldsymbol{\mu}, \boldsymbol{B}\boldsymbol{\Sigma}\boldsymbol{B}^T) \tag{2.56}$$

Compare with Example 2.1 on page 19.

Let $\boldsymbol{Z} = (Z_1, \ldots, Z_n)^T$ be a vector of independent $\mathrm{N}(0,1)$ random variables. *The (central)* χ^2 *distribution* with n *degrees of freedom* is obtained as the squared sum of n independent $\mathrm{N}(0,1)$ random variables, i.e.,

$$X^2 = \sum_{i=1}^{n} Z_i^2 = \boldsymbol{Z}^T\boldsymbol{Z} \in \chi^2(n) \tag{2.57}$$

From this it is clear that if $Y_1, \ldots, Y_n$ are independent $\mathrm{N}(\mu_i, \sigma_i^2)$ random variables, then

$$X^2 = \sum_{i=1}^{n} \left[\frac{Y_i - \mu_i}{\sigma_i}\right]^2 \in \chi^2(n), \tag{2.58}$$

since $Z_i = (Y_i - \mu_i)/\sigma_i$ is $\mathrm{N}(0,1)$ distributed.

For $\boldsymbol{Y} \in \mathrm{N}_n(\boldsymbol{\mu}, \boldsymbol{\Sigma})$ $(\boldsymbol{\Sigma} > 0)$, we have

$$X^2 = (\boldsymbol{Y} - \boldsymbol{\mu})^T\boldsymbol{\Sigma}^{-1}(\boldsymbol{Y} - \boldsymbol{\mu}) \in \chi^2(n). \tag{2.59}$$

This follows by using Theorem 2.5 and (2.57).

The non-central χ^2 *distribution* with n *degrees of freedom* and *non-centrality parameter* λ appears when considering the sum of squared normally distributed variables when the means are not necessarily zero. Hence,

$$X^2 = \boldsymbol{Y}^T\boldsymbol{\Sigma}^{-1}\boldsymbol{Y} \in \chi^2(n, \lambda), \tag{2.60}$$

where $\lambda = \frac{1}{2}\boldsymbol{\mu}^T\boldsymbol{\Sigma}^{-1}\boldsymbol{\mu}$. Compare (2.59) and (2.60).

Let $X_1^2, \ldots, X_m^2$ denote independent $\chi^2(n_i, \lambda_i)$ distributed random variables. Then the *reproduction property* of the χ^2 distribution is

$$\sum_{i=1}^{m} X_i^2 \in \chi^2\left[\sum_{i=1}^{m} n_i, \sum_{i=1}^{m} \lambda_i\right]. \tag{2.61}$$

If $\boldsymbol{\Sigma}$ is singular with rank $k < n$, then $\boldsymbol{Y}^T\boldsymbol{\Sigma}^{-}\boldsymbol{Y}$ is χ^2 distributed with k degrees of freedom and non-centrality parameter $\lambda = \frac{1}{2}\boldsymbol{\mu}^T\boldsymbol{\Sigma}^{-}\boldsymbol{\mu}$, where $\boldsymbol{\Sigma}^{-}$

denotes a generalized inverse (called *g*-inverse) for $\boldsymbol{\Sigma}$ (see, for instance, Rao (1973)).

The (student) t *distribution* with n *degrees of freedom* is obtained as

$$\mathrm{T} = \frac{Z}{(X^2/n)^{1/2}} \in \mathrm{t}(n), \tag{2.62}$$

where $Z \in \mathrm{N}(0,1)$, $X^2 \in \chi^2(n)$, and Z and X^2 are independent. *The non-central* t *distribution* is obtained from (2.62) if $Z \in \mathrm{N}(\mu, 1)$, and we write $\mathrm{T} \in \mathrm{t}(n, \mu)$.

The F *distribution* with (n, m) *degrees of freedom* appears as the following ratio

$$\mathrm{F} = \frac{X_1^2/n}{X_2^2/m} \in \mathrm{F}(n, m) \tag{2.63}$$

where $X_1^2 \in \chi^2(n)$, $X_2^2 \in \chi^2(m)$, and X_1^2 and X_2^2 are independent. It is clearly seen from (2.62) that $\mathrm{T}^2 \in \mathrm{F}(1, n)$.

The non-central F *distribution* with (n, m) *degrees of freedom* and *non-centrality parameter* λ is obtained from (2.63) if $X_1^2 \in \chi^2(n, \lambda)$, $X_2^2 \in \chi^2(m)$, and X_1^2 and X_2^2 are independent. The non-central F distribution is written $\mathrm{F} \in \mathrm{F}(n, m; \lambda)$.

2.8 Linear projections

This section contains the fundamental theorems used in, e.g., linear regression, where the independent variables are stochastic, as well as in linear predictions and the Kalman filter.

THEOREM 2.6 (LINEAR PROJECTION)
Let $\boldsymbol{Y} = (Y_1, \ldots, Y_m)^T$ and $\boldsymbol{X} = (X_1, \ldots, X_n)^T$ be random vectors, and let the $(m+n)$-dimensional vector $(\boldsymbol{Y}, \boldsymbol{X})^T$ have the mean

$$\begin{pmatrix} \boldsymbol{\mu}_Y \\ \boldsymbol{\mu}_X \end{pmatrix} \quad \textit{and covariance} \quad \begin{pmatrix} \boldsymbol{\Sigma}_{YY} & \boldsymbol{\Sigma}_{YX} \\ \boldsymbol{\Sigma}_{XY} & \boldsymbol{\Sigma}_{XX} \end{pmatrix}$$

Define the linear projection of $\boldsymbol{Y}$ on $\boldsymbol{X}$

$$\mathrm{E}[\boldsymbol{Y}|\boldsymbol{X}] = \boldsymbol{a} + \boldsymbol{B}\boldsymbol{X} \tag{2.64}$$

Then the projection—and the variance of the projection error—is given by

$$\mathrm{E}[\boldsymbol{Y}|\boldsymbol{X}] = \boldsymbol{\mu}_Y + \boldsymbol{\Sigma}_{YX}\boldsymbol{\Sigma}_{XX}^{-1}(\boldsymbol{X} - \boldsymbol{\mu}_X) \tag{2.65}$$

$$\mathrm{E}[\mathrm{Var}[\boldsymbol{Y}|\boldsymbol{X}]] = \boldsymbol{\Sigma}_{YY} - \boldsymbol{\Sigma}_{YX}\boldsymbol{\Sigma}_{XX}^{-1}\boldsymbol{\Sigma}_{YX}^T \tag{2.66}$$

Finally, the projection error, $\boldsymbol{Y} - \mathrm{E}[\boldsymbol{Y}|\boldsymbol{X}]$, and $\boldsymbol{X}$ are uncorrelated, i.e.,

$$\mathrm{C}[\boldsymbol{Y} - \mathrm{E}[\boldsymbol{Y}|\boldsymbol{X}], \boldsymbol{X}] = \boldsymbol{0} \tag{2.67}$$

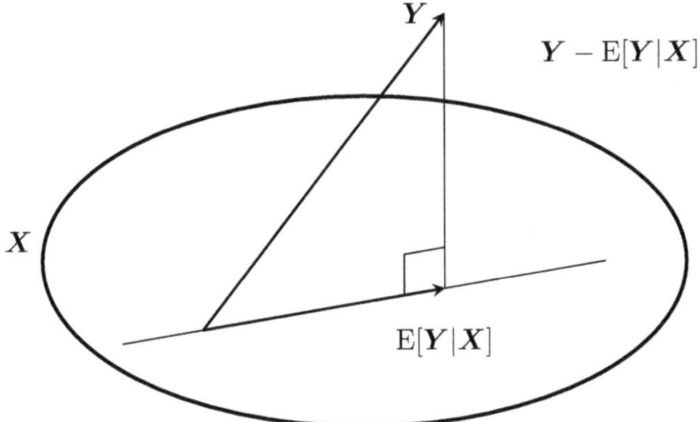

Figure 2.1: *The projection* $\mathrm{E}[\boldsymbol{Y}|\boldsymbol{X}]$ *of* $\boldsymbol{Y}$ *on* $\boldsymbol{X}$.

Proof From Theorem 2.4 on page 21:

$$\begin{aligned} \mathrm{C}[\boldsymbol{Y}, \boldsymbol{X}] &= \mathrm{E}[\mathrm{C}[\boldsymbol{Y}, \boldsymbol{X}|\boldsymbol{X}]] + \mathrm{C}[\mathrm{E}[\boldsymbol{Y}|\boldsymbol{X}], \mathrm{E}[\boldsymbol{X}|\boldsymbol{X}]] \\ &= \mathrm{E}[\boldsymbol{0}] + \mathrm{C}[\boldsymbol{a} + \boldsymbol{B}\boldsymbol{X}, \boldsymbol{X}] \\ &= \boldsymbol{B}\,\mathrm{Var}[\boldsymbol{X}] \end{aligned} \tag{2.68}$$

From this it is seen that $\boldsymbol{B} = \boldsymbol{\Sigma}_{\boldsymbol{YX}}\boldsymbol{\Sigma}_{\boldsymbol{XX}}^{-1}$

$$\mathrm{E}[\boldsymbol{Y}] = \mathrm{E}[\mathrm{E}[\boldsymbol{Y}|\boldsymbol{X}]] = \mathrm{E}[\boldsymbol{a} + \boldsymbol{B}\boldsymbol{X}] = \boldsymbol{a} + \boldsymbol{B}\,\mathrm{E}[\boldsymbol{X}],$$

which leads to

$$\boldsymbol{a} = \mathrm{E}[\boldsymbol{Y}] - \boldsymbol{B}\,\mathrm{E}[\boldsymbol{X}]$$

i.e.,

$$\boldsymbol{a} = \boldsymbol{\mu}_{\boldsymbol{Y}} - \boldsymbol{\Sigma}_{\boldsymbol{YX}}\boldsymbol{\Sigma}_{\boldsymbol{XX}}^{-1}\boldsymbol{\mu}_{\boldsymbol{X}} \tag{2.69}$$

Equation (2.65) is now obtained by using the values for $\boldsymbol{a}$ and $\boldsymbol{B}$ in (2.64).

Now

$$\begin{aligned} \mathrm{E}[\mathrm{Var}[\boldsymbol{Y}|\boldsymbol{X}]] &= \mathrm{Var}[\boldsymbol{Y} - \mathrm{E}[\boldsymbol{Y}|\boldsymbol{X}]] \\ &= \mathrm{Var}[\boldsymbol{Y} - \boldsymbol{a} - \boldsymbol{B}\boldsymbol{X}] \\ &= \boldsymbol{\Sigma}_{\boldsymbol{YY}} + \boldsymbol{B}\boldsymbol{\Sigma}_{\boldsymbol{XX}}\boldsymbol{B}^T - \boldsymbol{B}\boldsymbol{\Sigma}_{\boldsymbol{XY}} - \boldsymbol{\Sigma}_{\boldsymbol{YX}}\boldsymbol{B}^T \\ &= \boldsymbol{\Sigma}_{\boldsymbol{YY}} - \boldsymbol{\Sigma}_{\boldsymbol{YX}}\boldsymbol{\Sigma}_{\boldsymbol{XX}}^{-1}\boldsymbol{\Sigma}_{\boldsymbol{XY}} \end{aligned} \tag{2.70}$$

$$\begin{aligned} \mathrm{C}[\boldsymbol{Y} - \mathrm{E}[\boldsymbol{Y}|\boldsymbol{X}], \boldsymbol{X}] &= \mathrm{C}[\boldsymbol{Y} - \boldsymbol{a} - \boldsymbol{B}\boldsymbol{X}, \boldsymbol{X}] \\ &= \boldsymbol{\Sigma}_{\boldsymbol{YX}} - \boldsymbol{B}\boldsymbol{\Sigma}_{\boldsymbol{XX}} = \boldsymbol{0} \end{aligned} \tag{2.71}$$

∎

Referring to (2.67), we say that the error, $(\boldsymbol{Y} - \mathrm{E}[\boldsymbol{Y}|\boldsymbol{X}])$, and $\boldsymbol{X}$ are *orthogonal*—as illustrated in Figure 2.1. The linear projections above *give the minimum variance among all linear projections.* This is shown in the next theorem.

Since we are mostly interested in the variance, we assume for the moment that $\boldsymbol{\mu}_{\boldsymbol{Y}} = \mathbf{E}[\boldsymbol{Y}] = \mathbf{0}$ and $\boldsymbol{\mu}_{\boldsymbol{X}} = \mathrm{E}[\boldsymbol{X}] = \mathbf{0}$.

THEOREM 2.7 (PROJECTION THEOREM)
Let the random vectors $\boldsymbol{Y}$ *and* $\boldsymbol{X}$ *have zero mean. Then the linear projection*

$$\mathrm{E}[\boldsymbol{Y}|\boldsymbol{X}] = \boldsymbol{B}\boldsymbol{X} \tag{2.72}$$

which gives the minimum variance among all linear projections *is found by solving*

$$\mathrm{C}[\boldsymbol{Y} - \boldsymbol{B}\boldsymbol{X}, \boldsymbol{X}] = \mathbf{0} \tag{2.73}$$

Proof Assume that $\boldsymbol{B}\boldsymbol{X}$ satisfies (2.73). Let $\boldsymbol{\beta}\boldsymbol{X}$ denote any other linear projection. Then

$$\begin{aligned}
\mathrm{Var}[\boldsymbol{Y} - \boldsymbol{\beta}\boldsymbol{X}] &= \mathrm{Var}[\boldsymbol{Y} - \boldsymbol{B}\boldsymbol{X} + (\boldsymbol{B} - \boldsymbol{\beta})\boldsymbol{X}] = (\text{due to (2.73)}) \\
&= \mathrm{Var}[\boldsymbol{Y} - \boldsymbol{B}\boldsymbol{X}] + (\boldsymbol{B} - \boldsymbol{\beta})\,\mathrm{Var}[\boldsymbol{X}](\boldsymbol{B} - \boldsymbol{\beta})^T \\
&\geq \mathrm{Var}[\boldsymbol{Y} - \boldsymbol{B}\boldsymbol{X}]
\end{aligned}$$

■

The theorems above are based on a second order representation for the vector $(\boldsymbol{Y}^T, \boldsymbol{X}^T)^T$, and the projections are restricted to being linear projections.

Example 2.3 (Linear predictor)
Let $(Y_1, Y_2)^T$ be normally distributed with zero mean and covariance

$$\Sigma = \begin{pmatrix} \sigma_1^2 & \sigma_{12} \\ \sigma_{21} & \sigma_2^2 \end{pmatrix}.$$

The correlation between Y_1 and Y_2 is then $\rho = \sigma_{12}/\sigma_1\sigma_2$, cf. (2.32) on page 18.

Assume that Y_1 is observed. The projection theorem above can now be used to find the optimal linear predictor $\widehat{Y}_2 = aY_1$ by solving

$$\mathrm{Cov}[Y_2 - aY_1, Y_1] = 0$$

which leads to

$$\rho\sigma_1\sigma_2 - a\sigma_1^2 = 0$$

The optimal linear predictor is thus

$$\widehat{Y}_2 = \rho\frac{\sigma_2}{\sigma_1}Y_1$$

The variance of the prediction error is

$$\mathrm{Var}\left[Y_2 - \rho\frac{\sigma_2}{\sigma_1}Y_1\right] = \mathrm{Cov}\left[Y_2 - \rho\frac{\sigma_2}{\sigma_1}Y_1, Y_2\right] = \sigma_2^2(1-\rho^2)$$

The results can also be seen directly by using Theorem 2.8 below. Since $(Y_1, Y_2)^T$ is normally distributed, the optimal linear predictor is the optimal predictor.

In general $\mathrm{E}[\boldsymbol{Y}|\boldsymbol{X}]$ is not necessarily linear in $\boldsymbol{X}$. This is illustrated by the following example:

Example 2.4 (Non-linear $\mathrm{E}[\boldsymbol{Y}|\boldsymbol{X}]$)
Consider the density

$$f_{X,Y}(x,y) = \tfrac{1}{2}(x+y)\mathrm{e}^{-(x+y)}, \quad (x,y) \in (\mathbb{R}_+^2) \tag{2.74}$$

The marginal density for X is

$$f_X(x) = \tfrac{1}{2}(x+1)\mathrm{e}^{-x} \tag{2.75}$$

The conditional density for $Y|X = x$ is

$$f_{Y|X=x}(y) = \frac{f_{X,Y}(x,y)}{f_X(x)} = \frac{x+y}{1+x}\mathrm{e}^{-y} \tag{2.76}$$

Now the conditional expectation becomes

$$\begin{aligned}\mathrm{E}[Y|X=x] &= \int_0^\infty y\frac{x+y}{1+x}\mathrm{e}^{-y}dy \\ &= \frac{1}{1+x}\left(\int_0^\infty y^2\mathrm{e}^{-y}dy + x\int_0^\infty y\mathrm{e}^{-y}dy\right) \\ &= \frac{1}{1+x}(2+x) = \frac{2+x}{1+x}\end{aligned} \tag{2.77}$$

Hence, $\mathrm{E}[Y|X] = \frac{2+X}{1+X}$, which clearly is non-linear in X.

For normally distributed random variables, however, the conditional mean is linear. We have the following theorem.

THEOREM 2.8 (PROJECTIONS FOR NORMALLY DISTRIBUTED VARIABLES)
Let $(\boldsymbol{Y}^T, \boldsymbol{X}^T)^T$ be a normally distributed random variable with mean

$$\begin{pmatrix}\boldsymbol{\mu_Y}\\ \boldsymbol{\mu_X}\end{pmatrix} \quad \textit{and covariance} \quad \begin{pmatrix}\boldsymbol{\Sigma_{YY}} & \boldsymbol{\Sigma_{YX}}\\ \boldsymbol{\Sigma_{XY}} & \boldsymbol{\Sigma_{XX}}\end{pmatrix}$$

Then $\boldsymbol{Y}|\boldsymbol{X}$ *is normally distributed with mean*

$$\mathrm{E}[\boldsymbol{Y}|\boldsymbol{X}] = \boldsymbol{\mu_Y} + \boldsymbol{\Sigma_{YX}}\boldsymbol{\Sigma_{XX}^{-1}}(\boldsymbol{X} - \boldsymbol{\mu_X}) \tag{2.78}$$

and variance

$$\mathrm{Var}[\boldsymbol{Y}|\boldsymbol{X}] = \boldsymbol{\Sigma_{YY}} - \boldsymbol{\Sigma_{YX}}\boldsymbol{\Sigma_{XX}^{-1}}\boldsymbol{\Sigma_{YX}^{T}} \tag{2.79}$$

Furthermore, the error, $(\boldsymbol{Y} - \mathrm{E}[\boldsymbol{Y}|\boldsymbol{X}])$, *and* $\boldsymbol{X}$ *are* independent.

Proof Omitted—see, for instance, Jazwinski (1970). ■

▸ **Remark 2.4**
Without the assumption about normality, the error $(\boldsymbol{Y} - \mathrm{E}[\boldsymbol{Y}|\boldsymbol{X}])$ and $\boldsymbol{X}$ were uncorrelated; but if we consider normal distribution, they are also independent. ◂

▸ **Remark 2.5**
A more mathematical treatment of the projection theorem involves inner product spaces; see Madsen (1992) or Brockwell and Davis (1987). In the formulation used above, the inner product is defined by the covariance, i.e., $\langle \boldsymbol{X}, \boldsymbol{Y} \rangle = \mathrm{C}[\boldsymbol{X}, \boldsymbol{Y}]$. ◂

An application of the projection theorem is illustrated in the following important example.

Example 2.5 (Wiener filter—optimal linear reconstruction)
Consider two mutually correlated stochastic processes $\{X_t\}$ and $\{Y_t\}$, both with mean value zero. Assume that the autocovariance and cross-covariance functions are known (these functions, which are introduced formally in Chapter 5, give a total second moment representation for $\{X_t\}$ and $\{Y_t\}$).

We consider the problem of estimating the value of X_{t+k} by using observed values of Y_t at time $t_1, t_1+1, \dots, t_2$. Note, that no restriction on the value of k is given. In practice, however, we most often have $k = 0$ or $k > 0$.

The optimal linear projection

$$\widehat{X}_{t+k} = \sum_{i=t_1}^{t_2} h_{t,i} Y_i \tag{2.80}$$

is, due to the projection theorem, given as the solution to

$$\mathrm{Cov}\left[X_{t+k} - \sum_{i=t_1}^{t_2} h_{t,i} Y_i, Y_j\right] = 0, \quad j = t_1, \dots, t_2 \tag{2.81}$$

Note that in (2.81) the number of equations is equal to the number of unknown values of h_t.

The solution $\widehat{h}_{t,j}$ is known as the *Wiener filter* in discrete time and the *Wiener-Komogorow filter* in continuous time. See also Wiener (1949).

The variance of the reconstruction error is

$$\operatorname{Var}\left[X_{t+k} - \sum_{i=t_1}^{t_2} h_{t,i}Y_i\right] = \operatorname{Cov}\left[X_{t+k} - \sum_{i=t_1}^{t_2} h_{t,i}Y_i, X_{t+k}\right] \tag{2.82}$$

where the right hand side follows by using (2.81).

In communication theory a frequently used model is

$$Y_t = X_t + e_t \tag{2.83}$$

where Y_t is the measured signal, X_t is the true signal, and e_t is noise. Hence, the Wiener filter gives an estimate of the true signal.

Another very important example is the use of the linear projection theorem for deriving the Kalman filter, which is dealt with in Chapter 10.

2.9 Problems

Exercise 2.1

Let the two random variables X and Y, and their corresponding second order moment representation, be given by

$$\begin{aligned} &\mathrm{E}[X] = 5, &&\operatorname{Var}[X] = 1^2 \\ &\mathrm{E}[Y] = 10, &&\operatorname{Var}[Y] = 2^2, &&\rho[X,Y] = \tfrac{1}{2} \end{aligned}$$

Two new random variables H and L are now defined by

$$\begin{aligned} H &= 2X + 3Y \\ L &= -X + 2Y \end{aligned}$$

Calculate the second order moment representation of $\begin{pmatrix} H & L \end{pmatrix}^T$.

Exercise 2.2

Consider the linear model

$$Y = \alpha + \beta X + \varepsilon$$

where α and β are constants. X and ε are mutually uncorrelated random variables with mean values $\mathrm{E}[X] = \mu_X$ and $\mathrm{E}[\varepsilon] = 0$, variances $\operatorname{Var}[X] = \sigma_X^2$ and $\operatorname{Var}[\varepsilon] = \sigma_\varepsilon^2$, and correlations $\rho(X, Y) = \rho$.

Question 1 Find $\mathrm{E}[Y|X]$ and $\operatorname{Var}[Y|X]$.

Question 2 Find $\mathrm{E}[Y]$ and $\operatorname{Var}[Y]$.

Question 3 Establish a set of equations to determine the moment estimates of (α, β), where $\rho(X, Y)$ appears. (Hint: Moment estimates are found by expressing the moments of the random variables as a function of the parameters and then solving for the set of unknown parameters.)

Exercise 2.3

Let $X_1, X_2, \dots$ be a sequence of random variables. A new random variable is defined as

$$Y = X_1 + X_2 + \cdots + X_N$$

where N is a random variable with the sample space $\Omega_N = \{1, 2, \dots\}$ and

$$\mathrm{E}[N] = 20, \qquad \mathrm{Var}[N] = 2^2$$

It is also given that

$$\begin{aligned} \mathrm{E}[X_i|N = j] &= \mathrm{E}[X_i] = 2 \\ \mathrm{Var}[X_i|N = j] &= \mathrm{Var}[X_i] = \left(\tfrac{1}{8}\right)^2 \\ \mathrm{Cov}[X_r, X_s|N = j] &= \mathrm{Cov}[X_r, X_s] = 0\,, \quad r \neq s \end{aligned}$$

Question 1 Determine $\mathrm{E}[Y]$ and $\mathrm{Var}[Y]$.

Question 2 Determine the covariance

$$\mathrm{Cov}[Y, Z]$$

where

$$Z = X_1 + X_2 + \cdots + X_{\alpha N}\,, \quad \alpha \in \mathbb{N}$$

CHAPTER 3

Regression-based methods

In spite of the fact that a time series is an outcome of a stochastic process and, thus, an observation of a dynamical phenomena, methods, which are usually related to the analysis and modeling of static phenomena, are often applied. A class of such methods is closely related to the standard regression analysis.

3.1 The regression model

The classical regression model is used to describe a static relation between a dependent variable Y_t and p independent variables $X_{1t}, \ldots, X_{pt}$. In time series analysis, the observations occur successively in time and most frequently with an equidistant time distance. Therefore, an index t is introduced to denote the variable at time origin t—for other applications t denotes an arbitrary index.

In its most general form the *regression model* is written

$$Y_t = f(\boldsymbol{X}_t, t; \boldsymbol{\theta}) + \varepsilon_t \tag{3.1}$$

where $f(\boldsymbol{X}_t, t; \theta)$ is a known mathematical function of the $p+1$ independent variables $\boldsymbol{X}_t = (X_{1t}, \ldots, X_{pt})^T$ and t, but with unknown parameters $\boldsymbol{\theta} = (\theta_1, \ldots, \theta_m)^T$. The independent variable t is introduced to indicate that the model class described by (3.1) contains models where f is a function of t, e.g., $f(t, \theta) = \theta \sin(\omega t)$, and ε_t is a random variable with $\mathrm{E}[\varepsilon_t] = 0$ and $\mathrm{Var}[\varepsilon_t] = \sigma_t^2$. Furthermore, it is assumed that $\mathrm{Cov}[\varepsilon_{t_i}, \varepsilon_{t_j}] = \sigma^2 \boldsymbol{\Sigma}_{ij}$, i.e., $\sigma_t^2 = \sigma^2 \boldsymbol{\Sigma}_{ii}$. Finally, ε_t and $\boldsymbol{X}_t$ are assumed to be independent.

For the regression model (3.1) for Y_t given $\boldsymbol{X}_t = \boldsymbol{x}_t$, it holds:

$$\mathrm{E}[Y_t | \boldsymbol{X}_t = \boldsymbol{x}_t] = f(\boldsymbol{x}_t, t; \boldsymbol{\theta}) \tag{3.2}$$

$$\mathrm{Var}[Y_t | \boldsymbol{X}_t = \boldsymbol{x}_t] = \sigma_t^2 \tag{3.3}$$

$$\mathrm{Cov}[Y_{t_i}, Y_{t_j} | \boldsymbol{X}_t = \boldsymbol{x}_t] = \sigma^2 \boldsymbol{\Sigma}_{ij} \tag{3.4}$$

where $\sigma_{t_i}^2 = \sigma^2 \boldsymbol{\Sigma}_{ii}$. The total variation for Y_t depends on the variation of $\boldsymbol{X}_t$; see Theorem 2.4 on page 21.

In this chapter we shall restrict our attention to models with fixed and non-stochastic independent variables, i.e., $\boldsymbol{X}_t = \boldsymbol{x}_t$. However, most of the results are also valid for independent random variables.

Example 3.1 (Direct radiation in clear skies, part I)
In Sections 3.1 through 3.3 we will be following a case study to show how the methods are applied to solve a real problem. The purpose is to establish a stochastic model for direct radiation in clear skies as a function of the solar elevation. The radiation is known to depend on other variables as well, such as water vapor and aerosols, but these dependencies are ignored. However, by considering a stochastic approach, the model is able to give an indirect description of effects due to variations of the ignored variables.

In this example, only solar radiation in clear skies is considered. The observations are hourly values from February 1, 1966, through December 31, 1973, where the total number of hourly observations was 69,384. Only 1,183 observations of the total set correspond to clear sky during daytime. Figure 3.1 shows these observations as a function of the solar elevation. Furthermore, the figure shows two days with clear sky during the entire day. The direct radiation is actually not measured, but is calculated using observed, hourly measurements of global ($Y_{t,\text{glob}}$) and diffuse ($Y_{t,\text{dif}}$) radiation, along with the solar elevation, by isolating the direct radiation in the fundamental relationship

$$Y_{t,\text{glob}} = Y_{t,\text{dir}} \cdot \sin(h_t) + Y_{t,\text{dif}} \tag{3.5}$$

where $Y_{t,\text{dir}}$ is the measured direct radiation at time t, denoted as Y_t in the following.

A deterministic model, describing the relationship between direct radiation and solar elevation, has been suggested by Paltridge and Platt (1976):

$$f(h_t) = \theta_1 \left(1 - \mathrm{e}^{-\theta_2 h_t}\right) \tag{3.6}$$

where h_t is the solar elevation at time t, $f(h_t)$ is the corresponding direct radiation, and θ_1 and θ_2 are parameters of the model, suggested to be $1000\,\text{W/m}^2$ and $0.06\,\text{deg}^{-1}$, respectively. As mentioned above, the solar elevation alone is not able to describe the total variation. Thus, the model is expanded to the stochastic model

$$Y_t = Y_{t,\text{dir}} = f(h_t) + \varepsilon_t \tag{3.7}$$

where the deviation from the deterministic model (3.6) is described by the random variable ε_t.

Consider now the variance structure for ε_t. By isolating the direct radiation from (3.5) and assuming that the variance of the measurements of $Y_{t,\text{glob}}$ and $Y_{t,\text{dif}}$ are constant, a reasonable relationship for the variance is

$$\text{Var}[Y_t] = \text{Var}[\varepsilon_t] = \frac{\sigma^2}{\sin^2(h_t)}. \tag{3.8}$$

This tendency is recognized in Figure 3.1, where it is seen that the variance of the direct radiation decreases as the solar elevation increases.

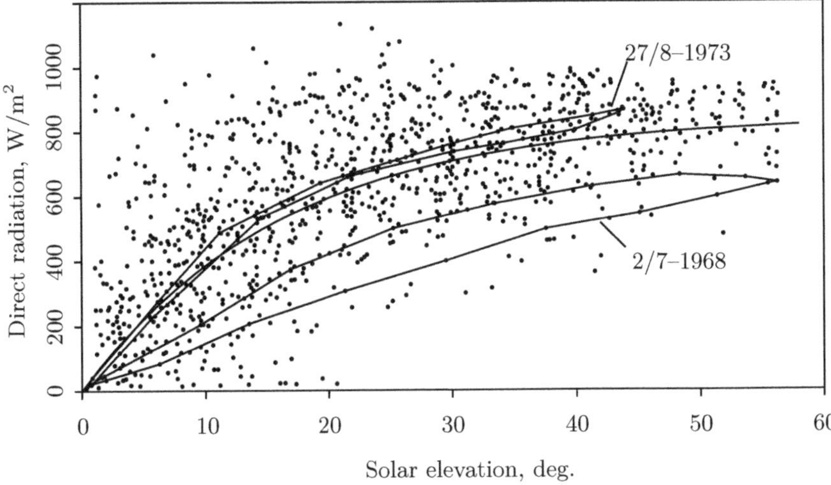

Figure 3.1: *Hourly observed direct radiation in clear skies as a function of solar elevation. Two days with clear sky for the entire day are shown, along with the final model.*

3.2 The general linear model (GLM)

The general linear model is a very important special case of the regression model (3.1) where the function f is linear in the parameters $\boldsymbol{\theta}$.

DEFINITION 3.1 (GENERAL LINEAR MODEL)
The general linear model (GLM) is a regression model with the following model structure

$$Y_t = \boldsymbol{x}_t^T \boldsymbol{\theta} + \varepsilon_t \tag{3.9}$$

where $\boldsymbol{x}_t = (x_{1t}, \ldots, x_{pt})^T$ is a known vector and $\boldsymbol{\theta} = (\theta_1, \ldots, \theta_p)^T$ are the unknown parameters. ε_t is a random variable with mean $\mathrm{E}[\varepsilon_t] = 0$ and covariance $\mathrm{Cov}[\varepsilon_{t_i}, \varepsilon_{t_j}] = \sigma^2 \boldsymbol{\Sigma}_{ij}$.

Let us consider some examples of a general linear model.

a) Constant mean model

$$Y_t = \theta_0 + \varepsilon_t \tag{3.10}$$

$$\boldsymbol{x}_t = 1, \qquad \boldsymbol{\theta} = \theta_0 \tag{3.11}$$

b) Simple linear regression model

$$Y_t = \theta_0 + \theta_1 z_t + \varepsilon_t \tag{3.12}$$

$$\boldsymbol{x}_t = (1, z_t)^T, \qquad \boldsymbol{\theta} = (\theta_0, \theta_1)^T \tag{3.13}$$

c) Quadratic model

$$Y_t = \theta_0 + \theta_1 z_t + \theta_2 z_t^2 + \varepsilon_t \tag{3.14}$$

$$\boldsymbol{x}_t = (1, z_t, z_t^2)^T, \qquad \boldsymbol{\theta} = (\theta_0, \theta_1, \theta_2)^T \tag{3.15}$$

d) Model with two independent variables

$$Y_t = \theta_0 + \theta_1 z_{1t} + \theta_2 z_{2t} + \varepsilon_t \tag{3.16}$$

$$\boldsymbol{x}_t = (1, z_{1t}, z_{2t})^T, \qquad \boldsymbol{\theta} = (\theta_0, \theta_1, \theta_2)^T \tag{3.17}$$

e) A mixed model

$$Y_t = \theta_0 + \theta_1 \sin(\omega t) + \theta_2 \xi_t + \theta_3 z_{1t} + \theta_4 z_{2t} + \theta_5 z_{1t} z_{2t} + \varepsilon_t \tag{3.18}$$

$$\xi_t = \begin{cases} 0 & t \leq 200 \\ 1 & t \geq 200 \end{cases} \quad \text{(step variable)} \tag{3.19}$$

$$\boldsymbol{x}_t = (1, \sin(\omega t), \xi_t, z_{1t}, z_{2t}, z_{1t} z_{2t})^T \tag{3.20}$$

$$\boldsymbol{\theta} = (\theta_0, \theta_1, \theta_2, \theta_3, \theta_4, \theta_5)^T \tag{3.21}$$

The model $Y_t = \theta_0 + \theta_1 (\sin(\omega t))^{\theta_2} + \varepsilon_t$ is not a linear model since it cannot be written on the form $Y_t = \boldsymbol{x}_t^T \boldsymbol{\theta} + \varepsilon_t$.

3.2.1 Least squares (LS) estimates

Let us assume that the parameters in the regression model (3.1) on page 31 or the general linear model (3.9) on the preceding page are unknown. Then they may be estimated from the following N observations of the dependent and independent variables

$$(y_1, \boldsymbol{x}_1), (y_2, \boldsymbol{x}_2), \ldots, (y_N, \boldsymbol{x}_N). \tag{3.22}$$

Based on these observations we want an estimate $\widehat{\boldsymbol{\theta}}$ of $\boldsymbol{\theta}$ such that $f(\boldsymbol{x}_t; \widehat{\boldsymbol{\theta}})$ describes the observations as well as possible, as defined by some measure of closeness. If the sum of squared errors (SSE) $\sum [y_t - f(\boldsymbol{x}_t; \boldsymbol{\theta})]^2$ is chosen then we have

---

DEFINITION 3.2 (LS ESTIMATES)
The *Least Squares (unweighted) estimates* are found from

$$\widehat{\boldsymbol{\theta}} = \arg\min_{\boldsymbol{\theta}} S(\boldsymbol{\theta}), \tag{3.23}$$

where

$$S(\boldsymbol{\theta}) = \sum_{t=1}^{N} [y_t - f(\boldsymbol{x}_t; \boldsymbol{\theta})]^2 = \sum_{t=1}^{N} \varepsilon_t^2(\boldsymbol{\theta}) \tag{3.24}$$

i.e., $\widehat{\boldsymbol{\theta}}$ is the $\boldsymbol{\theta}$ that minimizes the sum of squared residuals.

An estimate for the variance of $\widehat{\boldsymbol{\theta}}$ is

$$\operatorname{Var}[\widehat{\boldsymbol{\theta}}] = 2\widehat{\sigma}^2 \left[\frac{\partial^2}{\partial^2 \boldsymbol{\theta}} S(\boldsymbol{\theta}) \right]^{-1} \Bigg|_{\boldsymbol{\theta}=\widehat{\boldsymbol{\theta}}} \tag{3.25}$$

where $\widehat{\sigma}^2 = S(\widehat{\boldsymbol{\theta}})/(N-p)$.

▸ **Remark 3.1**

The term *unweighted* is used since no consideration is taken of residuals which may have a larger variance or residuals which may be correlated. Hence, unweighted least squares estimates are only reasonable if the variance of the residuals is constant (i.e., $\boldsymbol{\Sigma}_{ii} = 1$) and the residuals are mutually independent (i.e., $\boldsymbol{\Sigma}_{ij} = 0$ for $i \neq j$). Later in this section weighted least squares estimates will be considered. In order to distinguish between the two types of least squares estimates, unweighted least squares are often called *OLS estimates* (Ordinary Least Squares) and the weighted estimates are called *WLS estimates* (Weighted Least Squares). ◂

Since it is not possible in general to find an explicit expression for the LS estimator, numerical methods have to be used. A classical reference to non-linear parameter estimation is Bard (1974).

3.2.1.1 A simple numerical procedure

It is well known that for unconstrained optimization, the value $\boldsymbol{\theta}^\star$ which minimizes S, will satisfy the equation

$$\nabla_{\theta} S(\boldsymbol{\theta}^*) = \mathbf{0} \tag{3.26}$$

A Taylor expansion of $S(\boldsymbol{\theta})$ in the neighbourhood of $\boldsymbol{\theta} = \boldsymbol{\theta}_0$ is

$$\begin{aligned} S(\boldsymbol{\theta}) &= S(\boldsymbol{\theta}_0) + (\boldsymbol{\theta} - \boldsymbol{\theta}_0)^T \nabla_{\theta} S(\boldsymbol{\theta}_0) \\ &\quad + \frac{1}{2} (\boldsymbol{\theta} - \boldsymbol{\theta}_0)^T \boldsymbol{H}(\boldsymbol{\theta}_0)(\boldsymbol{\theta} - \boldsymbol{\theta}_0) + \cdots \end{aligned} \tag{3.27}$$

where $\boldsymbol{H}(\boldsymbol{\theta}_0)$ is the so-called *Hessian matrix*, given as

$$\boldsymbol{H}(\boldsymbol{\theta}_0) = \frac{\partial^2 S(\boldsymbol{\theta})}{\partial \theta_i \, \partial \theta_j} \Bigg|_{\boldsymbol{\theta}=\boldsymbol{\theta}_0} \tag{3.28}$$

The Hessian matrix is a symmetric matrix with dimension equal to the number of parameters in $\boldsymbol{\theta}$. It is assumed that the partial derivatives are continuous in $\boldsymbol{\theta}_0$.

An approximation for $S(\boldsymbol{\theta})$ in $\boldsymbol{\theta}_0$ is

$$\begin{aligned} S_{\text{app}}(\boldsymbol{\theta}) = S(\boldsymbol{\theta}_0) &+ (\boldsymbol{\theta} - \boldsymbol{\theta}_0)^T \nabla_\theta S(\boldsymbol{\theta}_0) \\ &+ \frac{1}{2}(\boldsymbol{\theta} - \boldsymbol{\theta}_0)^T \boldsymbol{H}(\boldsymbol{\theta}_0)(\boldsymbol{\theta} - \boldsymbol{\theta}_0) \end{aligned} \tag{3.29}$$

The minimum of the quadratic approximation is found by putting $\nabla_\theta S_{\text{app}}(\boldsymbol{\theta}) = 0$, that is

$$\nabla_\theta S(\boldsymbol{\theta}_0) + \boldsymbol{H}(\boldsymbol{\theta}_0)(\boldsymbol{\theta} - \boldsymbol{\theta}_0) = 0 \tag{3.30}$$

By putting the old value $\boldsymbol{\theta}_0 = \boldsymbol{\theta}_i$ and denoting the new value $\boldsymbol{\theta}_{i+1}$, we end up with the *Newton-Raphson method*

$$\boldsymbol{\theta}_{i+1} = \boldsymbol{\theta}_i - \boldsymbol{H}^{-1}(\boldsymbol{\theta}_i)\nabla_\theta S(\boldsymbol{\theta}_i), \tag{3.31}$$

which is an iterative method for finding $\boldsymbol{\theta}^*$, which is based on successive quadratic approximations. Other numerical procedures, e.g., the *Gauss-Newton method*, use the fact that the optimization is a least squares problem.

3.2.1.2 LS estimates for the general linear model

For the general linear model it is possible to find an explicit expression for the LS estimator. Considering the linear model (3.9) on page 33, the observations can be written

$$\begin{pmatrix} Y_1 \\ \vdots \\ Y_N \end{pmatrix} = \begin{pmatrix} \boldsymbol{x}_1^T \\ \vdots \\ \boldsymbol{x}_N^T \end{pmatrix} \boldsymbol{\theta} + \begin{pmatrix} \varepsilon_1 \\ \vdots \\ \varepsilon_N \end{pmatrix} \tag{3.32}$$

or

$$\boldsymbol{Y} = \boldsymbol{x\theta} + \boldsymbol{\varepsilon} \tag{3.33}$$

As assumed in the definition of the linear model (3.9), we have $\mathrm{E}[\boldsymbol{\varepsilon}] = \boldsymbol{0}$ and $\mathrm{Var}[\boldsymbol{\varepsilon}] = \mathrm{E}[\boldsymbol{\varepsilon\varepsilon}^T] = \sigma^2\boldsymbol{\Sigma}$, where $\boldsymbol{\Sigma} = [\Sigma_{ij}]$.

3.2.1.3 Unweighted LS estimates

For the case of no weights we have $\boldsymbol{\Sigma} = \boldsymbol{I}$.

THEOREM 3.1 (LS ESTIMATES)
Least squares estimates for $\boldsymbol{\theta}$ in the linear model (3.33) *are found by solving the normal equation*

$$\boldsymbol{x}^T\boldsymbol{x}\widehat{\boldsymbol{\theta}} = \boldsymbol{x}^T\boldsymbol{Y} \tag{3.34}$$

If $\boldsymbol{x}^T\boldsymbol{x}$ has full rank, we get

$$\widehat{\boldsymbol{\theta}} = (\boldsymbol{x}^T\boldsymbol{x})^{-1}\boldsymbol{x}^T\boldsymbol{Y} \tag{3.35}$$

Proof

$$\begin{aligned} S(\boldsymbol{\theta}) &= (\boldsymbol{Y} - \boldsymbol{x\theta})^T(\boldsymbol{Y} - \boldsymbol{x\theta}) \\ \nabla_\theta S(\boldsymbol{\theta}) &= -2\boldsymbol{x}^T(\boldsymbol{Y} - \boldsymbol{x\theta}) = \mathbf{0} \end{aligned}$$

i.e., $\boldsymbol{\theta}^*$ is found from $\boldsymbol{x}^T\boldsymbol{x\theta} = \boldsymbol{x}^T\boldsymbol{Y} \sim$ (3.34) and (3.35) follow directly. ∎

THEOREM 3.2 (PROPERTIES OF THE LS ESTIMATOR)
The LS estimator given by (3.35) *has the following properties,*

i) It is a linear function of the observations $\boldsymbol{Y}$.

ii) It is unbiased, i.e., $\mathrm{E}[\widehat{\boldsymbol{\theta}}] = \boldsymbol{\theta}$.

iii) $\mathrm{Var}[\widehat{\boldsymbol{\theta}}] = \mathrm{E}\left[(\widehat{\boldsymbol{\theta}} - \boldsymbol{\theta})(\widehat{\boldsymbol{\theta}} - \boldsymbol{\theta})^T\right] = \sigma^2(\boldsymbol{x}^T\boldsymbol{x})^{-1}$.

iv) $\widehat{\boldsymbol{\theta}}$ *is BLUE (Best Linear Unbiased Estimator), which means that it has the smallest variance among all estimators which is a linear function of the observations (the so-called Gauss-Markov theorem).*

Proof

i) From (3.35) we have that $\widehat{\boldsymbol{\theta}} = \boldsymbol{LY}$, where $\boldsymbol{L} = (\boldsymbol{x}^T\boldsymbol{x})^{-1}\boldsymbol{x}^T$

ii)

$$\begin{aligned} \mathrm{E}[\widehat{\boldsymbol{\theta}}] &= \mathrm{E}\left[(\boldsymbol{x}^T\boldsymbol{x})^{-1}\boldsymbol{x}^T\boldsymbol{Y}\right] \\ &= \mathrm{E}\left[(\boldsymbol{x}^T\boldsymbol{x})^{-1}\boldsymbol{x}^T(\boldsymbol{x\theta} + \boldsymbol{\varepsilon})\right] \\ &= (\boldsymbol{x}^T\boldsymbol{x})^{-1}(\boldsymbol{x}^T\boldsymbol{x})\boldsymbol{\theta} = \boldsymbol{\theta} \end{aligned} \tag{3.36}$$

iii) We have

$$\begin{aligned} \widehat{\boldsymbol{\theta}} - \boldsymbol{\theta} &= (\boldsymbol{x}^T\boldsymbol{x})^{-1}\boldsymbol{x}^T\boldsymbol{Y} - \boldsymbol{\theta} \\ &= (\boldsymbol{x}^T\boldsymbol{x})^{-1}\boldsymbol{x}^T(\boldsymbol{x\theta} + \boldsymbol{\varepsilon}) - \boldsymbol{\theta} \\ &= (\boldsymbol{x}^T\boldsymbol{x})^{-1}\boldsymbol{x}^T\boldsymbol{\varepsilon} \end{aligned}$$

which leads to

$$\begin{aligned} \mathrm{Var}[\widehat{\boldsymbol{\theta}}] &= \mathrm{E}\left[(\boldsymbol{x}^T\boldsymbol{x})^{-1}\boldsymbol{x}^T\boldsymbol{\varepsilon}\boldsymbol{\varepsilon}^T\boldsymbol{x}(\boldsymbol{x}^T\boldsymbol{x})^{-1}\right] \\ &= (\boldsymbol{x}^T\boldsymbol{x})^{-1}\boldsymbol{x}^T\,\mathrm{E}[\boldsymbol{\varepsilon}\boldsymbol{\varepsilon}^T]\boldsymbol{x}(\boldsymbol{x}^T\boldsymbol{x})^{-1} \\ &= \sigma^2(\boldsymbol{x}^T\boldsymbol{x})^{-1} \end{aligned}$$

iv) Omitted—cf. Kendall and Stuart (1983). ∎

Example 3.2 (Simple linear regression model)
For the stochastic process/random variable Y_t defined by

$$Y_t = \alpha + \beta x_t + \varepsilon_t$$

the observations $(Y_1, x_1), \ldots, (Y_N, x_N)$ are available. For all these observations the model is written

$$\begin{pmatrix} Y_1 \\ \vdots \\ Y_N \end{pmatrix} = \begin{pmatrix} 1 & x_1 \\ \vdots & \vdots \\ 1 & x_N \end{pmatrix} \begin{pmatrix} \alpha \\ \beta \end{pmatrix} + \begin{pmatrix} \varepsilon_1 \\ \vdots \\ \varepsilon_N \end{pmatrix}$$

According to (3.35) the LS estimator is

$$\widehat{\boldsymbol{\theta}} = \begin{pmatrix} \widehat{\alpha} \\ \widehat{\beta} \end{pmatrix} = (\boldsymbol{x}^T \boldsymbol{x})^{-1} \boldsymbol{x}^T \boldsymbol{Y} = \begin{pmatrix} N & \sum x_i \\ \sum x_i & \sum x_i^2 \end{pmatrix}^{-1} \begin{pmatrix} \sum Y_i \\ \sum x_i Y_i \end{pmatrix}$$

that is,

$$\begin{pmatrix} \widehat{\alpha} \\ \widehat{\beta} \end{pmatrix} = \begin{pmatrix} \dfrac{\sum x_i^2 \sum Y_i - \sum x_i \sum x_i Y_i}{N \sum x_i^2 - (\sum x_i)^2} \\ \dfrac{N \sum x_i Y_i - \sum x_i \sum Y_i}{N \sum x_i^2 - (\sum x_i)^2} \end{pmatrix}$$

The variance is

$$\operatorname{Var}[\widehat{\boldsymbol{\theta}}] = \sigma^2 (\boldsymbol{x}^T \boldsymbol{x})^{-1} = \frac{\sigma^2}{N \sum x_i^2 - (\sum x_i)^2} \begin{pmatrix} \sum x_i^2 & -\sum x_i \\ -\sum x_i & N \end{pmatrix}$$

3.2.1.4 Weighted LS estimates

Let us introduce weighted least squares estimates for $\boldsymbol{\theta}$ in the linear model

$$\boldsymbol{Y} = \boldsymbol{x}\boldsymbol{\theta} + \boldsymbol{\varepsilon} \tag{3.37}$$

Using $\mathrm{E}[\boldsymbol{\varepsilon}] = \mathbf{0}$ and $\operatorname{Var}[\boldsymbol{\varepsilon}] = \mathrm{E}[\boldsymbol{\varepsilon}\boldsymbol{\varepsilon}^T] = \sigma^2 \boldsymbol{\Sigma}$, where $\boldsymbol{\Sigma}$ is assumed known, the *weighted least squares estimates* for $\boldsymbol{\theta}$ are found by minimizing

$$S = (\boldsymbol{Y} - \boldsymbol{x}\boldsymbol{\theta})^T \boldsymbol{\Sigma}^{-1} (\boldsymbol{Y} - \boldsymbol{x}\boldsymbol{\theta}) \tag{3.38}$$

THEOREM 3.3 (WEIGHTED LS ESTIMATES)
The weighted least squares estimate for $\boldsymbol{\theta}$ in the linear model, is found by solving the normal equation

$$(\boldsymbol{x}^T \boldsymbol{\Sigma}^{-1} \boldsymbol{x}) \widehat{\boldsymbol{\theta}} = \boldsymbol{x}^T \boldsymbol{\Sigma}^{-1} \boldsymbol{Y} \tag{3.39}$$

If $\boldsymbol{x}^T \boldsymbol{\Sigma}^{-1} \boldsymbol{x}$ has full rank we get

$$\widehat{\boldsymbol{\theta}} = (\boldsymbol{x}^T \boldsymbol{\Sigma}^{-1} \boldsymbol{x})^{-1} \boldsymbol{x}^T \boldsymbol{\Sigma}^{-1} \boldsymbol{Y} \tag{3.40}$$

Proof Same as for Theorem 3.1 on page 36. However, when calculating the derivatives one should apply, for any symmetric $N \times N$-matrix, $\boldsymbol{A}$, it holds:

$$\begin{aligned}\boldsymbol{\alpha}^T \boldsymbol{A}\boldsymbol{\beta} &= \boldsymbol{\alpha}^T(\boldsymbol{\beta}^T \boldsymbol{A}^T)^T \\ &= ((\boldsymbol{\beta}^T \boldsymbol{A}^T)\boldsymbol{\alpha})^T \\ &= \boldsymbol{\beta}^T \boldsymbol{A}\boldsymbol{\alpha},\end{aligned}$$

when $\boldsymbol{\alpha}$ and $\boldsymbol{\beta}$ are N-dimensional vectors, i.e.,

$$S(\boldsymbol{\theta}) = (\boldsymbol{Y} - \boldsymbol{x}\boldsymbol{\theta})^T \boldsymbol{\Sigma}^{-1}(\boldsymbol{Y} - \boldsymbol{x}\boldsymbol{\theta}) \tag{3.41}$$

yields

$$\nabla_\theta S(\boldsymbol{\theta}) = -2\boldsymbol{x}^T \boldsymbol{\Sigma}^{-1}(\boldsymbol{Y} - \boldsymbol{x}\boldsymbol{\theta}) \tag{3.42}$$

From here the rest follows. ■

The variance of the estimator (3.40) is

$$\operatorname{Var}[\widehat{\boldsymbol{\theta}}] = \sigma^2(\boldsymbol{x}^T \boldsymbol{\Sigma}^{-1}\boldsymbol{x})^{-1} \tag{3.43}$$

Apart from this, the estimator has the properties mentioned in Theorem 3.2 on page 37.

It is also possible to find an estimator for the variance, σ^2.

THEOREM 3.4 (ESTIMATOR FOR THE VARIANCE)
An unbiased estimator for σ^2 is

$$\widehat{\sigma}^2 = \frac{S(\widehat{\boldsymbol{\theta}})}{N-p} = \frac{(\boldsymbol{Y} - \boldsymbol{x}\widehat{\boldsymbol{\theta}})^T \boldsymbol{\Sigma}^{-1}(\boldsymbol{Y} - \boldsymbol{x}\widehat{\boldsymbol{\theta}})}{N-p} \tag{3.44}$$

where N is the number of observations and p is the number of parameters.

Proof Omitted—see Kendall and Stuart (1983). ■

▸ **Remark 3.2**
A weighted least squares problem can be expressed as an unweighted least squares problem by a suitable transformation of the variables. The transformation is given by Theorem 2.5 on page 22, which shows that we can always find a linear transformation $\boldsymbol{T}$, determined by $\boldsymbol{\Sigma} = \boldsymbol{T}\boldsymbol{T}^T$, such that $\boldsymbol{\varepsilon}$ in (3.37) can be written as $\boldsymbol{\varepsilon} = \boldsymbol{T}\boldsymbol{e}$. That is,

$$\boldsymbol{Y} = \boldsymbol{x}\boldsymbol{\theta} + \boldsymbol{T}\boldsymbol{e}$$

where $\operatorname{Var}[\boldsymbol{e}] = \sigma^2 \boldsymbol{I}$. By multiplying both sides of the above equation by $\boldsymbol{T}^{-1}$, the weighted least squares problem is brought into an unweighted least squares problem. ◂

3.2.2 Maximum likelihood (ML) estimates

We shall now consider maximum likelihood estimators of the parameters $\boldsymbol{\theta}$ and σ^2 in the linear model $\boldsymbol{Y} = \boldsymbol{x}\boldsymbol{\theta} + \boldsymbol{\varepsilon}$. The ML estimates are based on an assumption of normality in $\boldsymbol{\varepsilon}$ and thus also in $\boldsymbol{Y}$.

THEOREM 3.5 (ML ESTIMATES)
Consider the N-dimensional random variable $\boldsymbol{Y}$ and let $\boldsymbol{Y} \in \mathrm{N}(\boldsymbol{x}\boldsymbol{\theta}, \sigma^2\boldsymbol{\Sigma})$ where $\boldsymbol{\Sigma}$ is assumed known. Then the maximum likelihood estimator *for $\boldsymbol{\theta}$ is equivalent to the least square estimator*

$$\widehat{\boldsymbol{\theta}} = (\boldsymbol{x}^T\boldsymbol{\Sigma}^{-1}\boldsymbol{x})^{-1}\boldsymbol{x}^T\boldsymbol{\Sigma}^{-1}\boldsymbol{Y} \tag{3.45}$$

Proof The density for $\boldsymbol{Y}$ is

$$f_Y(\boldsymbol{y}) = \frac{1}{\sqrt{(2\pi\sigma^2)^N \det \boldsymbol{\Sigma}}} \exp\left[-\frac{1}{2\sigma^2}(\boldsymbol{y} - \boldsymbol{x}\boldsymbol{\theta})^T\boldsymbol{\Sigma}^{-1}(\boldsymbol{y} - \boldsymbol{x}\boldsymbol{\theta})\right]$$

Since the likelihood function is equal to the simultaneous density for all the observations, the logarithm of the likelihood function, L, is

$$\begin{aligned}\log L(\boldsymbol{\theta}, \sigma^2; \boldsymbol{y}) &= -\frac{N}{2}\log(2\pi) - \frac{1}{2}\log(\det \boldsymbol{\Sigma}) - \frac{N}{2}\log(\sigma^2) \\ &\quad - \frac{1}{2\sigma^2}(\boldsymbol{y} - \boldsymbol{x}\boldsymbol{\theta})^T\Sigma^{-1}(\boldsymbol{y} - \boldsymbol{x}\boldsymbol{\theta}) \\ &= -\frac{N}{2}\log(\sigma^2) - \frac{1}{2\sigma^2}(\boldsymbol{y} - \boldsymbol{x}\boldsymbol{\theta})^T\boldsymbol{\Sigma}^{-1}(\boldsymbol{y} - \boldsymbol{x}\boldsymbol{\theta}) + c,\end{aligned} \tag{3.46}$$

where c is a constant. By differentiation with respect to $\boldsymbol{\theta}$ we obtain

$$\nabla_{\boldsymbol{\theta}} \log L(\boldsymbol{\theta}, \sigma^2; \boldsymbol{y}) = \frac{1}{\sigma^2}\boldsymbol{x}^T\boldsymbol{\Sigma}^{-1}(\boldsymbol{y} - \boldsymbol{x}\boldsymbol{\theta})$$

Setting $\nabla_{\boldsymbol{\theta}} \log L = 0$ implies that the maximum likelihood estimator (MLE) $\widehat{\boldsymbol{\theta}}$ is found by solving the normal equation

$$\boldsymbol{x}^T\boldsymbol{\Sigma}^{-1}\boldsymbol{x}\widehat{\boldsymbol{\theta}} = \boldsymbol{x}^T\boldsymbol{\Sigma}^{-1}\boldsymbol{Y}.$$

If $\boldsymbol{x}^T\boldsymbol{\Sigma}^{-1}\boldsymbol{x}$ has full rank, (3.45) is now obtained. ∎

THEOREM 3.6 (ML ESTIMATOR FOR THE VARIANCE)
The maximum likelihood estimator for σ^2 is given by

$$\widehat{\sigma}^2 = \frac{1}{N}(\boldsymbol{Y} - \boldsymbol{x}\widehat{\boldsymbol{\theta}})^T\boldsymbol{\Sigma}^{-1}(\boldsymbol{Y} - \boldsymbol{x}\widehat{\boldsymbol{\theta}}). \tag{3.47}$$

Proof From (3.46) it follows

$$\begin{aligned}\frac{\partial \log L(\widehat{\boldsymbol{\theta}}, \sigma^2; \boldsymbol{y})}{\partial \sigma^2} &= -\frac{N}{2}\frac{1}{\sigma^2} + \frac{1}{2\sigma^4}(\boldsymbol{y} - \boldsymbol{x}\widehat{\boldsymbol{\theta}})^T \boldsymbol{\Sigma}^{-1}(\boldsymbol{y} - \boldsymbol{x}\widehat{\boldsymbol{\theta}}) \\ &= -\frac{N}{2}\frac{1}{\sigma^4}\left[\sigma^2 - \frac{1}{N}(\boldsymbol{y} - \boldsymbol{x}\widehat{\boldsymbol{\theta}})^T \boldsymbol{\Sigma}^{-1}(\boldsymbol{y} - \boldsymbol{x}\widehat{\boldsymbol{\theta}})\right]\end{aligned}$$

Solving for σ, it can be seen that the estimate (3.47) implies that the partial derivative is 0. ∎

A comparison between (3.44) and (3.47) shows that the estimator given by (3.47) is biased.

THEOREM 3.7 (PROPERTIES OF THE ML ESTIMATOR)
The ML estimator given by (3.45) *has the following properties*

i) It is a linear function of the observations which now implies that it is normally distributed.

ii) It is unbiased, i.e., $\mathrm{E}[\widehat{\boldsymbol{\theta}}] = \boldsymbol{\theta}$.

iii) It has variance

$$\mathrm{Var}[\widehat{\boldsymbol{\theta}}] = \mathrm{E}[(\widehat{\boldsymbol{\theta}} - \boldsymbol{\theta})(\widehat{\boldsymbol{\theta}} - \boldsymbol{\theta})^T] = (\boldsymbol{x}^T \boldsymbol{\Sigma}^{-1} \boldsymbol{x})^{-1}\sigma^2.$$

iv) It is an efficient estimator.

Proof i), ii) and iii)—see the proof for Theorem 3.2 on page 37 and use that $\boldsymbol{Y}$ is now normally distributed. iv) omitted. ∎

Until now it has been assumed that $\boldsymbol{\Sigma}$ is known. *If* $\boldsymbol{\Sigma}$ *is unknown*, one possibility is to use the *relaxation algorithm* (Goodwin and Payne 1977, p. 49), which has the following steps.

Relaxation algorithm

a) Select a value for $\boldsymbol{\Sigma}$ (e.g., $\boldsymbol{\Sigma} = \boldsymbol{I}$).

b) Find the estimates for this value of $\boldsymbol{\Sigma}$, e.g., by solving the normal equations.

c) Consider the residuals $\{\widehat{\varepsilon}_t\}$ and calculate the correlation and variance structure of the residuals. Then select a new value for $\boldsymbol{\Sigma}$ which reflects that correlation and variance structure.

d) Stop if convergence; otherwise go to b).

Example 3.3 (Direct radiation in clear skies, part II)
We continue our analysis from Example 3.1 on page 32 with the suggested model which includes two parameters, θ_1 and θ_2. With observations for the direct radiation and the solar elevation, these parameters can be estimated, along with the variance σ^2, by using the maximum likelihood method. This implies that the loss function $S(\boldsymbol{\theta})$, as indicated in Definition 3.2 on page 34, is minimized with respect to $\boldsymbol{\theta}$, such that

$$\widehat{\boldsymbol{\theta}} = \arg\min_{\boldsymbol{\theta}} \left(\boldsymbol{Y} - \boldsymbol{f}(\boldsymbol{\theta})\right)^T \boldsymbol{\Sigma}^{-1} \left(\boldsymbol{Y} - \boldsymbol{f}(\boldsymbol{\theta})\right) \tag{3.48}$$

where for N observed values, the vectors of estimated and observed response are described by $\boldsymbol{f}(\boldsymbol{\theta}) = (f_1, \ldots, f_N)^T$ and $\boldsymbol{Y} = (Y_1, \ldots, Y_N)^T$, respectively. The vector $\boldsymbol{\theta} = (\theta_1, \theta_2)^T$ is the unknown parameters. Since the deterministic model is non-linear, there is no explicit solution to the minimization problem. Therefore an iterative procedure is applied and here the relaxation algorithm on the preceding page, along with the Newton-Raphson method on page 36, is used to find the estimates $\widehat{\boldsymbol{\theta}}$. The maximum likelihood estimator for the variance is similar to the linear case as indicated in (3.47), but for non-linear case it is

$$\widehat{\sigma}^2 = \frac{1}{N}\left(\boldsymbol{Y} - \boldsymbol{f}(\widehat{\boldsymbol{\theta}})\right)^T \boldsymbol{\Sigma}^{-1} \left(\boldsymbol{Y} - \boldsymbol{f}(\widehat{\boldsymbol{\theta}})\right). \tag{3.49}$$

To get reasonable estimates, the covariance matrix $\boldsymbol{\Sigma}$ has to be defined appropriately. Hence, both the variance and the correlation structure of ε_t must be described. By inspecting the trajectories of the direct radiation in Figure 3.1 on page 33, it is seen that time consecutive observations appear to be dependent. This indicates that the noise sequence $\{\varepsilon_t\}$ is strongly correlated in time. It is assumed that the correlation structure is an exponential decaying function of the time distance between two observations, i.e.,

$$\operatorname{Cor}\left[\varepsilon_{t_i}, \varepsilon_{t_j}\right] = \rho^{|t_i - t_j|} \tag{3.50}$$

where ρ is the hour-to-hour correlation. Furthermore, the residuals are assumed to be Gaussian distributed with mean zero and covariance matrix $\sigma^2\boldsymbol{\Sigma}$, where the $n \times n$-dimensional matrix $\boldsymbol{\Sigma}$, as suggested by (3.8) and (3.50), is given by

$$\{\Sigma_{ij}\} = \frac{\rho^{|t_i - t_j|}}{\sin(h_{t_i})\sin(h_{t_j})}. \tag{3.51}$$

If only the variance structure is considered, the matrix is given by

$$\{\Sigma_{ii}\} = \frac{1}{\sin^2(h_{t_i})}, \qquad \{\Sigma_{ij}\} = 0, \quad (i \neq j), \tag{3.52}$$

which indicates that deviations from $\boldsymbol{f}$ are assumed to be mutually independent, but the variance increases with decreasing solar elevation. If only the

Table 3.1: *The maximum likelihood estimates of the parameters in the model for direct radiation in clear skies with different covariance structure.*

Covariance given by	$\widehat{\theta}_1$ [W/m^2]	$\widehat{\theta}_2$ [deg^{-1}]	$\widehat{\sigma}^2_N$
$\boldsymbol{\Sigma} = \boldsymbol{I}$	798.6	0.0798	34631.8
Equation (3.52)	822.6	0.0706	3947.6
Equation (3.53)	827.8	0.0551	25180.1
Equation (3.51)	842.3	0.0614	2302.1

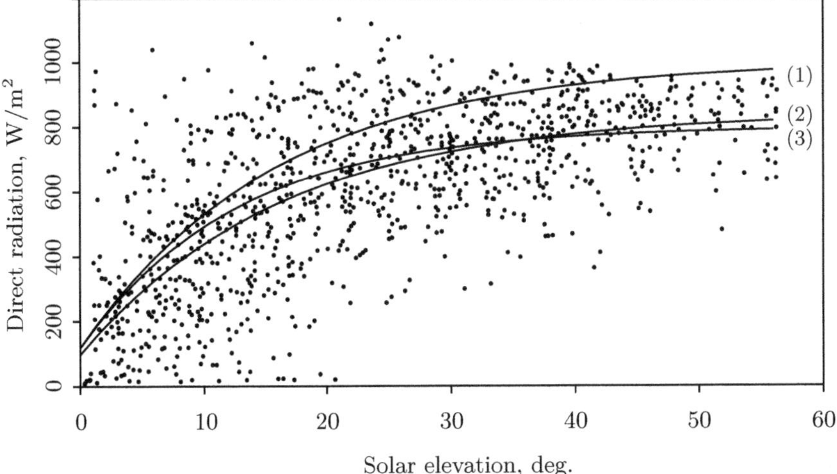

Figure 3.2: *Direct radiaton in clear skies. Final model (curve 2), model without consideration of variance and correlation structures (curve 3) and the model based on the coefficients suggested by Paltridge and Platt (1976) in Example 3.1.*

correlation structure is considered in the model, the matrix becomes

$$\{\Sigma_{ij}\} = \rho^{|t_i - t_j|} \tag{3.53}$$

and hence, the variance of the deviations from the model is assumed to be independent of the solar elevation but with strong correlation between hourly values. If no consideration is taken of either the variance or the correlation structure, the covariance structure is $\boldsymbol{\Sigma} = \boldsymbol{I}$.

The ML estimates $\widehat{\boldsymbol{\theta}}$ and $\widehat{\sigma}^2$ for the four different covariance matrices are shown in Table 3.1. For the model considering both variance and correlation structure, the estimated hour-to-hour correlation is $\widehat{\rho} = 0.859$, which indicates high hourly correlation in the deviation from the deterministic model.

In Figure 3.2 the model in equation (3.7), with two different covariance structures (curves 2 and 3), is compared with the model with the suggested coefficients from Example 3.1 (curve 1). A large difference is detected between the models, about 150 W/m^2, since the estimates for the models in Table 3.1 are very different from the suggested coefficients. The two models, one with the covariance structure as the identity matrix and the other one as Equation (3.51) implies, are very similar although the largest difference is about 50 Wm^2 occuring near solar elevation of 12°.

3.3 Prediction

Fundamentally, prediction is based on the following important theorem.

THEOREM 3.8
Let Y be a random variable with mean $\mathrm{E}[Y]$, *then the minimum of* $\mathrm{E}[(Y-a)^2]$ *is obtained for* $a = \mathrm{E}[Y]$.

Proof

$$\begin{aligned}
\mathrm{E}[(Y-a)^2] &= \mathrm{E}\left[(Y-\mathrm{E}[Y]+\mathrm{E}[Y]-a)^2\right] \\
&= \mathrm{E}\left[(Y-\mathrm{E}[Y])^2\right] + (\mathrm{E}[Y]-a)^2 \\
&\quad + 2\,\mathrm{E}\left[Y-\mathrm{E}[Y]\right](\mathrm{E}[Y]-a) \\
&= \mathrm{Var}[Y] + (\mathrm{E}[Y]-a)^2 \geq \mathrm{Var}[Y]
\end{aligned}$$

The equal sign holds for $a = \mathrm{E}[Y]$ and the result follows. ∎

This leads to an expression for the optimal prediction.

THEOREM 3.9 (OPTIMAL PREDICTION)
It holds that

$$\min_g \mathrm{E}\left[(Y-g(X))^2|X=x\right] = \mathrm{E}\left[(Y-g^*(x))^2|X=x\right] \tag{3.54}$$

where $g^*(x) = \mathrm{E}[Y|X=x]$.

Proof Follows immediately from Theorem 3.8. ∎

Theorem 3.9 is *very* important in time series analysis. The theorem says that if the criterion is to minimize the expected value of the squared prediction error, then *the optimal prediction is equal to the conditional expectation.*

3.3.1 Prediction in the general linear model

3.3.1.1 Known parameters

Consider the linear model with *known parameters.*

$$Y_t = \boldsymbol{X}_t^T \boldsymbol{\theta} + \varepsilon_t \tag{3.55}$$

where $\mathrm{E}[\varepsilon_t] = 0$ and $\mathrm{Var}[\varepsilon_t] = \sigma^2$ (i.e., constant). The prediction for a future value $Y_{t+\ell}$ given the independent variable $\boldsymbol{X}_{t+\ell} = \boldsymbol{x}_{t+\ell}$ is

$$\widehat{Y}_{t+\ell} = \mathrm{E}[Y_{t+\ell} | \boldsymbol{X}_{t+\ell} = \boldsymbol{x}_{t+\ell}] = \boldsymbol{x}_{t+\ell}^T \boldsymbol{\theta} \tag{3.56}$$

$$\mathrm{Var}[Y_{t+\ell} - \widehat{Y}_{t+\ell}] = \mathrm{Var}[\varepsilon_{t+\ell}] = \sigma^2 \tag{3.57}$$

Compare with Example 2.2.

3.3.1.2 Unknown parameters

It is often the case that the parameters are unknown but there may exist some estimates of θ. If the estimates are found by the LS estimator

$$\widehat{\boldsymbol{\theta}} = (\boldsymbol{x}^T \boldsymbol{x})^{-1} \boldsymbol{x}^T \boldsymbol{Y} \tag{3.58}$$

then the variance of the prediction error can be stated.

THEOREM 3.10 (PREDICTION IN THE GENERAL LINEAR MODEL)
Assume that the unknown parameters $\boldsymbol{\theta}$ in the linear model are estimated by using the least squares method (3.58), *then* the minimum variance prediction *is*

$$\widehat{Y}_{t+\ell} = \mathrm{E}[Y_{t+\ell} | \boldsymbol{X}_{t+\ell} = \boldsymbol{x}_{t+\ell}] = \boldsymbol{x}_{t+\ell}^T \widehat{\boldsymbol{\theta}} \tag{3.59}$$

The variance of the prediction error $e_{t+\ell} = Y_{t+\ell} - \widehat{Y}_{t+\ell}$ *becomes*

$$\mathrm{Var}[e_{t+\ell}] = \mathrm{Var}[Y_{t+\ell} - \widehat{Y}_{t+\ell}] = \sigma^2 [1 + \boldsymbol{x}_{t+\ell}^T (\boldsymbol{x}^T \boldsymbol{x})^{-1} \boldsymbol{x}_{t+\ell}] \tag{3.60}$$

Proof (3.59) follows immediately since the mean of the prediction error is 0.

$$\begin{aligned}
\mathrm{Var}[Y_{t+\ell} - \widehat{Y}_{t+\ell}] &= \mathrm{Var}[\boldsymbol{x}_{t+\ell}^T \boldsymbol{\theta} + \varepsilon_{t+\ell} - \boldsymbol{x}_{t+\ell}^T \widehat{\boldsymbol{\theta}}] \\
&= \mathrm{Var}[\boldsymbol{x}_{t+\ell}^T (\boldsymbol{\theta} - \widehat{\boldsymbol{\theta}}) + \varepsilon_{t+\ell}] \\
&= \boldsymbol{x}_{t+\ell}^T \mathrm{Var}[\widehat{\boldsymbol{\theta}}] \boldsymbol{x}_{t+\ell} + \sigma^2 + 2\,\mathrm{Cov}[\boldsymbol{x}_{t+\ell}^T (\boldsymbol{\theta} - \widehat{\boldsymbol{\theta}}), \varepsilon_{t+\ell}] \\
&= \sigma^2 + \boldsymbol{x}_{t+\ell}^T \mathrm{Var}[\widehat{\theta}] \boldsymbol{x}_{t+\ell}.
\end{aligned}$$

The result follows from the fact that $\mathrm{Var}[\widehat{\boldsymbol{\theta}}] = \sigma^2 (\boldsymbol{x}^T \boldsymbol{x})^{-1}$. ∎

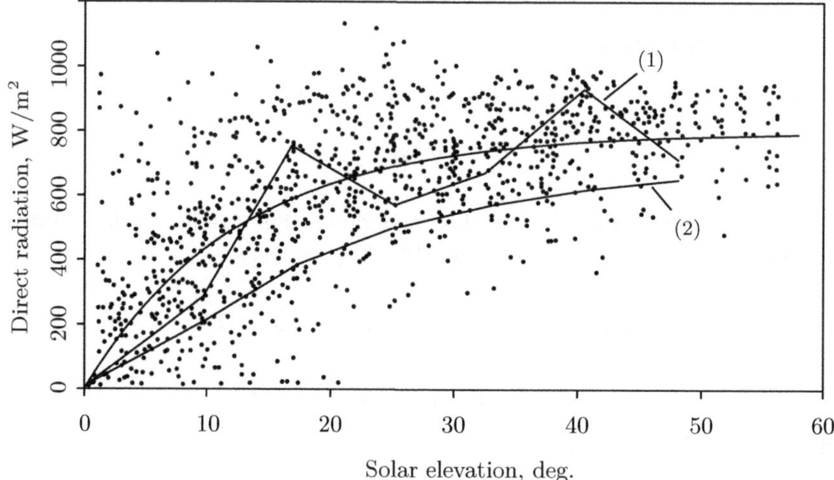

Figure 3.3: *Simulation of direct radiation in clear skies. Curve 1 has hour-to-hour correlation equal to zero; curve 2 has hour-to-hour correlation equal to* 0.859.

If we use an estimate for σ^2 (see Theorem 3.4 on page 39 applying $\boldsymbol{\Sigma} = \boldsymbol{I}$) then a $100(1-\alpha)\%$ *confidence interval for the future* $Y_{t+\ell}$ is given as

$$\begin{aligned}&\widehat{Y}_{t+\ell} \pm \mathrm{t}_{\alpha/2}(N-p)\sqrt{\mathrm{Var}[e_{t+\ell}]}\\ &\quad = \widehat{Y}_{t+\ell} \pm \mathrm{t}_{\alpha/2}(N-p)\widehat{\sigma}\sqrt{1 + \boldsymbol{x}_{t+\ell}^T(\boldsymbol{x}^T\boldsymbol{x})^{-1}\boldsymbol{x}_{t+\ell}}\end{aligned} \tag{3.61}$$

where $\mathrm{t}_{\alpha/2}(N-p)$ is the $\alpha/2$ quantile in the t distribution with $(N-p)$ degrees of freedom and N is the number of observations.

A confidence interval for a future value is also called a *prediction interval.*

Example 3.4 (Direct radiation in clear skies, part III)
Now it is illustrated that also for applications, such as simulation and prediction, it is also crucial to take into account both the variance and the correlation structure.

Figure 3.3 illustrates simulated direct radiation for two different covariance structures where the solid, long curve indicates the deterministic part of the model. Curve 1 is based on the covariance given in (3.52), i.e., with no consideration of the time correlation, and it is seen to be fluctuating around the deterministic part. This is clearly in conflict with the two specific days in Figure 3.1 on page 33. By considering the estimated correlation in $\boldsymbol{\Sigma}$ given by (3.51), the model is more able to reproduce the sequence of successive observations, as curve 2 indicates.

Another application showing importance of the covariance structure is prediction. In Figure 3.4, again the long curve is the deterministic part of

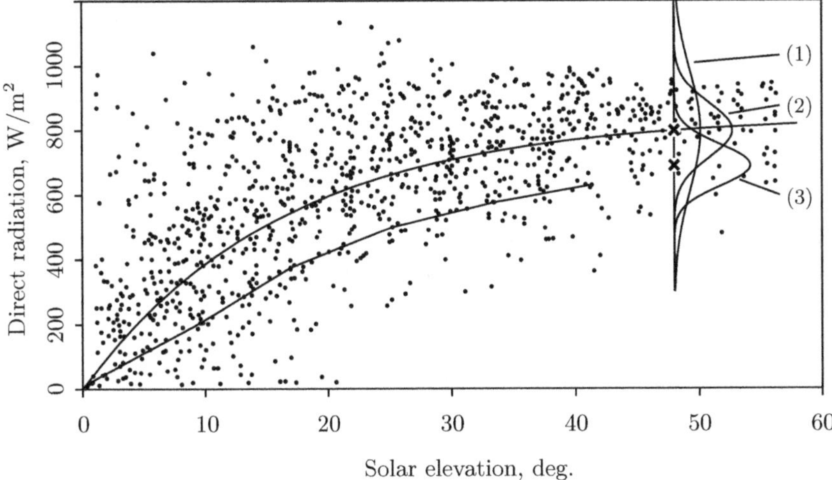

Figure 3.4: *Forecasting of direct radiation in clear skies. Curve 1 has no covariance structure; curve 2 only variance structure; and curve 3 both variance and correlation structure.*

the stochastic model and the short curve indicates observed direct radiation up to $h = 42°$, corresponding to 11 a.m. Now a forecast of the direct radiation is wanted at $h = 48°$, at 12 noon. Curves 1, 2, and 3 are the unit covariance matrix and the structures in Equations (3.52) and (3.51), respectively. Both curves, excluding the correlation structure, give forecasts equal to the point on the deterministic part corresponding to $h = 48°$, while the forecast considering the hour-to-hour correlation is found below the deterministic curve, as expected. The forecasts without consideration of the correlation structure are obviously about $108\,\text{W/m}^2$ too large. In Figure 3.4 the conditional distribution for the future values are depicted, corresponding to estimated variances in Table 3.1 on page 43. Comparing the curves shows the importance of the variance and correlation structure in forecasting.

This case study, of direct radiation in clear skies depending only on solar radiation, has revealed the importance of various methods given to this point in Chapter 3. The variance and correlation structures are shown to be very important in order to get correct values when we are making simulations and predictions.

3.4 Regression and exponential smoothing

In Section 3.2 we considered regression models relating the dependent variable Y to a set of independent variables $X_1, \ldots, X_p$. Such a model can be used for predicting the dependent variable Y if future values for the independent variables are specified. Another possibility, which we shall consider in this section, is that only observations of the dependent variable are available. In this situation both the modeling and the prediction have to be based on observations of Y only. Therefore, we consider models of the form

$$Y_t = f(t; \boldsymbol{\theta}) + \varepsilon_t \,. \tag{3.62}$$

Let us first introduce the relation between an exponential weighting together with the least squares method and *simple exponential smoothing* by considering a very simple model.

3.4.1 Predictions in the constant mean model

A very simple model belonging to the class of models in (3.62) is *the constant mean model*,

$$Y_t = \alpha + \varepsilon_t \tag{3.63}$$

where α is the constant and $\{\varepsilon_t\}$ is a sequence of independent, identically distributed (i.i.d) random variables with $\mathrm{E}[\varepsilon_t] = 0$ and $\mathrm{Var}[\varepsilon_t] = \sigma^2$. A process $\{\varepsilon_t\}$ with these characteristics is called *white noise* or the *innovation process.*

Given the observations $Y_1, \dots, Y_N$ the least squares estimator for α is

$$\widehat{\alpha} = \overline{Y} = \frac{1}{N}\sum_{t=1}^{N} Y_t \,. \tag{3.64}$$

We now introduce the step prediction $\widehat{Y}_{N+\ell|N}$ as the *prediction of* $Y_{N+\ell}$ *given the observations at time origin* N. Using Theorem 3.10 on page 45 the prediction is

$$\widehat{Y}_{N+\ell|N} = \widehat{\alpha} = \frac{1}{N}\sum_{t=1}^{N} Y_t. \tag{3.65}$$

The variance of the prediction error is

$$\mathrm{Var}\left[Y_{N+\ell} - \widehat{Y}_{N+\ell|N}\right] = \sigma^2\left(1 + \frac{1}{N}\right), \tag{3.66}$$

which is also found by using Theorem 3.10. Thus, the variance of the prediction error depends on the number of observations used for estimating α, and for $N \to \infty$ the variance approaches

$$\mathrm{Var}\left[Y_{N+\ell} - \widehat{Y}_{N+\ell|N}\right] \to \sigma^2.$$

Most frequently only the one-step prediction, i.e., $\widehat{Y}_{N+1|N}$, is of interest. Given the new observation Y_{N+1}, we are interested in the new one-step prediction, i.e., $\widehat{Y}_{N+2|N+1}$, which can be found by using Theorem 3.10. In many cases, however, it is much easier to use the updating equations given in the following theorem.

THEOREM 3.11 (UPDATING FORECASTS)
In the constant mean model (3.63), *the one-step prediction can be updated as*

$$\widehat{Y}_{N+2|N+1} = \frac{N}{N+1}\widehat{Y}_{N+1|N} + \frac{1}{N+1}Y_{N+1} \tag{3.67}$$

or

$$\widehat{Y}_{N+2|N+1} = \widehat{Y}_{N+1|N} + \frac{1}{N+1}(Y_{N+1} - \widehat{Y}_{N+1|N}) \tag{3.68}$$

Proof Using (3.64) and (3.65) at the time origin $N+1$:

$$\widehat{Y}_{N+2|N+1} = \frac{1}{N+1}\sum_{i=1}^{N+1} Y_i = \frac{1}{N+1}(Y_{N+1} + N\widehat{Y}_{N+1|N}) \quad \sim (3.67)$$

$$\begin{aligned}\widehat{Y}_{N+2|N+1} &= \frac{1}{N+1}(Y_{N+1} + N\widehat{Y}_{N+1|N})\\ &= \frac{1}{N+1}(Y_{N+1} - \widehat{Y}_{N+1|N} + (N+1)\widehat{Y}_{N+1|N})\\ &= \widehat{Y}_{N+1|N} + \frac{1}{N+1}(Y_{N+1} - \widehat{Y}_{N+1|N}) \quad \sim (3.68)\end{aligned}$$

■

▸ **Remark 3.3**
From (3.65) it is seen that the predictor $\widehat{Y}_{N+\ell|N}$ is equal to the estimate at time origin N, say $\widehat{\alpha}_N$, of the mean. This implies that (3.67) can be interpreted as the *recursive estimation*

$$\widehat{\alpha}_{N+1} = \frac{N}{N+1}\widehat{\alpha}_N + \frac{1}{1+N}Y_{N+1} = \widehat{\alpha}_N + \frac{1}{1+N}(Y_{N+1} - \widehat{\alpha}_N) \tag{3.69}$$

of the parameter α in the constant mean model. Note that the influence of new observations decreases as the number of observations increases. ◂

The equations above indicate that the one-step prediction at time origin $N+1$ can be expressed as a linear combination of the prediction from origin N and the most recent observation Y_{N+1} (3.67) or as a linear combination of the one-step prediction at the time origin t and the prediction error $(Y_{N+1} - \widehat{Y}_{N+1|N})$ (3.68). For the computation of the new forecast it is important that only the last observation and the most recent prediction error have to be stored. Since the mean α in the model is assumed constant, each observation contributes equally to the predictions.

3.4.2 Locally constant mean model and simple exponential smoothing

In practice the assumption of a global constant mean, as in the previous section, is often too restrictive. It might be obvious to allow for a slow variation in time of the mean. Heuristically, in such cases it would be more reasonable to give more weight to the most recent observation and less weight to the observations in the past. Let us consider weights which decrease geometrically with the age of the observations. Then the prediction of $Y_{N+\ell}$ at the time origin N is

$$\widehat{Y}_{N+\ell|N} = c \sum_{j=0}^{N-1} \lambda^j Y_{N-j} = c[Y_N + \lambda Y_{N-1} + \cdots + \lambda^{N-1} Y_1] \tag{3.70}$$

The constant λ ($|\lambda| < 1$) is called *the forgetting factor* or *the discount coefficient.* In Brown (1963) it is mentioned that a value between 0.7 and 0.95 will most often be reasonable. The normalizing constant c is chosen such that the sum of the weights is 1. Since

$$c[1 + \lambda + \cdots + \lambda^{N-1}] = \frac{c(1-\lambda^N)}{(1-\lambda)}, \tag{3.71}$$

the normalization constant is

$$c = \frac{1-\lambda}{1-\lambda^N}. \tag{3.72}$$

If N is large then $c \simeq (1-\lambda)$ and we obtain *an exponential weighting* of the past observations, i.e.,

$$\widehat{Y}_{N+\ell|N} = (1-\lambda) \sum_{j\geq 0} \lambda^j Y_{N-j}. \tag{3.73}$$

Successive one-step predictions are calculated by using

$$\begin{aligned}\widehat{Y}_{N+2|N+1} &= (1-\lambda) \sum_{j\geq 0} \lambda^j Y_{N+1-j} \\ &= (1-\lambda) Y_{N+1} + \lambda(1-\lambda) \sum_{j\geq 0} \lambda^j Y_{N-j} \\ &= (1-\lambda) Y_{N+1} + \lambda \widehat{Y}_{N+1|N} \end{aligned} \tag{3.74}$$

It is clearly seen that these results can be generalized to successive ℓ-step predictions, i.e.,

$$\widehat{Y}_{N+\ell+1|N+1} = (1-\lambda) Y_{N+1} + \lambda \widehat{Y}_{N+\ell|N} \tag{3.75}$$

The equation for the successive one-step prediction leads to the following definition.

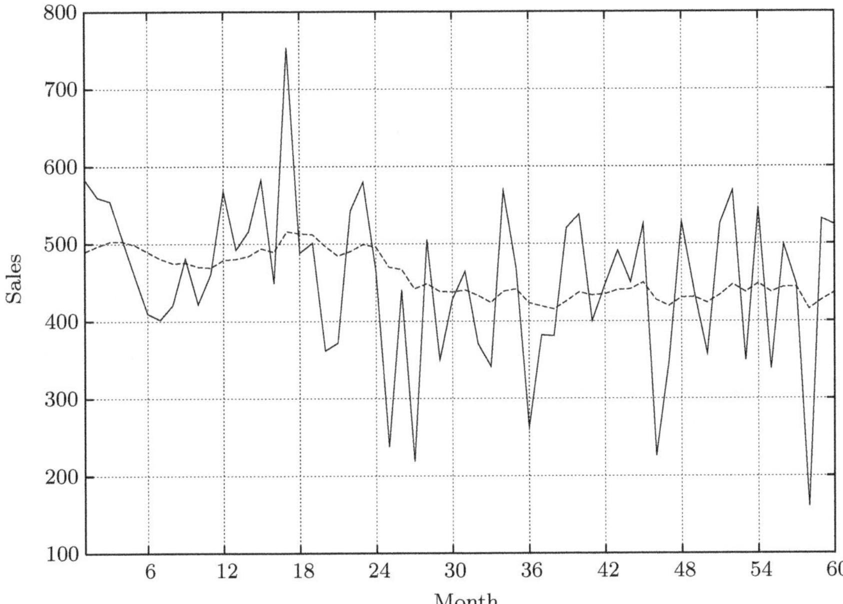

Figure 3.5: *Predicted sales figures using simple exponential smoothing ($\lambda = 0.9$).*

DEFINITION 3.3 (SIMPLE EXPONENTIAL SMOOTHING)
The sequence S_N defined as

$$S_N = (1 - \lambda)Y_N + \lambda S_{N-1} \tag{3.76}$$

is called *simple exponential smoothing* or *first order exponential smoothing.*

In the time series literature a *smoothing constant* $\alpha = (1-\lambda)$ is often considered instead of λ.

Example 3.5 (Simple exponential smoothing)
A Swedish company has used exponential smoothing for predicting the monthly sales figures for their product. The sales figures for a five-year period and the predicted values are shown in Figure 3.5. A forgetting factor of $\lambda = 0.9$ is used.

3.4.2.1 Choice of initial value

As an initial estimate for S_0 in (3.76), the simple arithmetic average of some recent historical data—or the first part of the considered series—can be used.

The influence of S_0 will decay rapidly since

$$S_N = (1-\lambda)[Y_N + \lambda Y_{N-1} + \cdots + \lambda^{N-1} Y_1] + \lambda^N S_0 \tag{3.77}$$

Alternatively the first observation can be used as the initial value. This initial value will be most reasonable if a level changes rapidly; see Makridakis and Wheelwright (1978).

3.4.2.2 Choice of smoothing constant

The smoothing constant $\alpha = (1-\lambda)$ determines to which extent the past observations influence the prediction. A small value for α results in a slow response to changes in the mean value of the process. A high value results in a rapid response to changes in the level, but also a high sensitivity against random irregular observations.

Let us consider how the smoothing constant α can be chosen. Given the observations $Y_1, \dots, Y_N$ and a value α, we can calculate the *one-step prediction error*,

$$e_t(\alpha) = Y_t - \widehat{Y}_{t|t-1}(\alpha) \tag{3.78}$$

and the sum of the squared one-step prediction errors (SSE),

$$S(\alpha) = \sum_{t=1}^{N} e_t^2(\alpha) \tag{3.79}$$

The smoothing constant which minimizes $S(\alpha)$ is then used. For this minimization a numerical procedure such as the Newton-Raphson procedure, see (3.31) on page 36, can be used.

3.4.3 Prediction in trend models

In this section the class of trend models is introduced. A trend model is a regression model in which certain functions of time are taken as independent variables.

A *trend model* is a regression model of the form

$$Y_{N+j} = \boldsymbol{f}^T(j)\boldsymbol{\theta} + \varepsilon_{N+j} \tag{3.80}$$

where $\boldsymbol{f}(j) = (f_1(j), \dots, f_p(j))^T$ is a vector of known *forecast functions* and $\boldsymbol{\theta} = (\theta_1, \dots, \theta_p)^T$ is a vector of parameters. The noise component, $\{\varepsilon_t, t = 0, \pm 1, \dots\}$, is a sequence of mutually independent, identically distributed random variables with $\mathrm{E}[\varepsilon_t] = 0$ and $\mathrm{Var}[\varepsilon_t] = \sigma^2$ (white noise).

Furthermore, it is assumed that $\boldsymbol{f}(j)$ satisfies the difference equation

$$\boldsymbol{f}(j+1) = \boldsymbol{L}\boldsymbol{f}(j), \quad \text{[start value } \boldsymbol{f}(0)] \tag{3.81}$$

where the *transition matrix* $\boldsymbol{L}$, is a fixed $p \times p$ nonsingular matrix.

For example, exponentials, polynomials, harmonic functions, and linear combinations of these functions satisfy (3.81).

▸ **Remark 3.4**

Equation (3.80) is a linear model in the class of the regression models given by (3.62) on page 48, which contain functions of time as independent variables. Note that the forecast functions are defined relative to time origin N. This parameterization turns out to be more convenient for updating and for making predictions. ◂

Let us introduce some important examples of trend models.

a) Constant mean model

$$Y_{N+j} = \theta_0 + \varepsilon_{N+j}, \tag{3.82}$$

is written in the form (3.80) by introducing $\boldsymbol{f}(j) = 1$, yielding $\boldsymbol{L} = \boldsymbol{f}(0) = 1$.

b) Linear trend model

$$Y_{N+j} = \theta_0 + \theta_1 j + \varepsilon_{N+j} \tag{3.83}$$

For this model, the form in (3.80) is obtained by introducing $\boldsymbol{f}(j) = (1, j)^T$. The transition matrix and the initial value of $\boldsymbol{f}$ are

$$\boldsymbol{L} = \begin{pmatrix} 1 & 0 \\ 1 & 1 \end{pmatrix}, \qquad \boldsymbol{f}(0) = \begin{pmatrix} 1 \\ 0 \end{pmatrix}.$$

c) Quadratic trend model

$$Y_{N+j} = \theta_0 + \theta_1 j + \theta_2 \frac{j^2}{2} + \varepsilon_{N+j} \tag{3.84}$$

Here, the form in (3.80) is obtained if $\boldsymbol{f}(j) = (1, j, j^2/2)^T$.

$$\boldsymbol{L} = \begin{pmatrix} 1 & 0 & 0 \\ 1 & 1 & 0 \\ 1/2 & 1 & 1 \end{pmatrix}, \quad \boldsymbol{f}(0) = \begin{pmatrix} 1 \\ 0 \\ 0 \end{pmatrix}.$$

The factor $j^2/2$ instead of j^2 yields a simpler difference equation.

d) k'th order polynomial trend

$$Y_{N+j} = \theta_0 + \theta_1 j + \theta_2 \frac{j^2}{2} + \cdots + \theta_k \frac{j^k}{k!} + \varepsilon_{N+j} \tag{3.85}$$

Here, we have $\boldsymbol{f}(j) = (1, j, \frac{j^2}{2}, \ldots, \frac{j^k}{k!})^T$ and the matrix $\boldsymbol{L}$ becomes

$$\boldsymbol{L} = \begin{pmatrix} 1 & 0 & \cdots & 0 \\ 1 & 1 & \cdots & 0 \\ 1/2 & 1 & \cdots & 0 \\ \vdots & \vdots & & \vdots \\ 1/k! & 1/(k-1)! & & 1 \end{pmatrix}, \qquad \boldsymbol{f}(0) = \begin{pmatrix} 1 \\ 0 \\ 0 \\ \vdots \\ 0 \end{pmatrix}.$$

$\boldsymbol{L}$ is thus a lower triangular matrix, having $\{L_{ij}\} = 1/(i-j)!$ for $i \geq j$.

e) Harmonic model with the period p

$$Y_{N+j} = \theta_0 + \theta_1 \sin\left(\frac{2\pi}{p}\right) j + \theta_2 \cos\left(\frac{2\pi}{p}\right) j + \varepsilon_{N+j} \tag{3.86}$$

i.e., $f(j) = (1, \sin\left(\frac{2\pi}{p}\right) j, \cos\left(\frac{2\pi}{p}\right) j)^T$ and

$$\boldsymbol{L} = \begin{pmatrix} 1 & 0 & 0 \\ 0 & \cos\left(\frac{2\pi}{p}\right) & \sin\left(\frac{2\pi}{p}\right) \\ 0 & -\sin\left(\frac{2\pi}{p}\right) & \cos\left(\frac{2\pi}{p}\right) \end{pmatrix}, \qquad \boldsymbol{f}(0) = \begin{pmatrix} 1 \\ 0 \\ 1 \end{pmatrix}.$$

The matrix $\boldsymbol{L}$ is derived from

$$\sin\left(\frac{2\pi}{p}\right)(j+1) = \cos\left(\frac{2\pi}{p}\right)\sin\left(\frac{2\pi}{p}\right) j + \sin\left(\frac{2\pi}{p}\right)\cos\left(\frac{2\pi}{p}\right) j$$
$$\cos\left(\frac{2\pi}{p}\right)(j+1) = -\sin\left(\frac{2\pi}{p}\right)\sin\left(\frac{2\pi}{p}\right) j + \cos\left(\frac{2\pi}{p}\right)\cos\left(\frac{2\pi}{p}\right) j$$

3.4.3.1 Estimation

Given the observations $Y_1, \ldots, Y_N$, the unknown parameters $\boldsymbol{\theta}$ can be estimated by the LS method. All N observations are written

$$\begin{pmatrix} Y_1 \\ Y_2 \\ \vdots \\ Y_N \end{pmatrix} = \begin{pmatrix} \boldsymbol{f}^T(-N+1) \\ \boldsymbol{f}^T(-N+2) \\ \vdots \\ \boldsymbol{f}^T(0) \end{pmatrix} \boldsymbol{\theta} + \begin{pmatrix} \varepsilon_1 \\ \varepsilon_2 \\ \vdots \\ \varepsilon_N \end{pmatrix} \tag{3.87}$$

or

$$\boldsymbol{Y} = \boldsymbol{x}_N \boldsymbol{\theta} + \varepsilon \tag{3.88}$$

Using Theorem 3.1 on page 36, the LS estimate is

$$\widehat{\boldsymbol{\theta}}_N = (\boldsymbol{x}_N^T \boldsymbol{x}_N)^{-1} \boldsymbol{x}_N^T \boldsymbol{Y}$$

or

$$\widehat{\boldsymbol{\theta}}_N = \boldsymbol{F}_N^{-1} \boldsymbol{h}_N \tag{3.89}$$

where $\boldsymbol{F}_N = \boldsymbol{x}_N^T \boldsymbol{x}_N$ and $\boldsymbol{h}_N = \boldsymbol{x}_N^T \boldsymbol{Y}$. The index N is introduced to stress that the estimate is based on the N observations, available at time N.

3.4.3.2 Prediction

Using Theorem 3.10 on page 45, the prediction $\widehat{Y}_{N+\ell|N}$, i.e., the *prediction of* $Y_{N+\ell}$ given the observations at time origin N, becomes

$$\widehat{Y}_{N+\ell|N} = \boldsymbol{f}^T(\ell)\widehat{\boldsymbol{\theta}}_N \tag{3.90}$$

and the variance of the prediction error, $e_N(\ell) = Y_{N+\ell} - \widehat{Y}_{N+\ell|N}$ is

$$\mathrm{Var}[e_N(\ell)] = \sigma^2 \left[1 + \boldsymbol{f}^T(\ell)\boldsymbol{F}_N^{-1}\boldsymbol{f}(\ell)\right] \tag{3.91}$$

Given an estimate $\widehat{\sigma}^2$ of σ^2, which can be obtained by using (3.44) on page 39, a $100(1-\alpha)\%$ *prediction interval* for the future value $Y_{N+\ell}$ is

$$\begin{aligned} &\widehat{Y}_{N+\ell|N} \pm \mathrm{t}_{\alpha/2}(N-p)\sqrt{\mathrm{Var}[e_N(\ell)]} \\ &\quad = \widehat{Y}_{N+\ell|N} \pm \mathrm{t}_{\alpha/2}(N-p)\widehat{\sigma}\sqrt{1 + \boldsymbol{f}^T(\ell)\boldsymbol{F}_N^{-1}\boldsymbol{f}(\ell)} \end{aligned} \tag{3.92}$$

3.4.3.3 Updating

When the next observation Y_{N+1} is available, the estimate can be updated by using the following theorem.

THEOREM 3.12 (UPDATING OF PARAMETERS IN TREND MODELS)
Given the next observation Y_{N+1}, the parameter estimate $\widehat{\boldsymbol{\theta}}_{N+1}$ is written

$$\widehat{\boldsymbol{\theta}}_{N+1} = \boldsymbol{F}_{N+1}^{-1}\boldsymbol{h}_{N+1} \tag{3.93}$$

where the updating formulas for $\boldsymbol{F}$ and $\boldsymbol{h}$ are

$$\boldsymbol{F}_{N+1} = \boldsymbol{F}_N + \boldsymbol{f}(-N)\boldsymbol{f}^T(-N) \tag{3.94}$$

and

$$\boldsymbol{h}_{N+1} = \boldsymbol{L}^{-1}\boldsymbol{h}_N + \boldsymbol{f}(0)Y_{N+1} \tag{3.95}$$

Proof (3.94) is a result of the definition of $\boldsymbol{F}$, and (3.95) follows from

$$\begin{aligned} \boldsymbol{h}_{N+1} &= \sum_{j=0}^{N} \boldsymbol{f}(-j)Y_{N+1-j} \\ &= \boldsymbol{f}(0)Y_{N+1} + \sum_{j=0}^{N-1} \boldsymbol{f}(-j-1)Y_{N-j} \\ &= \boldsymbol{f}(0)Y_{N+1} + \boldsymbol{L}^{-1}\sum_{j=0}^{N-1} \boldsymbol{f}(-j)Y_{N-j} \\ &= \boldsymbol{f}(0)Y_{N+1} + \boldsymbol{L}^{-1}\boldsymbol{h}_N \end{aligned}$$

where the difference equation $\boldsymbol{f}(j+1) = \boldsymbol{L}\boldsymbol{f}(j)$ is used. ■

3.4.4 Local trend and exponential smoothing

In the previous section *global* trend models were considered, i.e., models where the parameter $\boldsymbol{\theta}$ is constant in time, and each observation (recent or past) had the same weight. However, it is useful, as in Section 3.4.2, to allow for some time variation of the parameter $\boldsymbol{\theta}$. This means that observations in the past should be given less weight than recent observations in the least squares criterion. In this section the concept of *local* trend models is introduced by considering a *forgetting factor* or a *discount factor* in the least squares criterion. In the present derivation this is done simply by considering WLS (and an appropriate weight matrix $\boldsymbol{\Sigma}$) instead of LS (or OLS).

In *adaptive least squares* (or *discounted least squares*), the parameter estimates at time origin N, are

$$\widehat{\boldsymbol{\theta}}_N = \arg\min_{\boldsymbol{\theta}} S(\boldsymbol{\theta}; N) \tag{3.96}$$

where

$$S(\boldsymbol{\theta}; N) = \sum_{j=0}^{N-1} \lambda^j [Y_{N-j} - \boldsymbol{f}^T(-j)\boldsymbol{\theta}]^2 \tag{3.97}$$

The forgetting factor λ ($|\lambda| < 1$) determines the exponential discount of past observations.

Brown (1963) suggests choosing λ such that $0.70 < \lambda^p < 0.95$, where p is the number of parameters in the model. However, it is natural to choose a forgetting factor according to some criterion, e.g., the value which gives the minimum variance of the prediction error.

The estimation problem in (3.96) can be formulated as a WLS problem, since the model for all N observations is written as

$$\boldsymbol{Y} = \boldsymbol{x}_N\boldsymbol{\theta} + \boldsymbol{\varepsilon} \tag{3.98}$$

where $\boldsymbol{Y} = (Y_1, \ldots, Y_N)^T$, $\boldsymbol{x} = (\boldsymbol{f}^T(-N+1), \ldots, \boldsymbol{f}^T(0))^T$ and $\text{Var}[\boldsymbol{\varepsilon}] = \sigma^2\boldsymbol{\Sigma}$, and where $\boldsymbol{\Sigma} = \text{diag}[1/\lambda^{N-1}, \ldots, 1/\lambda, 1]$.

This WLS problem is solved by using Theorem 3.3 on page 38. The solution to (3.96) is

$$\widehat{\boldsymbol{\theta}}_N = \boldsymbol{F}_N^{-1}\boldsymbol{h}_N \tag{3.99}$$

where $\boldsymbol{F}_N = \boldsymbol{x}_N^T\boldsymbol{\Sigma}^{-1}\boldsymbol{x}_N$ and $\boldsymbol{h}_N = \boldsymbol{x}_N^T\boldsymbol{\Sigma}^{-1}\boldsymbol{Y}$.

Using (3.98) it is seen that

$$\boldsymbol{F}_N = \sum_{j=0}^{N-1} \lambda^j \boldsymbol{f}(-j)\boldsymbol{f}^T(-j), \qquad \boldsymbol{h}_N = \sum_{j=0}^{N-1} \lambda^j \boldsymbol{f}(-j)Y_{N-j}. \tag{3.100}$$

Using Theorem 3.10 the prediction of $Y_{N+\ell}$ given all observations at time N is

$$\widehat{Y}_{N+\ell|N} = \boldsymbol{f}^T(\ell)\widehat{\boldsymbol{\theta}}_N, \tag{3.101}$$

and the variance of the prediction error, $e_N(\ell) = Y_{N+\ell} - \widehat{Y}_{N+\ell|n}$, is

$$\mathrm{Var}[e_N(\ell)] = \sigma^2[1 + \boldsymbol{f}^T(\ell)\boldsymbol{F}_N^{-1}\boldsymbol{f}(\ell)], \tag{3.102}$$

where $\boldsymbol{F_N}$ is given by (3.100).

Note that there is no difference between the prediction equations for the global and local trend model. The difference lies in the estimation procedure.

THEOREM 3.13 (UPDATING IN LOCAL TREND MODELS)
When the next observation Y_{N+1} is available, the parameter estimate $\widehat{\boldsymbol{\theta}}_{N+1}$ is written

$$\widehat{\boldsymbol{\theta}}_{N+1} = \boldsymbol{F}_{N+1}^{-1}\boldsymbol{h}_{N+1} \tag{3.103}$$

where the recursive updating of $\boldsymbol{F}$ and $\boldsymbol{h}$ is

$$\boldsymbol{F}_{N+1} = \boldsymbol{F}_N + \lambda^N \boldsymbol{f}(-N)\boldsymbol{f}^T(-N)$$

and

$$\boldsymbol{h}_{N+1} = \lambda \boldsymbol{L}^{-1}\boldsymbol{h}_N + \boldsymbol{f}(0)Y_{N+1} \tag{3.104}$$

Proof Same as for Theorem 3.12. ■

For $N \to \infty$, $\lambda^N \boldsymbol{f}(-N)\boldsymbol{f}^T(-N) \to 0$ for nearly all functions (e.g., polynomials and harmonics). This implies that the following *stationary* or *steady state* value of $\boldsymbol{F}$ exists

$$\lim_{N\to\infty} \boldsymbol{F}_{N+1} = \boldsymbol{F} = \sum_{j\geq 0} \lambda^j \boldsymbol{f}(-j)\boldsymbol{f}^T(-j) \tag{3.105}$$

In the steady state case the procedure for updating the parameters is given by the following theorem.

THEOREM 3.14 (UPDATING IN STEADY STATE)
In steady state *the updating of the parameters in the locally constant trend model is*

$$\widehat{\boldsymbol{\theta}}_{N+1} = \boldsymbol{L}^T\widehat{\boldsymbol{\theta}}_N + \boldsymbol{F}^{-1}\boldsymbol{f}(0)[Y_{N+1} - \widehat{Y}_{N+1|N}] \tag{3.106}$$

where $\boldsymbol{F}$ is given in (3.105).

Proof See Abraham and Ledolter (1983, p. 103). ■

In Example 3.6 on the following page it is illustrated how the steady state value of $\boldsymbol{F}$ is calculated for a linear trend model.

In Section 3.4.2 the locally constant mean model was used to define simple exponential smoothing. Similarly, the much wider class of locally constant trend models can be used to define general exponential smoothing.

DEFINITION 3.4 (GENERAL EXPONENTIAL SMOOTHING)
The general exponential smoothing is related to a trend model with parameters $\boldsymbol{\theta} = (\theta_1, \ldots, \theta_p)$. For a given smoothing constant $\alpha = 1 - \lambda$ the p exponential smoothed values are defined as

$$\begin{pmatrix} S_N^{[1]} \\ \vdots \\ S_N^{[p]} \end{pmatrix} = \boldsymbol{A}(\lambda) \begin{pmatrix} \widehat{\theta}_{1,N} \\ \vdots \\ \widehat{\theta}_{p,N} \end{pmatrix},$$

where $\boldsymbol{A}(\lambda)$ is describing a mapping between the parameters of the smoothed values. This is illustrated in Example 3.6. $\widehat{\theta}_{i,N}$ is obtained from (3.106).

Using a k'th order polynomial trend model leads to a definition of the k'th *order exponential smoothing*:

$$S_N^{[k]} = (1 - \lambda) S_N^{[k-1]} + \lambda S_{N-1}^{[k]}, \quad S_N^{[0]} = Y_N \tag{3.107}$$

where $S_N^{[k]}$ is the k'th order exponential smoothing, which clearly is a simple exponential smoothing of $S_N^{[k-1]}$. The relation to $\widehat{\theta}$ is given by Definition 3.4, and for a k'th order polynomial trend, $\widehat{\theta}_{1,N}$ is the local level, $\widehat{\theta}_{2,N}$ is the local slope, $\widehat{\theta}_{3,N}$ the local curvature, etc.

Example 3.6 (Double exponential smoothing)
This example illustrates the relation between local linear trend and double exponential smoothing. Consider the linear trend model

$$Y_{N+j} = \theta_0 + \theta_1 j + \varepsilon_{N+j}$$

The steady state value of $\boldsymbol{F}$ is

$$\lim_{N \to \infty} \boldsymbol{F}_{N+1} = \boldsymbol{F} = \sum_{j \geq 0} \lambda^j \begin{pmatrix} 1 \\ -j \end{pmatrix} (1 \quad -j) = \sum_{j \geq 0} \lambda^j \begin{pmatrix} 1 & -j \\ -j & j^2 \end{pmatrix}$$

Since for $|\lambda| < 1$ it holds

$$\sum_{j=0}^{\infty} \lambda^j = \frac{1}{1 - \lambda}$$

$$\sum_{j=0}^{\infty} j \lambda^j = \frac{\lambda}{(1 - \lambda)^2}$$

$$\sum_{j=0}^{\infty} j^2 \lambda^j = \frac{\lambda(1 + \lambda)}{(1 - \lambda)^3}$$

the steady state value of $\boldsymbol{F}$ is

$$\boldsymbol{F} = \begin{pmatrix} \frac{1}{1-\lambda} & \frac{-\lambda}{(1-\lambda)^2} \\ \frac{-\lambda}{(1-\lambda)^2} & \frac{\lambda(1+\lambda)}{(1-\lambda)^3} \end{pmatrix}$$

and the parameter estimates are

$$\widehat{\boldsymbol{\theta}}_N = \boldsymbol{F}^{-1}\boldsymbol{h}_N = \begin{pmatrix} 1-\lambda^2 & (1-\lambda)^2 \\ (1-\lambda)^2 & \frac{(1-\lambda)^3}{\lambda} \end{pmatrix} \begin{pmatrix} \sum \lambda^j Y_{N-j} \\ -\sum j\lambda^j Y_{N-j} \end{pmatrix}$$

Or, since $\boldsymbol{\theta}_N = (\theta_{0,N}, \theta_{1,N})^T$

$$\widehat{\theta}_{0,N} = (1-\lambda^2)\sum \lambda^j Y_{N-j} - (1-\lambda)^2 \sum j\lambda^j Y_{N-j}$$

$$\widehat{\theta}_{1,N} = (1-\lambda)^2 \sum \lambda^j Y_{N-j} - \frac{(1-\lambda)^3}{\lambda} \sum j\lambda^j Y_{N-j}.$$

This gives the exponential weighted estimates of the local level, θ_0, and the local slope, θ_1.

Let us now illustrate the relation to the double (2nd order) exponential smoothing. As described by (3.107) the 1st and 2nd order exponential smoothed values are given as

$$S_N^{[1]} = (1-\lambda)Y_N + \lambda S_{N-1}^{[1]}$$

$$S_N^{[2]} = (1-\lambda)S_N^{[1]} + \lambda S_{N-1}^{[2]}.$$

By some (lengthy) calculations we find that

$$\widehat{\theta}_{0,N} = 2S_N^{[1]} - S_N^{[2]}$$

$$\widehat{\theta}_{1,N} = \frac{1-\lambda}{\lambda}(S_N^{[1]}) - (S_N^{[2]})$$

which clearly is an example of the relation in Definition 3.4.

Predictions in the local trend model are obtained by

$$\widehat{Y}_{N+\ell|N} = \widehat{\theta}_{0,N} + \widehat{\theta}_{1,N}\ell$$

$$= \left(2 + \frac{1-\lambda}{\lambda}\ell\right) S_N^{[1]} - \left(1 + \frac{1-\lambda}{\lambda}\ell\right) S_N^{[2]}$$

This prediction procedure is called *double exponential smoothing.*

3.5 Time series with seasonal variations

The trend models considered in Section 3.4.3 can be used to handle time series with seasonal variations (e.g., an annual variation). In this section some other methods based on the regression or exponential smoothing principles are considered. In later chapters dynamical models for describing time series with seasonal variations are introduced.

3.5.1 The classical decomposition

In the trend model the variation of the observations is decomposed into a trend and a random error, i.e., $Y_t = T_t + \varepsilon_t$, where T_t is the trend and ε_t is the random error (typically white noise). If the time series also shows a seasonal variation, the classical decomposition may be used. In the *classical decomposition* the variation is commonly decomposed in a *trend* T_t, a *seasonal* or *cyclic effect* S_t, and a *random error* ε_t. The trend is most frequently described by low order polynomials, and the seasonal effect by trigonometric functions or seasonal indicators (which will be introduced later). The random error is most frequently considered to be white noise.

If the effects are additive, then the *additive decomposition* can be used

$$Y_t = T_t + S_t + \varepsilon_t \tag{3.108}$$

This model might be appropriate if the seasonal effect is independent of the mean.

However, if the size of the seasonal effect appears to increase with the mean, it might be more reasonable to consider the *multiplicative decomposition*

$$Y_t = T_t \times S_t \times \varepsilon_t \tag{3.109}$$

By using a logarithmic transformation the multiplicative model is brought into an additive model. Note, however, that this transformation will stabilize the variance correctly only if the error term is also thought to be multiplicative. Another possibility is to use *mixed additive-multiplicative models.*

Traditionally the *trend component*, T_t, is described by polynomials in the time t

$$T_t = \theta_0 + \sum_{j=1}^{k} \theta_j \frac{t^j}{j!} \tag{3.110}$$

Usually we select $k = 0$, 1, or 2.

A *seasonal component* S_t with a period of s samples may be described by *seasonal indicators* (or *dummy variables*)

$$S_t = \sum_{j=1}^{s} \theta_j \delta_{tj}\,, \quad \sum_{j=1}^{s} \theta_j = 0 \tag{3.111}$$

where the special Kronecker's delta $\delta_{ti} = 1$ if t corresponds to the seasonal time point i, and otherwise $\delta_{ti} = 0$. For instance for monthly data with an annual seasonal period, if t corresponds to March ($i = 3$), then only δ_{t3} is one and the rest is zero. Unfortunately, for large values of s, the formulation of a model with a seasonal effect requires a large number of parameters which might conflict with the general idea of parsimony in model building.

Alternatively trigonometric functions can be used

$$S_t = \sum_{j=1}^{m} A_j \sin\left(\frac{j2\pi}{s}t + \varphi_j\right) \tag{3.112}$$

where the parameters are A_i and φ_i, which are the amplitude and the phase, respectively, of the harmonic component at the frequency $\omega_j = j2\pi/s$. The component (3.112) is not linear in the parameters; but using the identity $\sin(x+y) = \sin(x)\cos(y) + \cos(x)\sin(y)$, an alternative parameterization, which is linear in the parameters, is obtained.

The parameters of the above models can be estimated by using, for instance, the least squares method. The estimation procedure, as well as the prediction formulas follow from what we have seen previously in Section 3.2. See, e.g., Abraham and Ledolter (1983) or Harvey (1981) for a further discussion of the classical decomposition.

3.5.2 Holt-Winters procedure

For the locally constant trend models in Section 3.4.3, a single parameter (the smoothing constant $\alpha = 1 - \lambda$) is used to describe the variation in time of the parameters. However, the seasonal component may be more persistent than the trend component, and thus it is obvious to use different smoothing constants.

Holt (1957) formulated a double exponential smoothing with one smoothing constant for the level and another smoothing constant for the slope. Winters (1960) extended this procedure of using different smoothing constants to the additive model

$$Y_{N+j} = T_{N+j} + S_{N+j} + \varepsilon_{N+j} \tag{3.113}$$

where the trend is linear, i.e., $T_{N+j} = \mu_{N+j} = \mu_N + \beta_N j$, and where the seasonal component is described by seasonal indicators, i.e.,

$$S_t = \sum_{i=1}^{s} S_i \delta_{ti} \tag{3.114}$$

where $S_i = S_{i+s} = S_{i+2s} = \cdots$ for $i = 1, \ldots, s$ and $\sum_{i=1}^{s} S_i = 0$.

The procedure is usually referred to as *Holt-Winters procedure.* The procedure updates three smoothed statistics $\widehat{\mu}$, $\widehat{\beta}$, and $\widehat{S}$ recursively in order to provide adaptive estimates of the seasonal and trend components.

3.5.2.1 Estimation

An estimate of the *level* (or *mean*) of the time series at time origin $N+1$ is found by the recursive equation

$$\widehat{\mu}_{N+1} = \alpha_1(Y_{N+1} - \widehat{S}_{N+1-s}) + (1-\alpha_1)(\widehat{\mu}_N + \widehat{\beta}_N) \tag{3.115}$$

It is seen that the new estimate is a weighted average of the new observation Y_{N+1} adjusted for the seasonal component $\widehat{S}_{N+1-s}$ and the previous estimates for the level and the slope, $\widehat{\mu}_N$ and $\widehat{\beta}_N$. Thus, $(\widehat{\mu}_N + \widehat{\beta}_N)$ is the estimate of the level at time origin $N+1$ based on the observations available at time origin N.

Correspondingly, the estimate of the *slope* at time origin $N+1$ is found as

$$\widehat{\beta}_{N+1} = \alpha_2(\widehat{\mu}_{N+1} - \widehat{\mu}_N) + (1-\alpha_2)\widehat{\beta}_N \tag{3.116}$$

i.e., as a weighted average between the new observation of the slope $(\widehat{\mu}_{N+1} - \widehat{\mu}_N)$ and the previous estimate $\widehat{\beta}_N$.

Finally, the seasonal component is updated using

$$\widehat{S}_{N+1} = \alpha_3(Y_{N+1} - \widehat{\mu}_{N+1}) + (1-\alpha_3)\widehat{S}_{N+1-s} \tag{3.117}$$

which is a weight between the new observation of the seasonal component $(Y_{N+1} - \widehat{\mu}_{N+1})$ and the previous estimate $\widehat{S}_{N+1-s}$.

3.5.2.2 Prediction

Using the current estimates of the level, slope, and seasonal component, a prediction of $Y_{N+\ell}$, given the observations at time origin N, is

$$\widehat{Y}_{N+\ell|N} = \widehat{\mu}_N + \beta_N \ell + \widehat{S}_{N+\ell-s}\,, \quad \ell = 1, 2, \ldots, s \tag{3.118}$$

$$\widehat{Y}_{N+\ell|N} = \widehat{\mu}_N + \beta_N \ell + \widehat{S}_{N+\ell-2s}\,, \quad \ell = s+1, \ldots, 2s$$

etc.

3.5.2.3 Choice of smoothing constants

Several possible optimality criteria for choosing the set of smoothing constants (α_1, α_2 and α_3) exist. Abraham and Ledolter (1983) suggest that the smoothing constants are chosen so that the sum of squared *one-step prediction errors* is minimized, i.e.,

$$\sum_{t=1}^{N} [Y_t - \widehat{Y}_{t|t-1}]^2 \tag{3.119}$$

with respect to $(\alpha_1, \alpha_2, \alpha_3)$. The minimization is carried out by using, e.g., the Newton-Raphson procedure (3.31) from page 36.

3.6 Global and local trend model—an example

This section contains an example regarding a global and local linear trend model. Furthermore, the example illustrates in detail all the calculations involved, and the number of observations ($N = 6$) is kept low to ensure that you can follow the calculations without a computer. The available observations are shown in Figure 3.6.

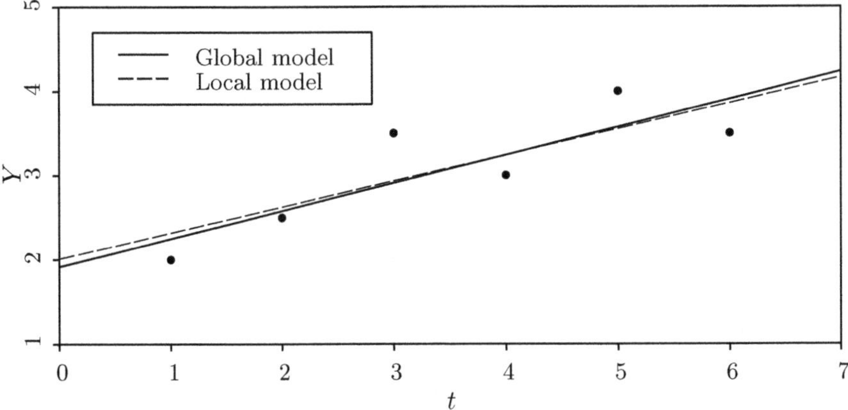

Figure 3.6: *Global and local linear trend.*

Example 3.7 (Global and local trend model)
Let us assume that we consider a linear trend model and assume that the 6 observations shown in Figure 3.6 are available. Let us first consider the global model:

a) Global linear trend model

$$Y_{N+j} = \theta_0 + \theta_1 j + \varepsilon_{N+j} \quad \Rightarrow \quad f(j) = (1 \quad j)^T$$

$\{\varepsilon_t, t = 0, \pm 1, \dots\}$ is white noise with zero mean and variance σ^2.

Using the above model for all six observations gives

$$\begin{pmatrix} 2.0 \\ 2.5 \\ 3.5 \\ 3.0 \\ 4.0 \\ 3.5 \end{pmatrix} = \begin{pmatrix} 1 & -5 \\ 1 & -4 \\ 1 & -3 \\ 1 & -2 \\ 1 & -1 \\ 1 & 0 \end{pmatrix} \begin{pmatrix} \theta_0 \\ \theta_1 \end{pmatrix} + \begin{pmatrix} \varepsilon_1 \\ \varepsilon_2 \\ \varepsilon_3 \\ \varepsilon_4 \\ \varepsilon_5 \\ \varepsilon_6 \end{pmatrix} \quad \Leftrightarrow \quad \boldsymbol{y} = \boldsymbol{x}_6 \boldsymbol{\theta} + \boldsymbol{\varepsilon}$$

that is

$$\boldsymbol{F}_6 = \boldsymbol{x}_6^T \boldsymbol{x}_6 = \begin{pmatrix} 6 & -15 \\ -15 & 55 \end{pmatrix}$$

$$\boldsymbol{h}_6 = \boldsymbol{x}_6^T \boldsymbol{y} = \begin{pmatrix} 18.5 \\ -40.5 \end{pmatrix}$$

$$\widehat{\boldsymbol{\theta}}_6 = \boldsymbol{F}_6^{-1} \boldsymbol{h}_6 = \begin{pmatrix} 0.5238 & 0.1429 \\ 0.1429 & 0.0571 \end{pmatrix} \begin{pmatrix} 18.5 \\ -40.5 \end{pmatrix} = \begin{pmatrix} 3.903 \\ 0.331 \end{pmatrix}$$

Using these estimates the solid line in Figure 3.6 is obtained. Based on the available observations ($N = 6$) the prediction of $Y_{6+\ell}$ is

$$\widehat{\boldsymbol{Y}}_{6+\ell|6} = \boldsymbol{f}^T(\ell)\widehat{\boldsymbol{\theta}}_6 = 3.903 + 0.331\ell$$

An unbiased estimate for σ^2 is

$$\begin{aligned}\widehat{\sigma}^2 &= (\boldsymbol{y} - \boldsymbol{x}_6\widehat{\boldsymbol{\theta}})^T(\boldsymbol{y} - \boldsymbol{x}_6\widehat{\boldsymbol{\theta}})/(6-2) \\ &= \frac{(-0.248)^2 + 0.079^2 + 0.590^2 + (-0.241)^2 + 0.428^2 + (-0.403)^2}{4} \\ &= 0.453^2\end{aligned}$$

Now an estimate of the variance of the prediction error $e_6(\ell) = Y_{6+\ell} - \widehat{Y}_{6+\ell|6}$ is

$$\begin{aligned}\widehat{\text{Var}}[e_6(\ell)] &= \widehat{\sigma}^2\left(1 + \boldsymbol{f}^T(\ell)\boldsymbol{F}_6^{-1}\boldsymbol{f}(\ell)\right) \\ &= 0.453^2\left(1 + (1 \quad \ell)\begin{pmatrix}0.5238 & 0.1429\\ 0.1429 & 0.0571\end{pmatrix}\begin{pmatrix}1\\ \ell\end{pmatrix}\right) \\ &= 0.453^2(1.5238 + 0.2858\ell + 0.0571\ell^2)\end{aligned}$$

As an example

$$\widehat{Y}_{7|6} = 4.234\,, \quad \widehat{\text{Var}}[e_6(1)] = 0.619^2$$

The 90% prediction interval for Y_7, given the observations at time origin $t = 6$, is

$$\widehat{Y}_{7|6} \pm \mathrm{t}_{0.05}(6-2)\sqrt{\widehat{\text{Var}}[e_6(1)]} = 4.234 \pm 1.320$$

Assume that *the next observation is* $y_7 = 3.5$.

Updating the parameters

$$\begin{aligned}\boldsymbol{F}_7 &= \boldsymbol{F}_6 + \boldsymbol{f}(-6)\boldsymbol{f}^T(-6) \\ &= \begin{pmatrix}6 & -15\\ -15 & 55\end{pmatrix} + \begin{pmatrix}1\\ -6\end{pmatrix}(1 \quad -6) \\ &= \begin{pmatrix}7 & -21\\ -21 & 91\end{pmatrix} \\ \boldsymbol{h}_7 &= \boldsymbol{L}^{-1}\boldsymbol{h}_6 + \boldsymbol{f}(0)Y_7 \\ &= \begin{pmatrix}1 & 0\\ 1 & 1\end{pmatrix}^{-1}\begin{pmatrix}18.5\\ -40.5\end{pmatrix} + \begin{pmatrix}1\\ 0\end{pmatrix}3.5 \\ &= \begin{pmatrix}22\\ -59\end{pmatrix} \\ \widehat{\boldsymbol{\theta}}_7 &= \begin{pmatrix}0.4643 & 0.1071\\ 0.1071 & 0.0357\end{pmatrix}\begin{pmatrix}22\\ -59\end{pmatrix} = \begin{pmatrix}3.896\\ 0.250\end{pmatrix}\end{aligned}$$

A tendency for a decrease in the slope is clearly seen.

b) Local linear trend model We choose the forgetting factor $\lambda = 0.9$.

The model is unchanged; but due to the discount of the past observations, we get

$$\boldsymbol{F}_6 = \sum_{j=0}^{5} \lambda^j \boldsymbol{f}(-j)\boldsymbol{f}^T(-j) = \begin{pmatrix} 4.6856 & -10.284 \\ -10.284 & 35.961 \end{pmatrix}$$

$$\boldsymbol{h}_6 = \sum_{j=0}^{5} \lambda^j \boldsymbol{f}(-j) Y_{6-j} = \begin{pmatrix} 14.902 \\ -28.580 \end{pmatrix}$$

$$\widehat{\boldsymbol{\theta}}_6 = \boldsymbol{F}_6^{-1}\boldsymbol{h}_6 = \begin{pmatrix} 0.5732 & 0.1639 \\ 0.1639 & 0.0747 \end{pmatrix} \begin{pmatrix} 14.902 \\ -28.580 \end{pmatrix} = \begin{pmatrix} 3.858 \\ 0.308 \end{pmatrix}$$

Compared with the globally constant trend model, we clearly see a smaller slope. This tendency for a smaller slope for the more recent observations is recognized in Figure 3.6.

Updating the parameters

Again the next observation is $y_7 = 3.5$. With the adaptive scheme the updating equations are

$$\begin{aligned} \boldsymbol{F}_7 &= \boldsymbol{F}_6 + \lambda^6 \boldsymbol{f}(-6)\boldsymbol{f}^T(-6) \\ &= \begin{pmatrix} 4.6856 & -10.284 \\ -10.284 & 35.961 \end{pmatrix} + 0.9^6 \begin{pmatrix} 1 \\ -6 \end{pmatrix} \begin{pmatrix} 1 & -6 \end{pmatrix} \\ &= \begin{pmatrix} 5.2170 & -13.473 \\ -13.473 & 55.093 \end{pmatrix} \\ \boldsymbol{h}_7 &= \lambda \boldsymbol{L}^{-1}\boldsymbol{h}_6 + \boldsymbol{f}(0) y_7 \\ &= 0.9 \begin{pmatrix} 1 & 0 \\ 1 & 1 \end{pmatrix}^{-1} \begin{pmatrix} 14.902 \\ -28.580 \end{pmatrix} + \begin{pmatrix} 1 \\ 0 \end{pmatrix} 3.5 \\ &= \begin{pmatrix} 16.912 \\ -39.134 \end{pmatrix} \end{aligned}$$

i.e., the adaptive estimated parameters are

$$\widehat{\boldsymbol{\theta}}_7 = \boldsymbol{F}_7^{-1}\boldsymbol{h}_7 = \begin{pmatrix} 0.5202 & 0.1272 \\ 0.1272 & 0.0493 \end{pmatrix} \begin{pmatrix} 16.912 \\ -39.134 \end{pmatrix} = \begin{pmatrix} 3.820 \\ 0.222 \end{pmatrix}$$

Again it is noted that the adaptive nature of this method implies a lower slope than for globally estimated parameters, which is in accordance with the tendency seen in Figure 3.6.

3.7 Problems

Exercise 3.1

Table 3.2 on the following page shows some observations of a stochastic process $\{Y_t\}$ and an independent variable.

Table 3.2: *Observations.*

t	1	2	3	4	5	6	7
y_t	1.0	0.5	2.0	2.0	3.5	3.0	4.0
x_t	4.0	4.0	3.5	4.0	2.0	2.5	1.5

Question 1 Formulate a model for the variations in Y_t and estimate the parameters.

Question 2 Knowing that $x_8 = 0.5$, find a 90% confidence interval for Y_8.

Question 3 Assume that we have observed only the time series $\{y_t\}$. Using this assumption, calculate a 90% confidence interval for Y_8.

Exercise 3.2
Consider the regression model

$$Y_t = \beta x_t + \varepsilon_t$$

where $\mathrm{E}[\{\varepsilon_t\}] = 0$. Suppose that N observations are given.

Question 1 Assume that $\mathrm{Var}[\varepsilon_t] = \sigma^2 / x_t^2$ but that the elements of the sequence $\{\varepsilon_t\}$ are mutually uncorrelated. Consider the unweighted least squares estimator $(\widehat{\beta}^*)$.

- Is the estimator unbiased?
- Calculate the variance of the estimator.

Question 2 Calculate the variance of the weighted least squares estimator $(\widehat{\beta})$.

Question 3 Compare the variances of $(\widehat{\beta}^*)$ and $(\widehat{\beta})$.

Question 4 Assume now that $\mathrm{Var}[\varepsilon_t] = \sigma^2$, but that the elements of the sequence $\{\varepsilon_t\}$ are mutually correlated. We assume that

$$\rho[\varepsilon_t, \varepsilon_{t-1}] = \rho\,, \quad \text{for } k = 1$$
$$\rho[\varepsilon_t, \varepsilon_{t-k}] = 0\,, \quad \text{for } k \geq 2$$

Consider the unweighted least square estimator.

- Is the estimator unbiased?
- Calculate the variance of the estimator.

Exercise 3.3
Suppose that $N = 2q + 1$ observations $Y_1, Y_2, \ldots, Y_N$ are given. Consider the model

$$Y_t = \mu + \alpha \cos(\omega t) + \beta \sin(\omega t) + \varepsilon_t$$

where $\{\varepsilon_t\}$ is a sequence of mutually uncorrelated random variables with $\mathrm{E}[\varepsilon_t] = 0$ and $\mathrm{Var}[\varepsilon_t] = \sigma^2$.

Question 1 State how (μ, α, β) can be determined using the least squares method and specify the corresponding matrices.

Question 2 Take $\omega = \omega_i = \frac{2\pi i}{N}$ $(i \in \mathbb{Z})$ and $(i \leq q)$ and find the least squares estimator for (μ, α, β). *Hint*:

$$\sum_{t=1}^{N} \cos^2(\omega_i t) = \frac{N}{2}$$

$$\sum_{t=1}^{N} \cos(\omega_i t) \sin(\omega_i t) = 0$$

$$\sum_{t=1}^{N} \sin^2(\omega_i t) = \frac{N}{2}$$

Question 3 The part of Y_t's variations which can be described by the periodic component ω_i is given by

$$I(\omega_i) = \sum_{t=1}^{N} \left(\widehat{\alpha} \cos(\omega_i t) + \widehat{\beta} \sin(\omega_i t)\right)^2$$

Show, this contribution can be written as

$$I(\omega_i) = (\widehat{\alpha}^2 + \widehat{\beta}^2)\frac{N}{2}$$

$I(\omega_i)$ is called the intensity (of the variations) in ω_i.

Exercise 3.4

In an airport near Copenhagen the, wind speed is measured every hour using a cup-anemometer. Suppose that now $(t = 8)$ the following observations of the wind speed are given [m/s]:

4.4, 3.4, 3.3, 2.5, 7.3, 4.9, 4.8, 4.4

Before the fifth measurement was performed, the anemometer was moved from the normal measuring location (2 m above ground level) to the roof of a building.

Question 1 Formulate a suitable model describing the variations in the observed wind speed.

Question 2 Estimate the parameters of the model and especially assess (model) the difference in wind speed at the two measurement locations.

Question 3 Predict the wind speed at the old measuring location at the next hour $(t = 9)$.

Exercise 3.5

Consider a model with local constant mean.

Question 1 Consider the formulas related to the local trend model. Show that the prediction of Y_{N+l}, given the observations $Y_1, Y_2, \ldots, Y_N$, is given by

$$\widehat{Y}_{N+l|N} = \frac{1-\lambda}{1-\lambda^N} \sum_{j=0}^{N-1} \lambda^j Y_{N-j}$$

Question 2 Show that for $\lambda \to 1$ (and fixed N) the prediction equation is reduced to

$$\widehat{Y}_{N+l|N} = \frac{1}{N} \sum_{j=0}^{N-1} Y_{N-j} = \overline{Y}$$

Compare with the result in Section 3.4.1.

Question 3 Now the case ($N \to \infty$) is considered (λ fixed).

Using Theorem 3.11 on page 49, show that updating of the one-step prediction is

$$\widehat{Y}_{N+2|N+1} = \widehat{Y}_{N+1|N} + (1-\lambda)[Y_{N+1} - \widehat{Y}_{N+1|N}]$$

and compare with the result in Section 3.4.2.

Exercise 3.6

Consider the numbers from Exercise 3.1, and suppose that only $\{Y_t\}$ can be observed. On the basis of a large data set, it is found acceptable to use a local linear trend with a forgetting factor $\lambda = 0.9$.

Question 1 Based on the above mentioned facts, calculate a 90% confidence interval for Y_8.

Question 2 Assume that the next observation is given as $y_8 = 5.0$. Update the parameter estimates and calculate a 90% confidence interval for Y_9.

CHAPTER 4

Linear dynamic systems

In the following we will consider signals (or processes) in *discrete time* $\{x_t; t = 0, \pm 1, \dots\}$ and in *continuous time* $\{x(t); -\infty \leq t \leq \infty\}$. In practical situations discrete signals are often a result of continuous signals being sampled, and this will be discussed later in this chapter.

A *system* converts an *input* $x(t)$ (possibly multidimensional) to a (possibly multidimensional) *output* $y(t)$:

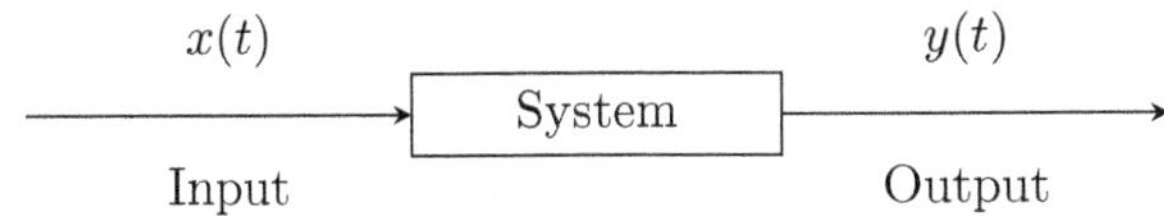

In mathematical terms the system is described by an operator $\mathcal{F}$—often called a *filter*. Within this framework we will limit ourselves to the following.

DEFINITION 4.1 (LINEAR SYSTEMS)
A system is said to be *linear* if

$$\mathcal{F}\left[\lambda_1 x_1(t) + \lambda_2 x_2(t)\right] = \lambda_1 \mathcal{F}\left[x_1(t)\right] + \lambda_2 \mathcal{F}\left[x_2(t)\right] \tag{4.1}$$

The class of linear systems can further be divided into, e.g., time-invariant and time-varying systems.

DEFINITION 4.2 (TIME-INVARIANT SYSTEMS)
A system is said to be *time-invariant* if

$$y(t) = \mathcal{F}\left[x(t)\right] \Rightarrow y(t-\tau) = \mathcal{F}\left[x(t-\tau)\right] \tag{4.2}$$

for all values of t and τ.

DEFINITION 4.3 (STABLE SYSTEMS)
A system is said to be *stable* if any constrained input implies a constrained output.

Let us consider some examples.

Example 4.1 (Non-linear system)
We consider the system $\{y_t\} = \mathcal{F}[\{x_t\}]$ (discrete time) defined by

$$y_t = (x_t)^t$$

This system is non-linear since $(\lambda_1 x_{1,t} + \lambda_2 x_{2,t})^t \neq \lambda_1 (x_{1,t})^t + \lambda_2 (x_{2,t})^t$. The system is not time-invariant since $y_t = (x_t)^t$ and $y_{t-\tau} = (x_{t-\tau})^{t-\tau} \neq (x_{t-\tau})^t$. Finally, the system is not stable since, e.g., $x_t = \text{constant} > 1$ implies that $y_t \to \infty$ for $t \to \infty$.

Example 4.2 (Linear system)
The system defined by

$$y_t = x_t + \theta x_{t-1}$$

is linear since

$$(\lambda_1 x_{1,t} + \lambda_2 x_{2,t}) + \theta(\lambda_1 x_{1,t-1} + \lambda_2 x_{2,t-1}) = \lambda_1 (x_{1,t} + \theta x_{1,t-1}) + \lambda_2 (x_{2,t} + \theta x_{2,t-1})$$

It is easily seen that the system is both time-invariant and stable.

4.1 Linear systems in the time domain

For linear and time-invariant systems we have the following fundamental result.

THEOREM 4.1 (IMPULSE RESPONSE FUNCTION)
For any linear and time-invariant system there exists a function h so that the output is obtained by applying the convolution integral

$$y(t) = \int_{-\infty}^{\infty} h(u) x(t-u)\, du \tag{4.3}$$

in continuous time and as the convolution sum

$$y_t = \sum_{k=-\infty}^{\infty} h_k x_{t-k} \tag{4.4}$$

in discrete time. Sometimes we write $y = h * x$, where $*$ is the convolution operator.

Proof Omitted. ∎

The weight function $h(u)$ is called the *impulse response function* because the output is $h(u)$ when the input is the *Dirac delta-function* $\delta(t)$. $\delta(t)$ is also called the *impulse function* and is defined by

$$\int_{-\infty}^{\infty} f(t)\delta(t - t_0)\, dt = f(t_0) \tag{4.5}$$

In discrete time the sequence $\{h_k\}$ is also called the impulse response function or just the impulse response, since the output is h_k when the input is *Kronecker's delta sequence* (sometimes referred to as the *impulse function*) and is defined by

$$\delta_k = \begin{cases} 1 & \text{for } k = 0 \\ 0 & \text{for } k = \pm 1, \pm 2, \dots \end{cases} \tag{4.6}$$

Instead of (4.3) and (4.4) the *convolution operator* $*$ is often used in both cases and the output is simply written as $y = h * x$.

THEOREM 4.2 (PROPERTIES OF THE CONVOLUTION OPERATOR)
The convolution operator has the following properties:

a) $h * g = g * h$ *(symmetric).*

b) $(h * g) * f = h * (g * f)$ *(associative).*

c) $h * \delta = h$, *where* δ *is the impulse function.*

Proof Left for the reader. ∎

▸ **Remark 4.1**
For a given (parameterized) system the impulse response function is often found most conveniently by simply putting $x = \delta$ and then calculating the response, $y = h$; cf. Theorem 4.1. This is illustrated in Example 4.3 on the following page. ◂

DEFINITION 4.4 (CAUSAL SYSTEMS)
A systems is said to be *physically feasible* or *causal* if the output at time t does not depend on future values of the input, i.e.,

$$h(u) = 0\,, \quad \text{for } u < 0 \tag{4.7}$$
$$h_k = 0\,, \quad \text{for } k < 0 \tag{4.8}$$

in continuous and discrete time, respectively.

Having introduced the impulse response function we introduce the following.

THEOREM 4.3 (STABILITY FOR LINEAR SYSTEMS)
A sufficient condition for a linear system being stable is that

$$\int_{-\infty}^{\infty} |h(u)|\, du < \infty \tag{4.9}$$

or

$$\sum_{k=-\infty}^{\infty} |h_k| < \infty \tag{4.10}$$

in continuous and discrete time, respectively.

Proof Omitted. ■

Example 4.3 (Calculation of h_k)
Consider the linear, time-invariant system

$$y_t - 0.8y_{t-1} = 2x_t - x_{t-1} \tag{4.11}$$

By putting $x = \delta$ we see that $y_k = h_k = 0$ for $k < 0$. For $k = 0$ we get

$$\begin{aligned} y_0 &= 0.8y_{-1} + 2\delta_0 - \delta_{-1} \\ &= 0.8 \times 0 + 2 \times 1 - 0 = 2 \end{aligned} \tag{4.12}$$

i.e., $h_0 = 2$. Continuing, we get

$$\begin{aligned} y_1 &= 0.8y_0 + 2\delta_1 - \delta_0 \\ &= 0.8 \times 2 + 2 \times 0 - 1 = 0.6 \\ y_2 &= 0.8y_1 = 0.48 \\ &\vdots \\ y_k &= 0.8^{k-1}0.6 \ (k > 0) \end{aligned}$$

Hence, the impulse response function is

$$h_k = \begin{cases} 0 & \text{for } k < 0 \\ 2 & \text{for } k = 0 \\ 0.8^{k-1}0.6 & \text{for } k > 0 \end{cases}$$

which clearly represents a causal system; cf. Definition 4.4. Furthermore, the system is stable since $\sum_0^\infty |h_k| = 2 + 0.6(1 + 0.8 + 0.8^2 + \cdots) = 5 < \infty$.

Based on the impulse response function, we define the *step response function* by

$$S(t) = \int_{-\infty}^{t} h(u)\, du \tag{4.13}$$

in continuous time and

$$S_t = \sum_{k \le t} h_k \tag{4.14}$$

in discrete time. The function is called the step response function because it equals the output when the input is a step.

4.2 Linear systems in the frequency domain

For some analysis it is more convenient to describe a system in the frequency domain, e.g., if the system is generating signals with periodicities.

For a linear time-invariant system we have seen that the relation between input and output for a system in *continuous time* is given by the convolution integral

$$y(t) = \int_{-\infty}^{\infty} h(u)x(t-u)\,du = (h * x)(t). \tag{4.15}$$

Since the output in the frequency domain is obtained by the *Fourier transform*

$$Y(\omega) = \int_{-\infty}^{\infty} y(t)\mathrm{e}^{-i\omega t}\,dt \tag{4.16}$$

where ω is the *frequency* [rad/s], we get by substitution of (4.15):

$$\begin{aligned} Y(\omega) &= \int_{-\infty}^{\infty}\int_{-\infty}^{\infty} h(u)x(t-u)\mathrm{e}^{-i\omega t}\,du\,dt \\ &= \int_{-\infty}^{\infty} h(u)\mathrm{e}^{-i\omega u}\left[\int_{-\infty}^{\infty} x(t-u)\mathrm{e}^{-i\omega(t-u)}\,dt\right]du \\ &= \int_{-\infty}^{\infty} h(u)\mathrm{e}^{-i\omega u}du \cdot \int_{-\infty}^{\infty} x(v)\mathrm{e}^{-i\omega v}\,dv \end{aligned} \tag{4.17}$$

where $v = t - u$. It is seen that the last integral is the Fourier transform $X(\omega)$ of the input signal $x(t)$; cf. (4.16). The first integral is the Fourier transform of the impulse response function, and is called the *frequency response function* (or sometimes the *transfer function*), and is denoted $\mathcal{H}(\omega)$. Thus, it is determined by

$$\mathcal{H}(\omega) = \int_{-\infty}^{\infty} h(u)\mathrm{e}^{-i\omega u}\,du\,, \quad -\infty < \omega < \infty \tag{4.18}$$

From (4.17), the following is seen.

THEOREM 4.4 (FREQUENCY RESPONSE FUNCTION)
For a linear, time-invariant system the output in the frequency domain is determined by multiplying *the input in the frequency domain with the frequency response function, i.e.,*

$$Y(\omega) = \mathcal{H}(\omega)X(\omega) \tag{4.19}$$

Proof Follows directly. ∎

For a system in *discrete time* we apply the *discrete Fourier transform*

$$Y(\omega) = \sum_{t=-\infty}^{\infty} y_t \mathrm{e}^{-i\omega t}, \quad -\pi \le \omega < \pi \tag{4.20}$$

From calculations equivalent to (4.17) it is seen that Theorem 4.4 is also valid for systems in discrete time where the frequency response function is determined by

$$\mathcal{H}(\omega) = \sum_{k=-\infty}^{\infty} h_k e^{-i\omega k}, \quad -\pi \le \omega < \pi \tag{4.21}$$

Based on a given frequency response function $\mathcal{H}(\omega)$, the impulse response function can be found by the *inverse Fourier transform*

$$h(t) = \frac{1}{2\pi}\int_{-\infty}^{\infty} \mathcal{H}(\omega)e^{i\omega t}\,d\omega, \quad \text{(continuous time)} \tag{4.22}$$

$$h_k = \frac{1}{2\pi}\int_{-\pi}^{\pi} \mathcal{H}(\omega)e^{i\omega k}\,d\omega, \quad \text{(discrete time)} \tag{4.23}$$

We observe, that for systems in discrete time, we consider ω only in the interval $[-\pi, \pi]$. This is due to *aliasing* of harmonic functions applied to equidistant observations. The problem will be discussed in Section 4.3 on page 78.

$\mathcal{H}(\omega)$ is in general complex since the impulse response function is not usually an even function. Thus, it is possible to split $\mathcal{H}(\omega)$ into a real and a complex part

$$\mathcal{H}(\omega) = |\mathcal{H}(\omega)|e^{i\arg\{\mathcal{H}(\omega)\}} = G(\omega)e^{i\phi(\omega)} \tag{4.24}$$

where $G(\omega)$ is the *amplitude* (*amplitude function*) or *gain* and $\phi(\omega)$ is the *phase* (*phase function*).

Let us consider two signals in the time domain $w(t)$ and $x(t)$, respectively, as well as the corresponding Fourier transforms, $W(\omega)$ and $X(\omega)$. The convolution in the frequency domain is given by

$$Y(\omega) = \frac{1}{2\pi}\int_{-\infty}^{\infty} W(\nu)X(\omega-\nu)\,d\nu = \frac{1}{2\pi}(W * X)(\omega) \tag{4.25}$$

And by taking an inverse Fourier transform we get

$$y(t) = w(t)x(t) \tag{4.26}$$

▸ Remark 4.2

In general it holds that convolution in the time domain corresponds to multiplication in the frequency domain. This is clear from (4.15) and (4.19) on the preceding page as well as from (4.25) and (4.26). ◂

It should be noted that different placement of the constant $1/(2\pi)$ for the pairs of *Fourier transforms* is often used in the literature. Alternatively, the pairs of Fourier transforms may be defined

$$Y(\omega) = \frac{1}{2\pi}\int_{-\infty}^{\infty} y(t)e^{-i\omega t}\,dt \tag{4.27}$$

$$y(t) = \int_{-\infty}^{\infty} Y(\omega) e^{i\omega t} \, d\omega \tag{4.28}$$

in continuous time and similarly in discrete time when changing the integration with summation and changing the limits according to (4.27) and (4.28). The pair (4.18) and (4.22) is most frequently used for frequency response functions, whereas the pair (4.27) and (4.28) is most frequently used in statistics (because the area under the spectra equals the total variance).

Furthermore, an alternative pair of Fourier transforms (Jenkins and Watts 1968) is

$$S(\nu) = \int_{-\infty}^{\infty} s(t) e^{-i2\pi\nu t} \, dt \tag{4.29}$$

$$s(t) = \int_{-\infty}^{\infty} S(\nu) e^{i2\pi\nu t} \, d\nu \tag{4.30}$$

where ν is called the frequency ($[\nu] = s^{-1} = \text{Hz}$). Since $\omega = 2\pi\nu$, we get the relation $S(\omega)d\omega = S(\nu)d\nu$ that $S(\omega)2\pi = S(\nu)$.

Let us consider the output that is obtained when the input is *a single harmonic signal*, i.e., the input is given by

$$x(t) = Ae^{i\omega t} = A\cos(\omega t) + iA\sin(\omega t) \tag{4.31}$$

Then the output becomes

$$\begin{aligned} y(t) &= \int_{-\infty}^{\infty} h(u) x(t-u) \, du \\ &= \int_{-\infty}^{\infty} h(u) A e^{i\omega(t-u)} \, du \\ &= Ae^{i\omega t} \int_{-\infty}^{\infty} h(u) e^{-i\omega u} \, du \\ &= \mathcal{H}(\omega) A e^{i\omega t} = G(\omega) A e^{i(\omega t + \phi(\omega))} \end{aligned} \tag{4.32}$$

From this is seen the following.

THEOREM 4.5 (SINGLE FREQUENCY RELATION)
A single harmonic input to a linear, time-invariant system will give an output having the same frequency ω. The amplitude of the output signal equals the amplitude of the input signal multiplied by $G(\omega)$. The change in phase from input to output is $\phi(\omega)$.

Proof Seen directly. ∎

Example 4.4 (An RC element)
Let us consider the electrical system in the figure below. The input voltage is denoted $x(t)$ and the output voltage $y(t)$.

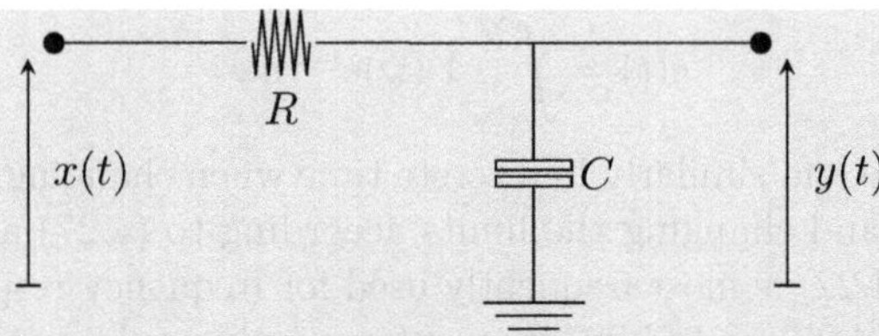

By considering the circuit, it is seen that

$$C\frac{dy(t)}{dt} = \frac{1}{R}(x(t) - y(t))$$

or

$$RC\frac{dy}{dt} + y = x(t). \tag{4.33}$$

The solution is

$$y(t) = \frac{1}{RC}\int_0^\infty \mathrm{e}^{-s/RC} x(t-s)\, ds. \tag{4.34}$$

We see directly, by comparing the convolution integral in Theorem 4.1 on page 70 that the impulse response function becomes

$$h(u) = \begin{cases} \frac{1}{RC}\mathrm{e}^{-u/RC} & \text{for } u \geq 0 \\ 0 & \text{for } u < 0 \end{cases} \tag{4.35}$$

The *time constant* is $\tau = RC$. The exponential decay and the time constant are illustrated in the figure below.

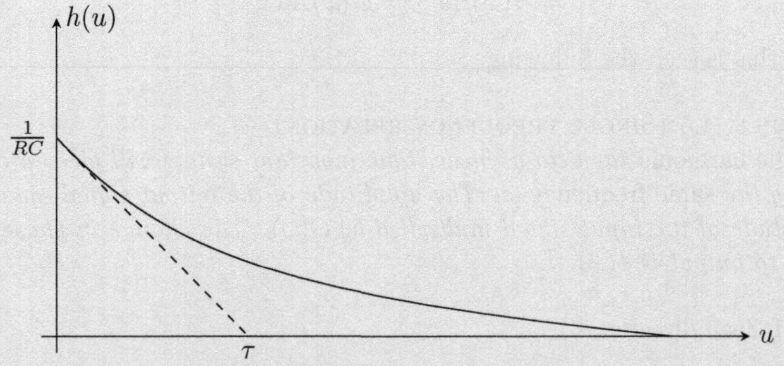

If the input is a pulse $\delta(t)$ we obtain the response $h(u)$ corresponding to the definition of $h(u)$.

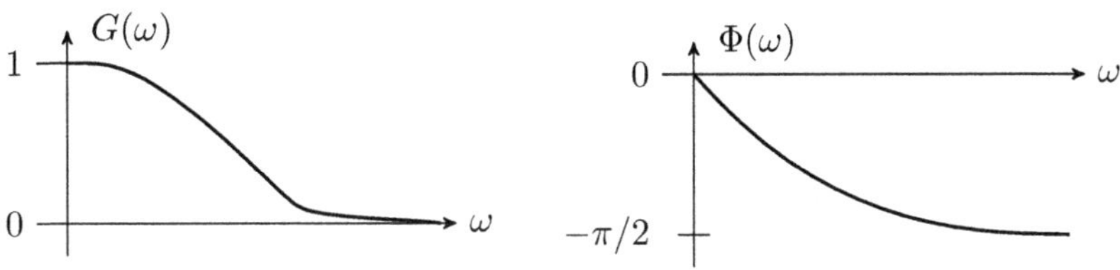

Figure 4.1: *Amplitude (gain) and phase for an RC-element.*

The frequency response function for the system becomes

$$\begin{aligned}\mathcal{H}(\omega) &= \int_{-\infty}^{\infty} h(u)\mathrm{e}^{-i\omega u}\,du \\ &= \frac{1}{RC}\int_0^{\infty} \mathrm{e}^{-u/RC}\mathrm{e}^{-i\omega u}\,du \\ &= \frac{1}{RC\left(\frac{1}{RC}+i\omega\right)}\int_0^{\infty} \mathrm{e}^{-(\frac{1}{RC}+i\omega)u}\,d\left(\frac{1}{RC}+i\omega\right)u\end{aligned}$$

i.e.,

$$\mathcal{H}(\omega) = \frac{1}{1+i\omega RC} \tag{4.36}$$

From this we calculate the amplitude,

$$G(\omega) = |\mathcal{H}(\omega)| = \frac{1}{\sqrt{1+\omega^2R^2C^2}} \tag{4.37}$$

and the phase,

$$\phi(\omega) = \arg\{\mathcal{H}(\omega)\} = \arctan(-RC\omega)\,. \tag{4.38}$$

The amplitude and the phase for the RC element is shown in Figure 4.1.

A filter (as described previously) where the slow variations (small values of ω) pass undisturbed, while the fast variations are damped (filtered), is called a *low-pass filter*. A filter, where the fast variations are kept intact, while the slow variations are filtered, is called a *high-pass filter*.

Example 4.5 (A time delay)
Let us consider a system where the input signal is delayed τ time units, i.e.,

$$y(t) = x(t-\tau)\,, \quad (\tau \text{ constant}) \tag{4.39}$$

The impulse response becomes

$$h(u) = \delta(u-\tau) \tag{4.40}$$

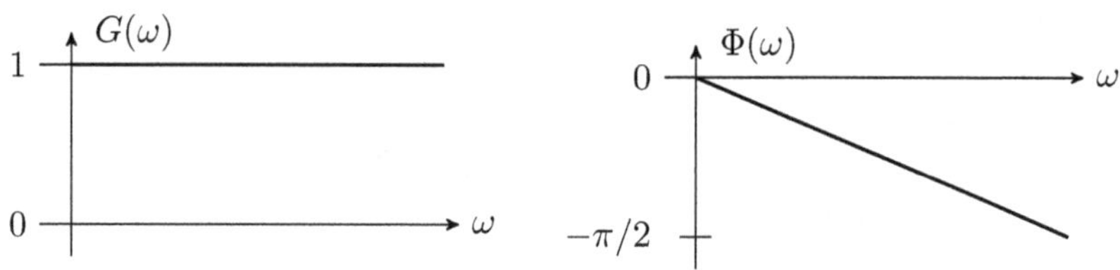

Figure 4.2: *Amplitude (gain) and phase of a time delay.*

and the frequency response function becomes

$$\mathcal{H}(\omega) = \mathrm{e}^{-i\omega\tau} \tag{4.41}$$

The amplitude of the system is $G(\omega) = 1$ and the phase is $\phi(\omega) = -\omega\tau$.

It is seen from Figure 4.2 that all frequencies pass undisturbed, but that the phase $\phi(\omega) \to -\infty$ for $\omega \to \infty$.

Example 4.6 (Design of low-pass filters)
An ideal low-pass filter is a filter, which cuts off frequencies higher than the threshold frequency ω_0 while frequencies lower than ω_0 pass undisturbed. We will now investigate how to select the weight function in the time domain so that we achieve such an *ideal low-pass filter*

$$\mathcal{H}_I(\omega) = \begin{cases} 1 & |\omega| \le \omega_0 \\ 0 & |\omega| > \omega_0 \end{cases} \tag{4.42}$$

We find

$$h_I(u) = \frac{1}{2\pi}\int_{-\omega_0}^{\omega_0} \mathrm{e}^{i\omega u}\, d\omega = \frac{1}{\pi}\int_0^{\omega_0} \cos(\omega u)\, d\omega = \frac{\sin(\omega_0 u)}{\pi u}. \tag{4.43}$$

It is not possible to construct this filter (not *causal*) since $h_I(u)$ is defined for all u. Even for "off-line" calculations we can only obtain an approximation to $\mathcal{H}_I(\omega)$, since $h_I(u)$ cannot be applied to the input signal in a non-finite time interval. An *ideal band-pass filter* is a filter where the signal passes undisturbed in the interval $[\omega_1, \omega_2]$, while it is cut off outside this interval.

4.3 Sampling

When a continuous signal is registered or acquired at discrete time instances, we say that we are sampling the signal. It is obvious that information is lost when a signal is sampled. Let us consider the case where a continuous signal $x(t)$ is sampled (or acquired) at *equidistant time instances* T (*sampling time*).

The sampled signal is the signal we obtain when the continuous signal is multiplied with a *series of δ functions* defined by

$$j(t) = \sum_{n=-\infty}^{\infty} \delta(t - nT) \tag{4.44}$$

Since $j(t)$ is a periodic function, it can be expanded in Fourier series; i.e., it can be written as

$$j(t) = \sum_{n=-\infty}^{\infty} F_n \mathrm{e}^{in\omega_0 t} \tag{4.45}$$

where $\omega_0 = 2\pi/T$ is the *sampling frequency*, and

$$\begin{aligned} F_n &= \frac{1}{T}\int_{-T/2}^{T/2} j(t)\mathrm{e}^{-in\omega_0 t}\, dt \\ &= \frac{1}{T}\int_{-T/2}^{T/2} \delta(t)\mathrm{e}^{-in\omega_0 t}\, dt = \frac{1}{T} \end{aligned}$$

i.e., the Fourier expansion of a series of δ functions becomes

$$j(t) = \frac{1}{T}\sum_{n=-\infty}^{\infty} \mathrm{e}^{in\omega_0 t} \tag{4.46}$$

The Fourier transformed becomes

$$\begin{aligned} J(\omega) &= \frac{1}{T}\int_{-\infty}^{\infty} \sum_{n=-\infty}^{\infty} \mathrm{e}^{in\omega_0 t}\mathrm{e}^{-i\omega t}\, dt \\ &= \frac{1}{T}\sum_{n=-\infty}^{\infty}\int_{-\infty}^{\infty} \mathrm{e}^{i(n\omega_0-\omega)t}\, dt \\ &= \frac{2\pi}{T}\sum_{n=-\infty}^{\infty} \delta(\omega - n\omega_0) \end{aligned} \tag{4.47}$$

i.e., again a series of δ functions. Let us consider the *sampled signal*

$$x_s(t) = x(t)j(t) \tag{4.48}$$

and let $x(t)$ be given in the frequency domain by $X(\omega)$. Since multiplication in the time domain corresponds to convolution in the frequency domain we get

$$\begin{aligned} X_s(\omega) &= \frac{1}{2\pi}\int_{-\infty}^{\infty} J(\nu)X(\omega-\nu)\, d\nu \\ &= \frac{1}{T}\int_{-\infty}^{\infty}\sum_{n=-\infty}^{\infty} \delta(\nu - n\omega_0)X(\omega-\nu)\, d\nu \end{aligned}$$

From this it is seen that

$$X_s(\omega) = \frac{1}{T} \sum_{n=-\infty}^{\infty} X(\omega - n\omega_0) \tag{4.49}$$

The sampled signal in the frequency domain is thus a periodic function with the period ω_0. Furthermore, it is seen that if $X(\omega)$ contains values outside the interval $[-\omega_0/2, \omega_0/2] = [-\pi/T, \pi/T]$ these values cannot be distinguished from values inside the interval. This phenomenon is known as *aliasing*. Values of $X(\omega)$ outside the interval $[-\pi/T, \pi/T]$ will thus be *added* with values inside the interval. The frequency $\omega = \pi/T$ is called the *Nyquist frequency*. If $X(\omega)$ contains values above the Nyquist frequency, we will encounter problems due to aliasing. Figure 4.3 shows a spectrum which we will consider later. However, if $X(\omega)$ is as shown in the figure, then $X_s(\omega)$ will be as shown.

The previous discussion can be used in the problem of choosing a suitable sampling time T, since the sampling frequency $2\pi/T$ should be at least twice as large as the limit frequency for $X(\omega)$. More precisely, we have the following.

THEOREM 4.6 (SHANNON'S THEOREM)
Let there be given a signal in continuous time $x(t)$ where $X(\omega) = 0$ for $|\omega| > \omega_c$ (i.e., the signal is limited to ω_c). This signal is completely described by the corresponding sampled signal x_t if the sampling frequency is at least twice as large as ω_c, i.e., $2\pi/T > 2\omega_c \Leftrightarrow T < \pi/\omega_c$.

Proof Follows directly. ∎

Alternatively, Shannon's theorem can be formulated: Any continuous signal, $x(t)$, can be reproduced from the corresponding sampled signal, $x_s(t)$, if, and only if, it does not contain frequencies above the Nyquist frequency.

Aliasing can result in faulty conclusions since a peak in $X(\omega)$ over the Nyquist frequency might be strong enough to appear as a peak at another frequency for the sampled signal. A practical solution to this problem is to choose a shorter sampling time than what was initially thought of and then choose a filter which cuts off the variations approximately outside the desired sampling interval. The most simple, but not always the most ideal, method is to apply subsampling in terms of the mean of k successive values. This filter is of course not adequate when it is important to restore the high frequency variation.

4.4 The z-transform

The z-transform is a useful way to describe dynamical systems in discrete time.

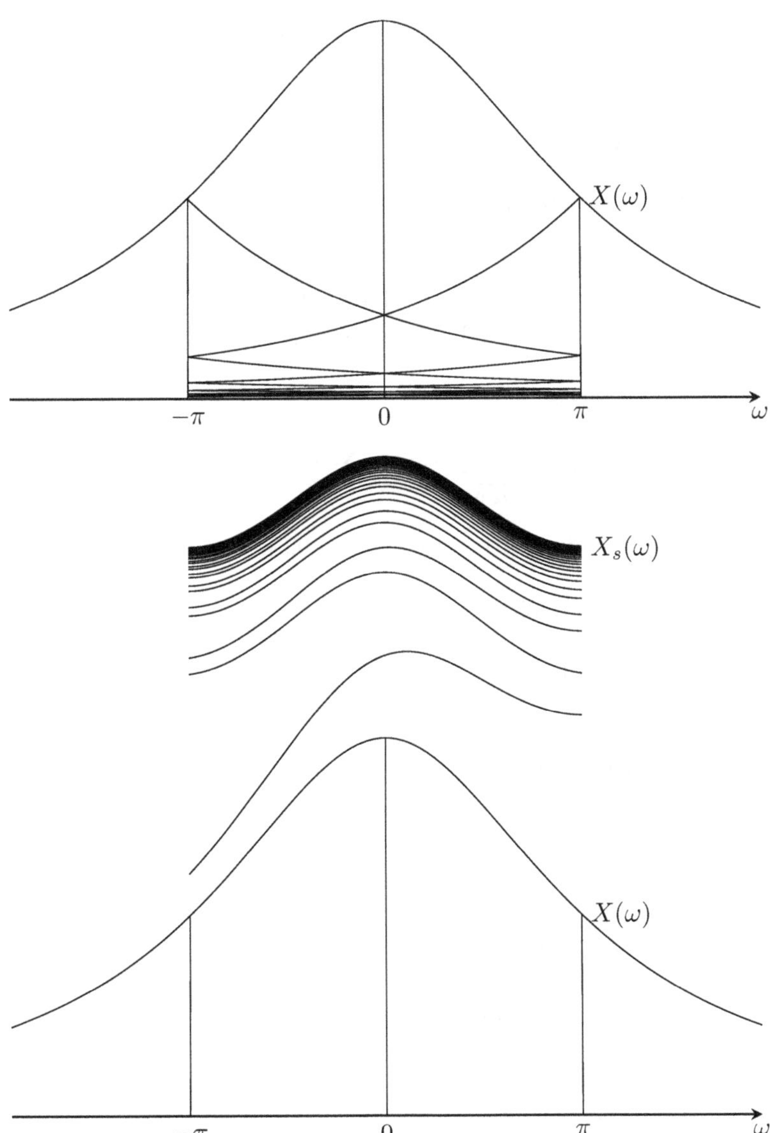

Figure 4.3: *Aliasing. From continuous to sampled signal.*

DEFINITION 4.5 (THE z-TRANSFORM)
For a sequence $\{x_t\}$ the z-transform of $\{x_t\}$ is defined as the complex function

$$Z(\{x_t\}) = X(z) = \sum_{t=-\infty}^{\infty} x_t z^{-t}. \tag{4.50}$$

The z-transform is defined for the complex variables z for which the Laurent series (4.50) is convergent.

Example 4.7 (z-transform of a sequence)
For the sequence $\{x_t\}$ defined as

$$x_t = \begin{cases} 0 & \text{for } t < 0 \\ 2^{-t} & \text{for } t \geq 0 \end{cases}$$

the z-transform becomes

$$Z(\{x_t\}) = X(z) = \sum_{t=0}^{\infty} 2^{-t} z^{-t} = \sum_{t=0}^{\infty} (2z)^{-t} = \frac{1}{1 - 0.5z^{-1}}$$

(with the region of convergence given as $|z| > 0.5$).

In the example we exploit that

$$\frac{1}{1-x} = 1 + x + x^2 + \cdots = \sum_{k=0}^{\infty} x^k \tag{4.51}$$

and that this geometric series is convergent for $|x| < 1$.

The z-transform is a linear operator, i.e.,

$$Z\left(\lambda_1\{x_t\} + \lambda_2\{y_t\}\right) = \lambda_1 Z\left(\{x_t\}\right) + \lambda_2 Z\left(\{y_t\}\right) \tag{4.52}$$

which is readily seen from the definition (4.50).

Of special importance is the backward shift operator.

DEFINITION 4.6 (BACKWARD SHIFT OPERATOR)
The backward shift operator z^{-1} is defined by

$$\begin{aligned} Z(\{x_{t-1}\}) &= \sum_{t=-\infty}^{\infty} x_{t-1} z^{-t} = z^{-1} \sum_{t=-\infty}^{\infty} x_{t-1} z^{-(t-1)} \\ &= z^{-1} X(z) = z^{-1} Z(\{x_t\}) \end{aligned} \tag{4.53}$$

Similarly, we have $Z(\{x_{t-k}\}) = z^{-k} X(z)$ and the *forward shift operator* $Z(\{x_{t+k}\}) = z^k X(z)$.

Example 4.8
It is seen immediately that the z-transform of the impulse function δ_k is

$$Z(\{\delta_k\}) = 1 \tag{4.54}$$

Example 4.9 (Solution of a difference equation)
Consider the linear time-invariant system defined by the difference equation

$$y_{t+1} - 0.9y_t = \delta_t\,, \quad y_0 = 0$$

applying the z-transform we obtain

$$zY(z) - 0.9Y(z) = 1 \Leftrightarrow Y(z) = \frac{1}{z-0.9} = \frac{z^{-1}}{1-0.9z^{-1}},$$

From this it is seen that the solution to the difference equation is

$$Y_t = 0.9^{t-1}\,, \quad t = 1, 2, \ldots$$

The advantage of using the z-transform for linear time-invariant systems is derived from the following.

THEOREM 4.7 (CONVOLUTION IN THE TIME DOMAIN CORRESPONDS TO MULTIPLICATION IN THE z DOMAIN)
Consider the sequences $\{h_k\}$ and $\{x_t\}$, and a sequence $\{y_t\}$ defined by the convolution

$$y_t = \sum_{k=-\infty}^{\infty} h_k x_{t-k} \tag{4.55}$$

then (4.55) *corresponds to*

$$Y(z) = H(z)X(z) \tag{4.56}$$

with $Y(z) = \sum_{t=-\infty}^{\infty} y_t z^{-t}$, $H(z) = \sum_{t=-\infty}^{\infty} h_t z^{-t}$, *and* $X(z) = \sum_{t=-\infty}^{\infty} x_t z^{-t}$.

Proof Use Definition 4.5. ■

$H(z)$ in (4.56) is called the *transfer function* of the system and it can be written as

$$H(z) = \frac{Y(z)}{X(z)} = \sum_{t=-\infty}^{\infty} h_t z^{-t} \tag{4.57}$$

where h_t is the impulse response function.

THEOREM 4.8 (THE LINEAR DIFFERENCE EQUATION)
The difference equation

$$y_t + a_1 y_{t-1} + \cdots + a_p y_{t-p} = b_0 x_{t-\tau} + b_1 x_{t-\tau-1} + \cdots + b_q x_{t-\tau-q} \tag{4.58}$$

represents a linear time-invariant system with input $\{x_t\}$ *and output* $\{y_t\}$. τ *is a constant positive integer-valued* time delay *from input to output.*

Proof The system is linear because the difference equation is linear. Furthermore, the system is time-invariant because the coefficients and the time delay are constant. ∎

Using the z-transform on both sides of (4.58) yields

$$(1 + a_1 z^{-1} + \cdots + a_p z^{-p})Y(z) = z^{-\tau}(b_0 + b_1 z^{-1} + \cdots + b_q z^{-q})X(z),$$

From (4.57) it is seen that the transfer function is

$$H(z) = \frac{(b_0 + b_1 z^{-1} + \cdots + b_q z^{-q})z^{-\tau}}{(1 + a_1 z^{-1} + \cdots + a_p z^{-p})} \tag{4.59}$$

Besides the time delay operator $z^{-\tau}$ the numerator contains the following polynomial in z

$$\begin{aligned} B(z) &= (b_0 + b_1 z^{-1} + \cdots + b_q z^{-q}) \\ &= z^{-q}(b_0 z^q + b_1 z^{q-1} + \cdots + b_q) \\ &= z^{-q}(z - n_1)(z - n_2)\cdots(z - n_q)b_0 \\ &= (1 - n_1 z^{-1})(1 - n_2 z^{-1})\cdots(1 - n_q z^{-1})b_0 \end{aligned} \tag{4.60}$$

where we have used the fact that any polynomial can be factorized. The roots $n_1, n_2, \ldots, n_q$, which may be complex, are called the *zeros of the system* and are found as the solution to $z^q B(z) = 0$.

For the denominator we introduce

$$A(z) = (1 + a_1 z^{-1} + \cdots + a_p z^{-p}) \tag{4.61}$$

$$= (1 - \lambda_1 z^{-1})(1 - \lambda_2 z^{-1})\cdots(1 - \lambda_p z^{-1}) \tag{4.62}$$

where the complex numbers $\lambda_1, \lambda_2, \ldots, \lambda_p$ are called the poles of the system.

Using the introduced polynomials the transfer function for the system defined by the difference equation (4.58) can be written

$$H(z) = \frac{B(z)z^{-\tau}}{A(z)} \tag{4.63}$$

THEOREM 4.9 (STABILITY)
The system described by the difference equation (4.58) *is stable if all the poles* $\lambda_1, \ldots, \lambda_p$ *lie within the unit circle in the complex plane.*

Proof Omitted—see, however, Appendix A, "The solution to difference equations." ■

THEOREM 4.10 (FREQUENCY RESPONSE AND TRANSFER FUNCTIONS)
For a linear time-invariant and stable system, the frequency response function is found by evaluating the transfer function on the unit circle $\{z \in \mathbb{C} \mid |z| = 1\}$*, i.e., by putting* $z = \mathrm{e}^{i\omega}$*. Hence,*

$$\mathcal{H}(\omega) = H(\mathrm{e}^{i\omega}) \tag{4.64}$$

Proof Follows by a comparison between (4.21) on page 74 and (4.57) on page 83. ■

Example 4.10 (A simple system)
Consider the linear time-invariant system defined as

$$y_t = x_t - 0.9x_{t-1}$$

Using the z-transform, we find the transfer function

$$H(z) = 1 - 0.9z^{-1}$$

Using the z-transform, the frequency response function is obtained

$$\mathcal{H}(\omega) = H(\mathrm{e}^{i\omega}) = 1 - 0.9\mathrm{e}^{-i\omega}$$

The gain is

$$|\mathcal{H}(\omega)| = \sqrt{\mathcal{H}(\omega)\overline{\mathcal{H}}(\omega)} = \sqrt{1.81 - 1.8\cos(\omega)}$$

and finally the phase

$$\phi(\omega) = \arg\{\mathcal{H}(\omega)\} = \arctan\left[\frac{0.9\sin(\omega)}{1 - 0.9\cos(\omega)}\right]$$

THEOREM 4.11 (SYSTEMS IN SERIES AND IN PARALLEL)
For systems in series

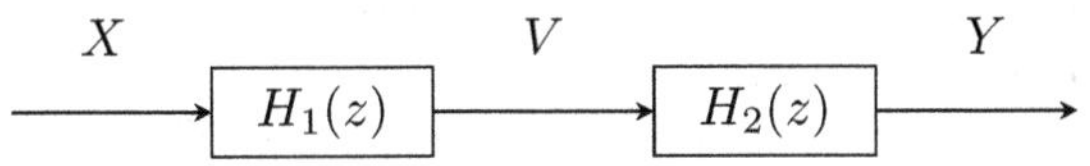

$$\xrightarrow{\;X\;} \boxed{H_1(z)} \xrightarrow{\;V\;} \boxed{H_2(z)} \xrightarrow{\;Y\;}$$

we have the transfer function from X to Y,

$$H(z) = H_1(z)H_2(z) \tag{4.65}$$

and for systems in parallel

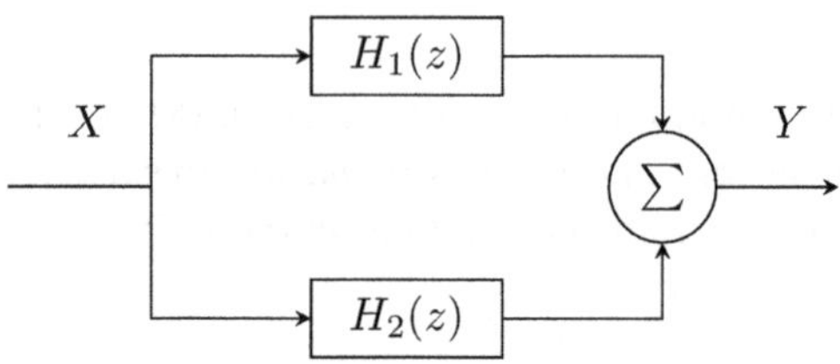

the transfer function from X to Y is

$$H(z) = H_1(z) + H_2(z) \tag{4.66}$$

Proof Seen immediately. ∎

THEOREM 4.12 (SYSTEM WITH A FEEDBACK LOOP)
For the feedback *system*

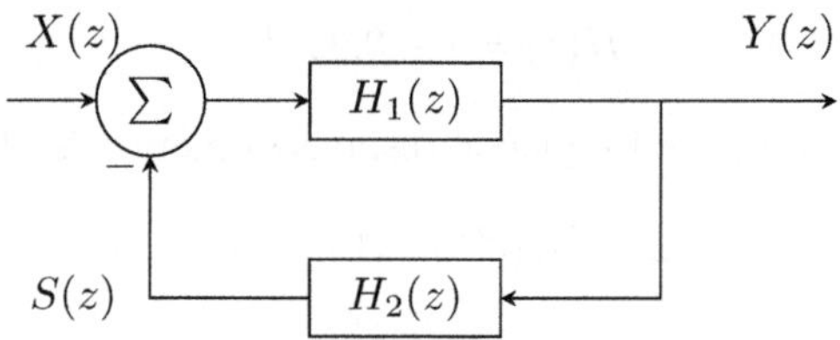

the transfer function from X to Y is

$$H(z) = \frac{H_1(z)}{1 + H_1(z)H_2(z)} \tag{4.67}$$

Proof $S(z) = H_2(z)Y(z)$, and hence $E(z) = X(z) - H_2(z)Y(z)$. Furthermore, we have $Y(z) = H_1(z)E(z)$, and then $Y(z) = H_1(z)X(z) - H_1(z)H_2(z)Y(z)$. By isolating $Y(z)/X(z)$, (4.67) follows. ∎

For a more complex system the total transfer function can be found by using (4.65), (4.66), and (4.67) repeatedly. As an alternative, *Mason's rule* can be applied, as described in Jensen, Høholdt, and Nielsen (1992).

4.5 Frequently used operators

In the previous section the backward shift operator was introduced in the z domain. In this section we will introduce a number of operators which directly apply to the time series $\{x_t\}$ or the stochastic process $\{X_t\}$ in the time domain. Without loss of generality the sampling time is assumed to be 1.

In the following we shall use extensively the *backward shift operator* B defined by

$$\mathrm{B}x_t = x_{t-1} \tag{4.68}$$

If B is used j times we write

$$\mathrm{B}(\mathrm{B}(\cdots(\mathrm{B}x_t)\cdots)) = \mathrm{B}^j x_t = x_{t-j} \tag{4.69}$$

The control theory literature most frequently uses q^{-1} instead of B. By using q^{-1} a more direct correlation with a z-transform is obtained since it is able to just exchange z^{-1} with q^{-1}, thereby changing $X(z)$ to X_t. Within econometrics the backward shift operator B is most frequently denoted L (for *lag*).

Correspondingly the *forward shift operator* F is defined by

$$\mathrm{F}\,x_t = x_{t+1} \tag{4.70}$$

$$\mathrm{F}^j\,x_t = x_{t+j} \tag{4.71}$$

It is clearly seen that F is the inverse operator for B and vice versa, i.e., $\mathrm{B}^{-1} = \mathrm{F}$ and $\mathrm{F}^{-1} = \mathrm{B}$.

The (backward) *difference operator* ∇ is defined as

$$\nabla x_t = x_t - x_{t-1} = 1x_t - \mathrm{B}x_t = (1-\mathrm{B})x_t \tag{4.72}$$

where the unit operator 1 is defined by $1x_t = x_t$.

It is seen from (4.72) that

$$\nabla = 1 - \mathrm{B} \tag{4.73}$$

The *summation operator* S is given by

$$\mathrm{S}\,x_t = (1 + \mathrm{B} + \cdots + \mathrm{B}^i + \cdots)x_t \tag{4.74}$$

Since

$$\nabla^{-1} = \frac{1}{1-\mathrm{B}} = 1 + \mathrm{B} + \cdots + \mathrm{B}^i + \cdots = \mathrm{S} \tag{4.75}$$

it is seen that S and ∇ are the inverse operators of each other.

Finally the *s-season difference operator* ∇_s is defined by

$$\nabla_s x_t = x_t - x_{t-s} = (1 - \mathrm{B}^s)x_t \tag{4.76}$$

All the defined operators are linear, i.e.,

$$\mathrm{H}[\lambda_1 x_t + \lambda_2 y_t] = \lambda_1 \,\mathrm{H}[x_t] + \lambda_2 \,\mathrm{H}[y_t] \tag{4.77}$$

where H is any of the introduced operators. Furthermore the operators are commutative, i.e.,

$$\mathrm{H}_1\,\mathrm{H}_2\, x_t = \mathrm{H}_2\,\mathrm{H}_1\, x_t \tag{4.78}$$

For each of the above defined operators, it is possible to define new *combined operators.* For the power series

$$a(z) = \sum_{i=0}^{\infty} a_i z^i \tag{4.79}$$

a new operator H is defined by

$$a(\mathrm{H}) = \sum_{i=0}^{\infty} a_i \,\mathrm{H}^i \tag{4.80}$$

If $a_i = 0$ for all i higher than some value, then an *operator polynomial* is obtained. For example we have

$$\theta(\mathrm{B}) = (1 + \theta_1 \,\mathrm{B} + \cdots + \theta_q \,\mathrm{B}^q) \tag{4.81}$$

a q-order polynomial in B.

▸ **Remark 4.3**

By the definition (4.80), an isomorphism between the class of combined operators and the class of power series is obtained such that the calculation rules which are used for power series also are valid for the combined operators. In particular we operate with the combined operator polynomials exactly as we operate with ordinary polynomials. Using the above definition a very strong tool for description of linear time-invariant systems in discrete time is obtained without considering the z-transform. ◂

The similarity with power series is illustrated in the following.

THEOREM 4.13

For the operator H *the following operators are given*

$$\lambda(\mathrm{H}) = \sum_{i=0}^{\infty} \lambda_i \,\mathrm{H}^i, \quad \psi(\mathrm{H}) = \sum_{i=0}^{\infty} \psi_i \,\mathrm{H}^i, \quad \pi(\mathrm{H}) = \sum_{i=0}^{\infty} \pi_i \,\mathrm{H}^i$$

such that

$$\lambda(\mathrm{H})\psi(\mathrm{H}) = \pi(\mathrm{H}) \tag{4.82}$$

Then λ_1, ψ_1 and π_i satisfy the equations

$$\{\lambda_i\} * \{\psi_i\} = \{\pi_i\}, \quad (i \geq 0) \tag{4.83}$$

that is,

$$\begin{aligned}
\pi_0 &= \lambda_0 \psi_0 \\
\pi_1 &= \lambda_1 \psi_0 + \lambda_0 \psi_1 \\
&\vdots \\
\pi_i &= \lambda_i \psi_0 + \lambda_{i-1} \psi_1 + \cdots + \lambda_0 \psi_i \\
&\vdots
\end{aligned}$$

Proof Use the well-known relations for the series $\lambda(z)$, $\psi(z)$ and $\pi(z)$. ■

Theorem 4.13 can be used, e.g., to find the inverse combined operators. The inverse $\lambda^{-1}(\mathrm{B})$ for $\lambda(\mathrm{B})$ is found by putting $\psi(\mathrm{B}) = \lambda^{-1}(\mathrm{B})$ and $\pi(\mathrm{B}) = 1$.

Example 4.11 (Box-Jenkins transfer function)
Let us consider the difference equation

$$y_t + \varphi_1 y_{t-1} + \cdots + \varphi_p y_{t-p} = \omega_0 x_{t-\tau} + \omega_1 x_{t-\tau-1} + \cdots + \omega_q x_{t-\tau-q} \tag{4.84}$$

Using the backward shift operator B, (4.84) is written

$$(1 + \varphi_1 \mathrm{B} + \cdots + \varphi_p \mathrm{B}^p) y_t = (\omega_0 + \omega_1 \mathrm{B} + \cdots + \omega_q \mathrm{B}^q) \mathrm{B}^\tau x_t \tag{4.85}$$

By introducing the operator polynomials

$$\varphi(\mathrm{B}) = (1 + \varphi_1 \mathrm{B} + \cdots + \varphi_p \mathrm{B}^p) \tag{4.86}$$

$$\omega(\mathrm{B}) = (\omega_0 + \omega_1 \mathrm{B} + \cdots + \omega_q \mathrm{B}^q) \tag{4.87}$$

then (4.85) is rewritten

$$\varphi(\mathrm{B}) y_t = \omega(\mathrm{B}) \mathrm{B}^\tau x_t \tag{4.88}$$

and finally by isolating y_t

$$y_t = \varphi^{-1}(\mathrm{B}) \omega(\mathrm{B}) \mathrm{B}^\tau x_t = h(\mathrm{B}) x_t = \left[\sum_{i=0}^{\infty} h_i \mathrm{B}^i \right] x_t = \sum_{i=0}^{\infty} h_i x_{t-i} \tag{4.89}$$

According to Box and Jenkins (1970/1976), the model (4.88) is called a *transfer function model* and $h(\mathrm{B}) = \varphi^{-1}(\mathrm{B}) \omega(\mathrm{B}) \mathrm{B}^\tau$ is called the *transfer function* for the system which clearly is based on the impulse response function, $\{h_k\}$. Compare (4.89) with (4.57).

4.6 The Laplace transform

The Laplace transform is a useful tool for description of dynamical systems in continuous time.

DEFINITION 4.7 (THE LAPLACE TRANSFORM)
For a given function in continuous time, $x(t)$, the Laplace transform is defined as

$$\mathcal{L}\{x(t)\} = X(s) = \int_{-\infty}^{\infty} \mathrm{e}^{-st} x(t)\, dt \tag{4.90}$$

provided the integral exists.

Example 4.12 (Laplace transform of a function)
For $x(t)$ defined by

$$x(t) = \begin{cases} 0 & \text{for } t < 0 \\ 2^{-t} & \text{for } t \geq 0 \end{cases}$$

the Laplace transform is

$$\begin{aligned} \mathcal{L}\{x(t)\} &= \int_0^{\infty} \mathrm{e}^{-st} 2^{-t}\, dt = \int_0^{\infty} \mathrm{e}^{-(s+\ln 2)t}\, dt \\ &= \frac{-1}{s+\ln 2} \left[\mathrm{e}^{-(s+\ln 2)t} \right]_0^{\infty} \\ &= \frac{1}{s+\ln 2}, \quad \text{for } \Re(s) > -\ln 2 \end{aligned}$$

where $\Re(s)$ denotes the real part of s. Compare with Example 4.7 on page 82 for the z-transform—also for the region of convergence.

▸ **Remark 4.4**
The Laplace transform is often used on causal systems where $f(t) = 0$ for $t < 0$; therefore, an alternative definition is

$$\mathcal{L}_u\{x(t)\} = X(s) = \int_0^{\infty} x(t) \mathrm{e}^{-st}\, dt.$$

This is called the *unilateral Laplace transform*, and (4.90) is called the *bilateral Laplace transformation.* ◂

From the example above it is seen that more generally

$$\mathcal{L}_u\{c\mathrm{e}^{at}\} = \frac{c}{s-a} \tag{4.91}$$

for $\Re(s) > a$. The index u denotes the unilateral Laplace transform.

The Laplace operator is a *linear operator*, i.e.,

$$\mathcal{L}\left\{\lambda_1 x_1(t) + \lambda_2 x_2(t)\right\} = \lambda_1 \mathcal{L}\left\{x_1(t)\right\} + \lambda_2 \mathcal{L}\left\{x_2(t)\right\} \tag{4.92}$$

which immediately follows from the definition.
Of special importance is the following.

THEOREM 4.14 (DIFFERENTIATION)
For the bilateral Laplace transform

$$\mathcal{L}\left\{\frac{dx(t)}{dt}\right\} = s\mathcal{L}\left\{x(t)\right\} = sX(s) \tag{4.93}$$

and for the unilateral

$$\mathcal{L}_u\left\{\frac{dx(t)}{dt}\right\} = sX(s) - x(0) \tag{4.94}$$

Proof

$$\begin{aligned}\int_0^\infty \dot{x}(t)\mathrm{e}^{-st}dt &= \left[x(t)\mathrm{e}^{-st}\right]_0^\infty + s\int_0^\infty x(t)\mathrm{e}^{-st}dt \\ &= -x(0) + sX(s)\end{aligned}$$

■

THEOREM 4.15 (TIME DELAY)
For a time delay we have

$$\mathcal{L}\left\{x(t-\tau)\right\} = X(s)\mathrm{e}^{-\tau s} \tag{4.95}$$

Proof

$$\begin{aligned}\mathcal{L}\{x(t-\tau)\} &= \int_{-\infty}^{\infty} x(t-\tau)\mathrm{e}^{-st}\,dt \\ &= \mathrm{e}^{-s\tau}\int_{-\infty}^{\infty} x(t-\tau)\mathrm{e}^{-s(t-\tau)}\,d(t-\tau) \\ &= \mathrm{e}^{-s\tau}\mathcal{L}\left\{x(t)\right\} = \mathrm{e}^{-s\tau}X(s).\end{aligned}$$

■

The Laplace transform, combined with tables for the inverse Laplace transform, is a strong tool for solving differential equations.

Example 4.13 (RC element)
Consider the simple electrical system from Example 4.4 on page 75 given as

$$RC\frac{dy(t)}{dt} + y(t) = x(t), \qquad y(t) = 0\,, \quad \text{for } t \le 0$$

By setting the input voltage equal to the Dirac-delta-function $\delta(t)$ and using the Laplace transform (since $\mathcal{L}\{\delta(t)\} = 1$),

$$(RC)sY(s) - y(0) + Y(s) = 1$$

or

$$Y(s) = \frac{1}{1 + (RC)s} = \frac{1/RC}{s + 1/RC}$$

From (4.91) it is seen that the inverse Laplace transform gives

$$y(t) = \frac{1}{RC} e^{-t/RC}$$

which obviously is the impulse response function found previously in Example 4.1.

THEOREM 4.16 (CONVOLUTION IN THE TIME DOMAIN CORRESPONDS TO MULTIPLICATION IN THE s-DOMAIN)
Consider the functions $h(t)$ and $x(t)$ and a function $y(t)$ defined by the convolution integral

$$y(t) = \int_{-\infty}^{\infty} h(u)x(t-u)\, du \tag{4.96}$$

Then (4.96) *implies that*

$$Y(s) = H(s)X(s) \tag{4.97}$$

where $Y(s) = \mathcal{L}\{y(t)\}$, $H(s) = \mathcal{L}\{h(t)\}$ and $X(s) = \mathcal{L}\{x(t)\}$.

Proof As previously for the Fourier transform, see Theorem 4.4 on page 73. ∎

$H(s)$ is the *transfer function* for the continuous time system, and it is clear that

$$H(s) = \frac{Y(s)}{X(s)} = \int_{-\infty}^{\infty} h(t) e^{-st}\, dt \tag{4.98}$$

where $h(t)$ is the impulse response function.

THEOREM 4.17 (THE LINEAR DIFFERENTIAL EQUATION)
The differential equation

$$\begin{aligned} \frac{d^p y}{dt^p}(t) + a_1 \frac{d^{p-1} y}{dt^{p-1}}(t) + a_p y(t) = b_0 \frac{d^q x}{dt^q}(t-\tau) \\ + b_1 \frac{d^{q-1} x}{dt^{q-1}}(t-\tau) + \cdots + b_q x(t-\tau) \end{aligned} \tag{4.99}$$

represents a linear time-invariant system. (τ is a time delay from the input $x(t)$ to the output $y(t)$.)

Proof The system is linear because the differential equation is linear and time-invariant because the coefficients and the time delay are constant. ∎

Using the Laplace transform on both sides of (4.99) gives

$$(s^p + a_1 s^{p-1} + \cdots + a_p)Y(s) = (b_0 s^q + b_1 s^{q-1} + \cdots + b_q)\mathrm{e}^{-\tau s} X(s) \quad (4.100)$$

The *transfer function* for the system described by differential equation (4.99) becomes

$$H(s) = \frac{b_0 s^q + b_1 s^{q-1} + \cdots + b_q}{s^p + a_1 s^{p-1} + \cdots + a_p} \cdot \mathrm{e}^{-\tau s} \quad (4.101)$$

The numerator contains a polynomial

$$B(s) = b_0 s^q + b_1 s^{q-1} + \cdots + b_q \quad (4.102)$$
$$= (s - n_1)(s - n_2) \cdots (s - n_q) b_0 \quad (4.103)$$

and $n_1, n_2, \ldots, n_q$ are the zeros of the system.

Similarly the denominator contains the polynomial

$$A(s) = s^p + a_1 s^{p-1} + \cdots + a_p \quad (4.104)$$
$$= (s - \lambda_1)(s - \lambda_2) \cdots (s - \lambda_p) \quad (4.105)$$

where $\lambda_1, \lambda_2, \ldots, \lambda_p$ are called the poles of the system.

Using the introduced polynomials, the transfer function for the continuous time system is written

$$H(s) = \frac{B(s)}{A(s)} \cdot \mathrm{e}^{-\tau s} \quad (4.106)$$

Theorem 4.18 (Stability)
The linear system described by the differential equation (4.99) *is stable provided that all poles lie in the left part of the complex s-plane, i.e.,* $\Re\{\lambda_i\} < 0$ *for all* i.

Proof Omitted. ∎

Theorem 4.19 (Frequency response and transfer functions)
For a linear time-invariant and stable system, described by the transfer function (4.106)*, the frequency response function is found as the transfer function evaluated on the imaginary axis, i.e., by putting* $s = i\omega$*. Hence,*

$$\mathcal{H}(\omega) = H(i\omega) \quad (4.107)$$

where H *is given by* (4.106).

Proof Follows immediately by a comparison between (4.18) on page 73 and (4.98) on page 92. ■

Section 4.4 on page 80 describes how the total transfer function is obtained for systems in series or parallel and for feedback systems. Exactly the same calculation rules can be used for continuous time systems.

4.7 A comparison between transformations

Assume $x(t)$ is a function which is sampled at the time instants $\{\dots, -T, 0, T, 2T, \dots\}$. In Section 4.3 on page 78, it was shown that the sampled signal $x_s(t)$ can be written

$$x_s(t) = \sum_{n=-\infty}^{\infty} x(t)\delta(t-nT) \tag{4.108}$$

By using $x(t)\delta(t) = x(0)\delta(t)$, we have

$$x_s(t) = \sum_{n=-\infty}^{\infty} x(nT)\delta(t-nT) \tag{4.109}$$

Then using the Laplace transform gives

$$x_s(s) = \sum_{n=-\infty}^{\infty} x(nT)\mathcal{L}\{\delta(t-nT)\} \tag{4.110}$$

Since $\mathcal{L}\{\delta(t)\} = 1$, using Theorem 4.16 on page 92, we have

$$\mathcal{L}\{\delta(t-nT)\} = \mathrm{e}^{-snT}\mathcal{L}\{\delta(t)\} = \mathrm{e}^{-snT}$$

thus, (4.110) can be written

$$X_s(s) = \sum_{n=-\infty}^{\infty} x(nT)\mathrm{e}^{-snT} \tag{4.111}$$

Finally, using the substitution

$$z = \mathrm{e}^{sT} \tag{4.112}$$

(4.111) is written

$$X_s(s)\big|_{z=\mathrm{e}^{st}} = \sum_{n=-\infty}^{\infty} x(nT)z^{-n} = X(z) \tag{4.113}$$

where $X(z)$ is the z-transform of $\{x(nT), n = 0, \pm 1, \dots\}$.

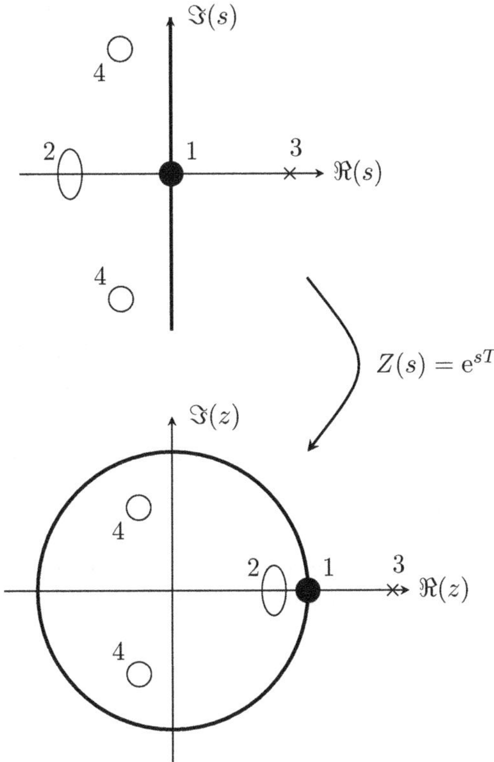

Figure 4.4: *From s-plane to z-plane (a sampled continuous signal). $\Re$ denotes real part, and $\Im$ the imaginary part of the complex number s or z.*

It is clearly seen that the z-transform corresponds to the Laplace transform of the sampled signal. Using the substitution (4.112) it is seen that when the transfer function from x to y is written as a rational function in the Laplace operator, s (i.e., (4.101) without time delay), then the transfer function for the sampled system is written as a rational function in e^{-sT} (see (4.59) on page 84).

The choice of sampling time T imposes the mapping defined by $z(s) = \mathrm{e}^{sT}$, which is a mapping of the complex s-plane onto the complex z-plane as illustrated in Figure 4.4. It maps the imaginary axis, $\Re(s) = 0$, onto the unit circle, $|z| = 1$, in the complex z-plane. Furthermore, the left half of the complex s-plane, ($\Re(s) < 0$), falls inside the unit circle in the complex z-plane. To ensure uniqueness, only s-plane values in the region: $-\pi < \Im(s) \leq \pi$ are considered, since the mapping will put all the values $s \pm 2p\pi\,, p \in \mathbb{N}$ into the same value in the z-plane.

The mapping is particularly useful for a transformation of the location of poles and zeros from discrete to continuous time and vice versa.

If we choose $s = i\omega$, so that s is on the imaginary axis, then we obtain X_s (we also set $T = 1$)

$$X_s(i\omega) = \sum_{n=-\infty}^{\infty} x(n)\mathrm{e}^{-in\omega} \tag{4.114}$$

which obviously is the Fourier transform (see (4.21) on page 74).

4.8 Problems

Exercise 4.1
A system in discrete time can be described by the following difference equation:

$$y_t - 1.2y_{t-1} + 0.61y_{t-2} = x_t - 0.8x_{t-1}$$

Question 1 Determine the poles and zeros of the system. Is the system stable?

Question 2 Determine the impulse response function (to and including $k = 5$).

Question 3 Determine the frequency response function and sketch the amplitude function. Compare the amplitude function with the pole placement—comment!

Exercise 4.2
Consider a system which can be described by the following differential equation:

$$\frac{d^2y(t)}{dt^2} + 5\frac{dy(t)}{dt} + 4y(t) = x(t)$$

Question 1 Determine the transfer function for the system. Is the system stable?

Question 2 Determine the frequency response function for the system.

Question 3 Suppose we want to sample the signals. Specify a suitable sampling time T.

Question 4 Determine the poles for the sampled system.

CHAPTER 5

Stochastic processes

5.1 Introduction

A *time series* $\{x_t\,, t = 0, \pm 1, \ldots\}$ is a *realization* of a *stochastic process* $\{X_t\,, t = 0, \pm 1, \ldots\}$ and an important application of modern time series analysis is to model, e.g., a physical phenomenon as a stochastic process.

This book provides only a brief introduction to stochastic processes. Its main purpose is to provide only the most essential notation and definitions and to introduce a class of useful stochastic processes. For a rather detailed introduction to stochastic processes, we refer to Grimmit and Stirzaker (1992), Doob (1953), Yaglom (1962), Parzen (1962), and Cox and Miller (1968). More applied oriented introductions are found in, e.g., Papoulis (1983) and Davis and Vinter (1985).

5.2 Stochastic processes and their moments

A stochastic process is defined as a family of random variables $\{X(t)\}$ or $\{X_t\}$, where t belongs to an *index set*. In the first case we are considering a process in *continuous time* while in the latter a process in *discrete time*. In this section only processes in continuous time will be described due to the notation, although the results and definitions can easily be formulated in discrete time.

A stochastic process is a function having two arguments, $\{X(t, \omega)\,, \omega \in \Omega\}$, where Ω is the *sample space*, or the *ensemble*, of the process. This implies that Ω is the set of all the possible time series that can be generated by the process. For a fixed t we say that $X(t, \cdot)$ is a *random variable*, and for fixed $\omega \in \Omega$ we call it a *realization* of the process, i.e., a *time series*, $X(\cdot, \omega)$; cf. Figure 5.1 on the next page.

Most often only one realization of the process is available, and in such cases it is necessary to assume several properties for the process in order to postulate distributional characteristics about the process.

Since a stochastic process is a family of random variables, its distributional specifications are more or less equivalent to what we have seen for multidimensional random variables.

The function, $f_{X(t_1),\ldots,X(t_n)}(x_1, \ldots, x_n)$, is called the n-dimensional probability distribution function for the process $\{X(t)\}$. The family of all these

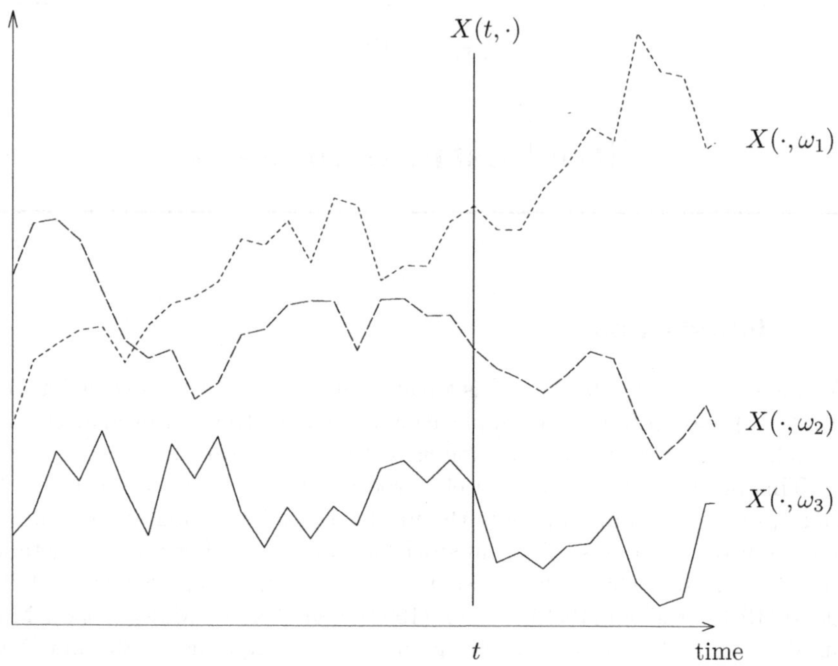

Figure 5.1: *Realizations of a stochastic process.*

probability distribution functions, for $n = 1, 2, \dots$ and for all values of t, forms the *family of finite-dimensional probability distribution functions* for the process $\{X(t)\}$. The stochastic process is completely characterized by the family of finite-dimensional probability distribution functions.

Just as it is adequate to characterize random variables by their moments, it is suitable to characterize stochastic processes by their *moment functions.*

The simplest moment is determined by the *mean value* (or the *mean value function*):

$$\mu(t) = \mathrm{E}[X(t)] = \int_{-\infty}^{\infty} x f_{X(t)}(x)\, dx, \tag{5.1}$$

which is a function of t. The expression (5.1) is used in case the sample space for $X(t)$ is continuous, while the corresponding summation is used when the sample space is discrete. The mean value is also referred to as the *first moment* of the process.

In a similar fashion we define the *variance* (or the *variance function*) of the process

$$\sigma^2(t) = \mathrm{Var}[X(t)] = \mathrm{E}\left[(X(t) - \mu(t))^2\right]. \tag{5.2}$$

An important concept is the *autocovariance function*:

DEFINITION 5.1 (AUTOCOVARIANCE FUNCTION)
The autocovariance function is given by

$$\begin{aligned}\gamma_{XX}(t_1,t_2) = \gamma(t_1,t_2) &= \mathrm{Cov}\left[X(t_1), X(t_2)\right] \\ &= \mathrm{E}\big[(X(t_1)-\mu(t_1))(X(t_2)-\mu(t_2))\big].\end{aligned} \tag{5.3}$$

As a special case, we have $\sigma^2(t) = \gamma(t,t)$. The *second moment* is given by the autocovariance function. The mean value and the autocovariance function form the *second order moment representation* of the process.

In a similar manner we can define moments of higher order. For example,

$$\begin{aligned}&\mathrm{K}(t_1, t_2, \ldots, t_n) \\ &\quad = \mathrm{E}\left[(X(t_1)-\mu(t_1))(X(t_2)-\mu(t_2))\cdots(X(t_n)-\mu(t_n))\right]\end{aligned} \tag{5.4}$$

is called the *moment function of n'th order*. For more detailed information on higher order moments, see Brillinger (1981) or Papoulis (1983).

In Section 5.2.2 we will consider the autocovariance functions more closely.

5.2.1 Characteristics for stochastic processes

In this section we will introduce a number of useful characteristics for stochastic processes. This gives us a possibility to limit our attention to stationary and ergodic processes.

5.2.1.1 Stationary processes

DEFINITION 5.2 (STRONG STATIONARITY)
A process $\{X(t)\}$ is said to be *strongly stationary* if all finite-dimensional distributions are invariant for changes in time, i.e., for every n, any set $(t_1, t_2, \ldots, t_n)$ and for any h it holds

$$f_{X(t_1),\ldots,X(t_n)}(x_1,\ldots,x_n) = f_{X(t_1+h),\ldots,X(t_n+h)}(x_1,\ldots,x_n) \tag{5.5}$$

DEFINITION 5.3 (WEAK STATIONARITY)
A process $\{X(t)\}$ is said to be *weakly stationary of order k* if all the first k moments are invariant to changes in time. A weakly stationary process of order 2 is simply called *weakly stationary.*

THEOREM 5.1
A weakly stationary process is characterized by the fact that both the mean value and the variance are constant, while the autocovariance function depends only on the time difference, i.e., by $(t_1 - t_2)$.

Proof Follows directly from Definition 5.3. ∎

In the remainder of the book we shall use the term *stationary* to denote a weakly stationary process.

DEFINITION 5.4 (ERGODICITY)
A stationary process is said to be ergodic if its ensemble averages equal appropriate time averages.

By this definition we mean that any statistic of $X(t)$ can be determined from a single realization $x(t)$. This is a general requirement. In most applications we consider only specific statistics, as, e.g., the mean value.

A stationary process is said to be mean-ergodic if

$$\mathrm{E}[X(t)] = \int_\Omega x(t,\omega) f(\omega)\, d\omega = \lim_{T\to\infty} \frac{1}{2T} \int_{-T}^{T} x(t,\omega)\, dt, \tag{5.6}$$

i.e., if the *mean of the ensemble* equals the *mean over time* (see also Figure 5.1 on page 98).

5.2.1.2 Normal processes

DEFINITION 5.5 (NORMAL PROCESS)
A process $\{X(t)\}$ is said to be a *normal process* (or a *Gaussian process*) if all the n-dimensional distribution functions $f_{X(t_1),\ldots,X(t_n)}(x_1,\ldots,x_n)$ for any n are (multidimensional) normal distributions.

A normal process is completely specified by its mean value function

$$\mu(t_i) = \mathrm{E}[X(t_i)] \tag{5.7}$$

and autocovariance function

$$\gamma(t_i, t_j) = \mathrm{Cov}[X(t_i), X(t_j)]. \tag{5.8}$$

Introducing the vector $\boldsymbol{\mu} = (\mu(t_1), \mu(t_2), \ldots, \mu(t_n))^T$ and the variance matrix $\boldsymbol{\Sigma} = \{\gamma(t_i, t_j)\}$, the joint distribution for $\boldsymbol{X} = (X(t_1), X(t_2), \ldots, X(t_n))^T$ is given by

$$f_{\boldsymbol{X}}(\boldsymbol{x}) = \frac{1}{2\pi^{n/2}\sqrt{\det \boldsymbol{\Sigma}}} \exp\left(-\frac{1}{2}(\boldsymbol{x}-\boldsymbol{\mu})^T \boldsymbol{\Sigma}^{-1}(\boldsymbol{x}-\boldsymbol{\mu})\right) \tag{5.9}$$

where $\boldsymbol{\Sigma}$ is assumed to be regular. For a stationary normal process the mean value is constant and the autocovariance function only depends on $(t_i - t_j)$.

Weak stationarity and strong stationarity are equivalent for normal processes, since a normal process is completely characterized by the first two moments.

5.2.1.3 Markov processes

DEFINITION 5.6 (MARKOV PROCESS)
A process $\{X(t)\}$ is called a *Markov process*, if for $t_1 < t_2 < \cdots < t_n$, the distribution of $X(t_n)$ given $(X(t_1), \ldots, X(t_{n-1}))$ is the same as the distribution of $X(t_n)$ given $X(t_{n-1})$. This implies that

$$\mathrm{P}\{X(t_n) \leq x | X(t_{n-1}), \ldots, X(t_1)\} = \mathrm{P}\{X(t_n) \leq x | X(t_{n-1})\} \tag{5.10}$$

A Markov process is thus characterized by the fact that all information about $X(t_n)$ from past observations of $\{X(t)\}$ is contained in the previous observation $X(t_{n-1})$. Due to the fact that only the most recent observation is needed, the process is also called a first order Markov process.

Example 5.1 (AR(1) process, part I)
Let $\{\varepsilon_t\}$ be a sequence of uncorrelated normally distributed variables with $\mathrm{E}[\varepsilon_t] = 0$ and $\mathrm{Var}[\varepsilon_t] = \sigma^2$, i.e., $\{\varepsilon_t\}$ is white noise. Let $\{\varepsilon_t\}$ be the input to a dynamical system defined by the difference equation

$$Y_t = \phi Y_{t-1} + \varepsilon_t, \tag{5.11}$$

which defines a new stochastic process $\{Y_t\}$. By successively substituting $Y_{t-1} = \phi Y_{t-2} + \varepsilon_{t-1}$, $Y_{t-2} = \phi Y_{t-3} + \varepsilon_{t-2}$, ... on the right hand side of (5.11), it is seen that Y_t can be written

$$Y_t = \varepsilon_t + \phi \varepsilon_{t-1} + \phi^2 \varepsilon_{t-2} + \cdots + \phi^i \varepsilon_{t-i} + \cdots . \tag{5.12}$$

We find

$$\mu_Y = \mathrm{E}[Y_t] = 0 \tag{5.13}$$

and

$$\sigma_Y^2 = \mathrm{Var}[Y_t] = (1 + \phi^2 + \phi^4 + \cdots + \phi^{2i} + \cdots)\sigma^2 = \frac{\sigma^2}{1 - \phi^2} \tag{5.14}$$

conditioned that $|\phi| < 1$. For $|\phi| \geq 1$ the variance is unbounded.

Finally, we have the covariance function for $(t_1 > t_2)$

$$\begin{aligned}
\gamma(t_1, t_2) &= \mathrm{Cov}[Y_{t_1}, Y_{t_2}] \\
&= \mathrm{Cov}[\varepsilon_{t_1} + \phi \varepsilon_{t_1 - 1} + \cdots + \phi^{t_1 - t_2} \varepsilon_{t_2} + \cdots, \varepsilon_{t_2} + \phi \varepsilon_{t_2 - 1} + \cdots] \\
&= \phi^{t_1 - t_2}(1 + \phi^2 + \phi^4 + \cdots)\sigma^2 \\
&= \phi^{t_1 - t_2} \sigma_Y^2
\end{aligned}$$

since $\text{Cov}[\varepsilon_{t_i}, \varepsilon_{t_2} + \phi\varepsilon_{t_2-1} + \cdots] = 0$ for all $t_i > t_2$. Similarly for $t_1 < t_2$,

$$\gamma(t_1, t_2) = \phi^{t_2-t_1}\sigma_Y^2.$$

We observe, that $\gamma(t_1, t_2)$ depends only on the time difference $t_1 - t_2$, i.e.,

$$\gamma(t_1, t_2) = \gamma(t_1 - t_2) = \phi^{|t_1-t_2|}\sigma_Y^2. \tag{5.15}$$

Since the mean value and the variance are constant for $|\phi| < 1$, and the autocovariance function depends only on the time difference, $\{Y_t\}$ is a *weakly stationary* process for $|\phi| < 1$.

It is well known that a sum of normally distributed random variables is normally distributed. Hence $\{Y_t\}$ is a *normal process* since ε_t is normally distributed. The process is thus *strongly stationary* for $|\phi| < 1$.

From (5.11) it is seen that

$$\text{P}\{Y_t \le y | Y_{t-1}, Y_{t-2}, \dots\} = \text{P}\{Y_t \le y | Y_{t-1}\}, \tag{5.16}$$

and hereby that $\{Y_t\}$ is a Markov process.

$\{Y_t\}$ is called an AutoRegressive process of first order (in short AR(1)). The name indicates that new values of Y are obtained by regression on old values of Y, and the order is given by the order of the difference equation.

5.2.1.4 Deterministic processes

DEFINITION 5.7 (DETERMINISTIC PROCESSES)
A process is said to be deterministic if it can be predicted *without uncertainty* from past observations.

Example 5.2
A process $\{Y_t\}$ defined by

$$Y_t = A_1 \cos(\omega t) + A_2 \sin(\omega t), \tag{5.17}$$

where A_1 and A_2 are random variables, is a deterministic process.

Example 5.3
The process $\{Y_t\}$ defined by

$$Y_t = \phi_1 Y_{t-1} - \phi_2 Y_{t-2}, \quad (Y_0 = A_1; Y_1 = A_2) \tag{5.18}$$

where ϕ_1 and ϕ_2 are constants, is a deterministic process. Here A_1 and A_2 are random variables.

The process in Example 5.3 is an example of a subset of the deterministic processes where all the values of the process can be expressed as a linear combination of a finite number of variables.

5.2.1.5 Purely stochastic processes

DEFINITION 5.8 (PURELY STOCHASTIC PROCESS)
A process is said to be (purely) stochastic, if it can be written in the form

$$X_t = \varepsilon_t + \psi_1 \varepsilon_{t-1} + \psi_2 \varepsilon_{t-2} + \cdots$$

where $\{\varepsilon_t\}$ is a sequence of uncorrelated stochastic variables, where $\mathrm{E}[\varepsilon_t] = 0$, $\mathrm{Var}[\varepsilon_t] = \sigma^2$, and $\sum_{i=0}^{\infty} \psi_i^2 < \infty$.

An important theorem for stationary processes follows.

THEOREM 5.2 (WOLD'S DECOMPOSITION)
Any stationary process in discrete time $\{X_t\}$ can be written in the form

$$X_t = S_t + D_t \,,$$

where $\{S_t\}$ is a purely stochastic process and $\{D_t\}$ is a purely deterministic process.

Proof See, e.g., Cox and Miller (1968). ∎

5.2.2 Covariance and correlation functions

For a stochastic process, $\{X(t)\}$, we have the *autocovariance function*

$$\gamma_{XX}(t_1, t_2) = \gamma(t_1, t_2) = \mathrm{Cov}[X(t_1), X(t_2)], \tag{5.19}$$

where the variance for the process is $\sigma^2(t) = \gamma_{XX}(t, t)$.

Similarly the *autocorrelation function* (ACF) is defined as

$$\rho_{XX}(t_1, t_2) = \rho(t_1, t_2) = \frac{\gamma_{XX}(t_1, t_2)}{\sqrt{\sigma^2(t_1)\sigma^2(t_2)}}. \tag{5.20}$$

If the process is stationary, then (5.20) is only a function of the time difference $t_2 - t_1$. Denoting the time difference by τ, we have for *stationary processes* the *autocovariance function*

$$\gamma_{XX}(\tau) = \mathrm{Cov}[X(t), X(t+\tau)], \tag{5.21}$$

and the *autocorrelation function*

$$\rho_{XX}(\tau) = \frac{\gamma_{XX}(\tau)}{\gamma_{XX}(0)} = \frac{\gamma_{XX}(\tau)}{\sigma_X^2}, \tag{5.22}$$

where $\sigma_X^2(t)$ is the variance of the process. Please note that $\rho_{XX}(0) = 1$.

Let us consider the stochastic processes, $\{X(t)\}$ and $\{Y(t)\}$. The covariance between these two stochastic processes is described by the *cross-covariance function*

$$\gamma_{XY}(t_1, t_2) = \mathrm{Cov}[X(t_1), Y(t_2)] \tag{5.23}$$
$$= \mathrm{E}\big[(X(t_1) - \mu_X(t_1))(Y(t_2) - \mu_Y(t_2))\big] \tag{5.24}$$

where $\mu_X(t) = \mathrm{E}[X(t)]$ and $\mu_Y(t) = \mathrm{E}[Y(t)]$.

Similarly to (5.20), we define the *cross-correlation function* (CCF)

$$\rho_{XY}(t_1, t_2) = \frac{\gamma_{XY}(t_1, t_2)}{\sqrt{\sigma_X^2(t_1)\sigma_Y^2(t_2)}}, \tag{5.25}$$

where $\sigma_X^2(t) = \mathrm{Var}[X(t)]$ and $\sigma_Y^2(t) = \mathrm{Var}[Y(t)]$.

Thus, the cross-covariance[1] and cross-correlation functions for *stationary processes* become

$$\gamma_{XY}(\tau) = \mathrm{Cov}[X(t), Y(t + \tau)], \tag{5.26}$$

$$\rho_{XY}(\tau) = \frac{\gamma_{XY}(\tau)}{\sqrt{\gamma_{XX}(0)\gamma_{YY}(0)}} = \frac{\gamma_{XY}(\tau)}{\sigma_X \sigma_Y}. \tag{5.27}$$

The autocovariance function for a stationary process has the following properties.

THEOREM 5.3 (PROPERTIES OF THE AUTOCOVARIANCE FUNCTION)
Let $\{X(t)\}$ be a stationary process having the autocovariance function $\gamma(\tau)$. It then holds that

i) $\gamma(\tau) = \gamma(-\tau)$.

ii) $|\gamma(\tau)| \le \gamma(0)$.

iii) The quadratic form of z given $\sum_{i=1}^{n}\sum_{j=1}^{n} z_i z_j \gamma(t_i - t_j)$, is for any z, n and times t_i, non-negative definite.

iv) If $\gamma(\tau)$ is continuous for $\tau = 0$ then it is continuous everywhere.

Proof

i)
$$\begin{aligned}\gamma(\tau) &= \mathrm{Cov}[X(t), X(t + \tau)]\\ &= \mathrm{Cov}[X(t + \tau), X(t)]\\ &= \mathrm{Cov}[X(t), X(t - \tau)]\\ &= \gamma(-\tau).\end{aligned}$$

---

[1] Actually, the bivariate process $(X(t), Y(t))^T$ must be stationary.

ii) Since $\text{Var}[\lambda_1 X(t) + \lambda_2 X(t+\tau)] \geq 0$ it follows that

$$(\lambda_1^2 + \lambda_2^2)\gamma(0) + 2\lambda_1\lambda_2\gamma(\tau) \geq 0,$$

since $\{X(t)\}$ is stationary, and hereby $\text{Var}[X(t)] = \text{Var}[X(t+\tau)] = \gamma(0)$.

If we put $\lambda_1 = \lambda_2 = 1$ we obtain $\gamma(\tau) \geq -\gamma(0)$, and if we put $\lambda_1 = 1$, and $\lambda_2 = -1$, we obtain $\gamma(\tau) \leq \gamma(0)$. In conclusion, $|\gamma(\tau)| \leq \gamma(0)$.

iii) For any choice of z, n, and time t_i, it holds that

$$\text{Var}\left[\sum_{i=1}^{n} z_i X(t_i)\right] \geq 0$$

which implies

$$\sum_{i=1}^{n}\sum_{j=1}^{n} z_i z_j \gamma(t_i, t_j) \geq 0.$$

iv) The proof is as follows:

$$\begin{aligned} |\gamma(\tau+h) - \gamma(\tau)| &= \left|\text{Cov}\left[X(t+\tau+h) - X(t+\tau), X(t)\right]\right| \\ &= \left|\text{Cov}\left[Y(t), X(t)\right]\right| \quad \text{where} \\ &\qquad Y(t) = X(t+\tau+h) - X(t+\tau) \\ &\leq \sqrt{\text{Cov}\left[Y(t), Y(t)\right]\text{Cov}\left[X(t), X(t)\right]} \\ &= \sqrt{2\left(\gamma(0) - \gamma(h)\right)\gamma(0)} \to 0\,, \quad \text{for } h \to 0 \end{aligned}$$

if $\gamma(\tau)$ is continuous in 0. ■

It follows from Theorem 5.3 and (5.22) on page 103 that the autocorrelation function for a stationary stochastic process is an even function. This implies that $\rho(-\tau) = \rho(\tau)$. It should be noted that from ii) it follows that $|\rho(\tau)| \leq 1$.

Since the autocovariance and the autocorrelation functions are symmetric around the lag $k = 0$, these functions are often plotted only for non-negative lags as illustrated in Figure 5.2 on the next page. The plot of the autocorrelation function is sometimes called a *correlogram*.

Similarly, we have the following properties.

THEOREM 5.4 (PROPERTIES OF THE CROSS-COVARIANCE FUNCTION)

i) $\gamma_{XY}(\tau) = \gamma_{YX}(-\tau)$.

ii) $|\gamma_{XY}(\tau)|^2 \leq \gamma_{XX}(0)\gamma_{YY}(0)$.

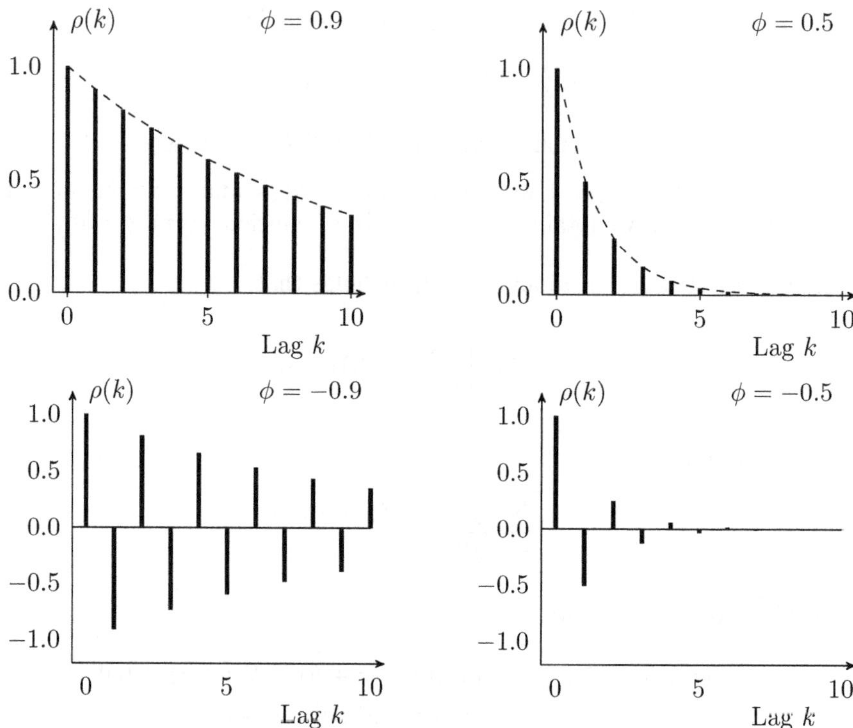

Figure 5.2: *Examples of autocorrelation functions for AR(1) processes.*

Proof

i)
$$\begin{aligned}\gamma_{XY}(\tau) &= \text{Cov}[X(t), Y(t+\tau)] \\ &= \text{Cov}[Y(t+\tau), X(t)] \\ &= \text{Cov}[Y(t), X(t-\tau)] \\ &= \gamma_{YX}(-\tau).\end{aligned}$$

ii) Since $\text{Var}[\lambda_1 X(t) + \lambda_2 Y(t+\tau)] \geq 0$, we obtain
$$\lambda_1^2 \gamma_{XX}(0) + \lambda_2^2 \gamma_{YY}(0) + 2\lambda_1\lambda_2\gamma_{XY}(\tau) \geq 0.$$
Taking $\lambda_1 = 1/\sqrt{\gamma_{XX}(0)}$, $\lambda_2 = 1/\sqrt{\gamma_{YY}(0)}$, and subsequently $\lambda_1 = 1/\sqrt{\gamma_{XX}(0)}$, $\lambda_2 = -1/\sqrt{\gamma_{YY}(0)}$ the result follows. ∎

It follows from Theorem 5.4 on the preceding page and (5.27) on page 104 that for the cross-correlation function, it holds that $\rho_{XY}(\tau) = \rho_{YX}(-\tau)$, and that $|\rho_{XY}(\tau)| \leq 1$. Finally, it should be noted, that while for the autocorrelation it holds that $\rho_{XX}(0) = 1$, it holds for the cross-correlation function that $|\rho_{XY}(0)| \leq 1$.

The cross-correlation function is not symmetric, and therefore it should be noticed that the cross-correlation functions is a measure of not only the strength of a relation between to stochastic process but also its direction. To give a full description of the relationship it is then important to examine the cross-correlation function for both positive ($k > 0$) and negative ($k < 0$) lags. A plot of the cross-correlation function as a function of the lag k is sometimes called the *cross-correlogram.*

Let us illustrate the previous fact by example. This example concerns the same stochastic process as in Example 5.1 on page 101.

Example 5.4 (AR(1) process, part II)
Consider the same process as in (5.11) on page 101. The process is stationary for $|\phi| < 1$. We find for ($k > 0$)

$$\begin{aligned}\gamma(k) = \gamma(-k) &= \mathrm{Cov}[Y_t, Y_{t-k}] \\ &= \mathrm{Cov}[\phi Y_{t-1} + \varepsilon_t, Y_{t-k}] \\ &= \phi\,\mathrm{Cov}[Y_{t-1}, Y_{t-k}] \\ &= \phi\gamma(k-1) = \phi^2\gamma(k-2) = \cdots\end{aligned}$$

Hereby we obtain that $\gamma(k) = \phi^k\gamma(0)$. Since $\gamma(k)$ is an even function it follows that

$$\gamma(k) = \phi^{|k|}\gamma(0),$$

and the autocorrelation function becomes

$$\rho(k) = \phi^{|k|}, \quad |\phi| < 1. \tag{5.28}$$

The argument k is often referred to as the *lag*, i.e., time distance.

The coefficient ϕ determines the memory of the process. For ϕ close to 1, there is a long memory, while the memory is short for small values of ϕ. Finally, $\rho(k)$ will oscillate for $\phi < 0$. This is illustrated in Figure 5.2.

It should be noted that for $\phi < 0$ there exists no embedded first order autoregressive process in continuous time. This corresponds to the fact that the equation $z = \mathrm{e}^{sT}$ cannot be solved with respect to s in case z is negative.

5.3 Linear processes

This section is concerned with linear stochastic processes in the time domain. Subsequently in Section 5.4 stochastic processes in the frequency domain will be considered.

5.3.1 Processes in discrete time

In general, a linear process $\{Y_t\}$ can be interpreted as the output from a linear system, where the input is white noise.

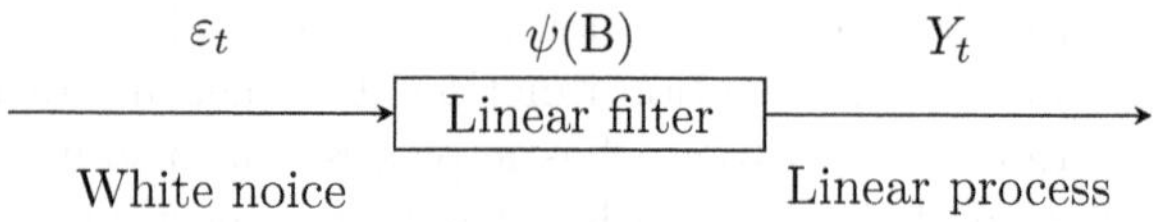

Figure 5.3: *Illustration of a linear process as the output from a linear filter having white noise as input.*

DEFINITION 5.9 (WHITE NOISE)
A process $\{\varepsilon_t\}$ is said to be *a completely random process* or *white noise*, if $\{\varepsilon_t\}$ is a sequence of mutually uncorrelated identically distributed random variables with mean value 0 and constant variance σ_ε^2. This implies that

$$\mu_t = \mathrm{E}[\varepsilon_t] = 0, \qquad \sigma_t^2 = \mathrm{Var}[\varepsilon_t] = \sigma_\varepsilon^2, \tag{5.29}$$

$$\gamma_\varepsilon(k) = \mathrm{Cov}[\varepsilon_t, \varepsilon_{t+k}] = 0\,, \quad \text{for } k \neq 0. \tag{5.30}$$

For white noise we obtain the autocorrelation function

$$\rho_\varepsilon(k) = \begin{cases} 1 & k = 0 \\ 0 & k = \pm 1, \pm 2, \ldots \end{cases} \tag{5.31}$$

The name white noise is due to the fact that the spectral density for $\{\varepsilon_t\}$ is constant, i.e., $f(\omega) = \sigma_\varepsilon^2/2\pi$. In other words the variations of $\{\varepsilon_t\}$ can be described by a uniform proportion of all frequencies. This will be discussed in Section 5.4 on page 113.

As discussed in Chapter 4, the output from a linear, time-invariant system can be written as a convolution of the input by a weight function (the impulse response function).

DEFINITION 5.10 (THE LINEAR PROCESS)
A (general) linear process $\{Y_t\}$ is a process that can be written in the form

$$Y_t - \mu = \sum_{i=0}^{\infty} \psi_i \varepsilon_{t-i}, \tag{5.32}$$

where $\{\varepsilon_t\}$ is white noise and μ is the mean value of the process.

The sequence $\{\psi_k\}$ is often referred to as the ψ *weights*, and (5.32) is referred to as the *random shock form*. *Please notice that without loss of generality we will assume that $\mu = 0$ in the following.*

It is common to scale $\{\varepsilon_t\}$ so that $\psi_0 = 1$. By introducing the linear operator

$$\psi(\mathrm{B}) = 1 + \sum_{i=1}^{\infty} \psi_i \mathrm{B}^i, \tag{5.33}$$

(5.32) can be formulated (for $\mu = 0$)

$$Y_t = \psi(\mathrm{B})\varepsilon_t. \tag{5.34}$$

Here $\psi(\mathrm{B})$ is referred to as the *transfer function* of the process.

If there exists an inverse operator $\pi(\mathrm{B})$ so that

$$\pi(\mathrm{B})\psi(\mathrm{B}) = 1 \Leftrightarrow \pi(\mathrm{B}) = \psi^{-1}(\mathrm{B}), \tag{5.35}$$

the linear process (5.34) can be written in the form

$$\pi(\mathrm{B})Y_t = \varepsilon_t, \tag{5.36}$$

where

$$\pi(\mathrm{B}) = 1 + \sum_{i=1}^{\infty} \pi_i \mathrm{B}^i. \tag{5.37}$$

Given that $\pi(\mathrm{B})$ exists we can determine (5.37) using a Taylor series expansion of $\psi^{-1}(\mathrm{B})$, and the sequence $\{\pi_k\}$ is then referred to as the π *weights* of the process. Equation (5.36) is called the *inverse form.*

The autocovariance function for the linear process is given by

$$\begin{aligned} \gamma_{YY}(k) &= \mathrm{Cov}[Y_t, Y_{t+k}] \\ &= \mathrm{Cov}\left[\sum_{i=0}^{\infty} \psi_i \varepsilon_{t-i}, \sum_{i=0}^{\infty} \psi_i \varepsilon_{t+k-i}\right] \\ &= \sigma_\varepsilon^2 \sum_{i=0}^{\infty} \psi_i \psi_{i+k} \,. \end{aligned} \tag{5.38}$$

As a special case the variance becomes

$$\sigma_Y^2 = \gamma_{YY}(0) = \sigma_\varepsilon^2 \sum_{i=0}^{\infty} \psi_i^2. \tag{5.39}$$

The autocovariance and in particular the variance obviously only exist when the sums on the right hand side in (5.38) and (5.39) exist.

Often the autocovariance function is determined using the *autocovariance generating function*

$$\Gamma(z) = \sum_{k=-\infty}^{\infty} \gamma(k) z^{-k}, \tag{5.40}$$

which is the z-transformation of the autocovariance function.

By substitution of (5.38) we obtain (since $\psi_i = 0$ for $i < 0$)

$$\begin{aligned}\Gamma(z) &= \sigma_\varepsilon^2 \sum_{k=-\infty}^{\infty}\sum_{i=0}^{\infty} \psi_i\psi_{i+k} z^{-k}\\ &= \sigma_\varepsilon^2 \sum_{i=0}^{\infty}\psi_i z^i \sum_{j=0}^{\infty}\psi_j z^{-j}\\ &= \sigma_\varepsilon^2 \psi(z^{-1})\psi(z).\end{aligned}$$

Furthermore, since we have $\pi(z)\psi(z) = 1$, we get

$$\Gamma(z) = \sigma_\varepsilon^2\psi(z^{-1})\psi(z) = \sigma_\varepsilon^2\pi^{-1}(z^{-1})\pi^{-1}(z). \tag{5.41}$$

Please note that one can get from $\psi(z) = \sum \psi_i z^{-i}$ to $\psi(\mathrm{B}) = \sum \psi_i \mathrm{B}^i$ by replacing z^{-i} with B^i (z^{-1} corresponds to B). From (5.40) it is seen that $\Gamma(e^{i\omega})$ is the Fourier transform of the autocovariance function, which implies, as we shall see, that (5.41) becomes very useful in determining the spectrum of a linear process.

The convergence of the series in (5.39) ensures that the variance of the process is finite. In Section 5.2.2 we have shown important properties for the autocovariance function for a stationary process. For linear processes this can be summarized.

THEOREM 5.5 (STATIONARITY FOR LINEAR PROCESSES)
The linear process $Y_t = \psi(\mathrm{B})\varepsilon_t$ is stationary if

$$\psi(z) = \sum_{i=0}^{\infty} \psi_i z^{-i} \tag{5.42}$$

converges for $|z| \geq 1$.

Proof Omitted. Since stationary processes are characterized by the fact that the variance is finite, one can interpret the result by comparison with (5.39). ∎

▸ **Remark 5.1**
The claim in Theorem 5.5 ensures that the influence from past observations of the noise variable ε_t goes sufficiently fast to 0. ◂

Similarly, it is convenient to claim that the π weights belonging to the alternative formulation of the process, $\pi(\mathrm{B})Y_t = \varepsilon_t$, go sufficiently fast to 0. Thus, we define the following.

DEFINITION 5.11 (INVERTIBILITY FOR LINEAR PROCESSES)
The linear process $\pi(\mathrm{B})Y_t = \varepsilon_t$ is said to be invertible if

$$\pi(z) = \sum_{i=0}^{\infty} \pi_i z^{-i} \tag{5.43}$$

converges for $|z| \geq 1$.

▸ **Remark 5.2**
The previous definition ensures that the influence from past observations of Y_t is going sufficiently fast to 0 for invertible processes. Furthermore, as we shall see later, the claim of the process being invertible ensures identifiability for certain linear processes (moving average processes which we will introduce shortly). ◂

5.3.2 Processes in continuous time

As we have done with discrete processes in the previous paragraph, we will define a linear process in continuous time as the output from a linear system having "continuous white noise" as input. Therefore, we will first introduce the term "continuous white noise."

In many practical situations we are working with a process $\{X(t)\}$ for which values $X(t)$ at given times t_1 and t_2 are independent if $|t_1 - t_2| > \xi$, where ξ is small and less than the smallest time interval of physical interest. This would be the case where the process describes the forces that work on a particle in a Brownian motion. A mathematical description of this motion will lead to a definition of stochastic processes with independent increments.

DEFINITION 5.12 (A PROCESS WITH INDEPENDENT INCREMENTS)
Consider the times $t_1 < t_2 < \cdots < t_n$. A process $\{X(t)\}$ is said to be a *process with independent* (or orthogonal) *increments* if the random variables

$$X(t_1), X(t_2) - X(t_1), \ldots, X(t_n) - X(t_{n-1})$$

are mutually independent. If the variables only are mutually uncorrelated then the process $\{X(t)\}$ is said to be a *process with mutually uncorrelated increments.*

The most important process in the class of processes with independent increments is the Wiener process.

DEFINITION 5.13 (THE WIENER PROCESS)
A process $\{X(t), 0 \leq t \leq \infty\}$ is said to be a *Wiener process* (or *Brownian motion process*) if

i) $P\{X(0) = 0\} = 1$.
ii) The increments $X(t_1) - X(t_0), X(t_2) - X(t_1), \ldots, X(t_n) - X(t_{n-1})$ are mutually independent for arbitrary time points $0 \leq t_0 < t_1 < \cdots < t_n$.
iii) For arbitrary t and $h > 0$ the increment $X(t+h) - X(t)$ is normally distributed

$$\mathrm{E}[X(t+h) - X(t)] = 0 \tag{5.44}$$

$$\mathrm{Var}[X(t+h) - X(t)] = \sigma^2 h \tag{5.45}$$

where σ^2 is called the variance.

Furthermore, it can be shown that all the finite dimensional distributions for a Wiener process are multidimensional normal distributions (cf., e.g., Sobczyk (1985)).

Another important process belonging to the family of processes with independent incremenents, but defined on a discrete sample space ($\mathbb{N}_0$), is the Poisson process, as described in Grimmit and Stirzaker (1992). For a Poisson process the probability of n events in the time interval t is

$$P(n, t) = \frac{(\lambda t)^n}{n!} e^{\lambda t} \tag{5.46}$$

where λ is the intensity of the Poisson process.

In mathematical terms it is more difficult to define continuous white noise as opposed to discrete white noise; see also the discussion in Øksendal (1995). The definition of white noise is based on the assumption that the spectral density is constant, i.e., $f(\omega) = c$, $-\infty < \omega < \infty$. But since the variance of the process equals the integral of $f(\omega)$ (cf. next section), the variance becomes infinite. Formally, these difficulties can be avoided by introducing continuous white noise as a generalized stochastic process from a generalized time derivative of the Wiener process.

DEFINITION 5.14 (CONTINUOUS WHITE NOISE)
Continuous white noise is formally defined as a generalized process $\{\varepsilon(t)\}$ with autocovariance function

$$\gamma(\tau) = \sigma_\varepsilon^2 \delta(\tau), \tag{5.47}$$

where $\delta(\tau)$ is the Dirac delta function (see Section 4.1).

The definition of white noise introduces an application, which has many similarities to the impulse response function for linear systems.

DEFINITION 5.15 (LINEAR PROCESSES IN CONTINUOUS TIME)
A linear process in continuous time $\{Y(t)\}$ is a process that can be written in the form

$$Y(t) - \mu = \int_0^\infty \psi(\tau)\varepsilon(t-\tau)\, d\tau, \tag{5.48}$$

where $\{\varepsilon(t)\}$ is continuous white noise. Here $\{Y(t) - \mu\}$ is the output corresponding to the input $\{\varepsilon(t)\}$ from a linear system having the impulse response function $\psi(\tau)$.

The linear process $\{Y(t)\}$ defined by (5.48) has the mean value

$$\mathrm{E}[Y(t)] = \mu \tag{5.49}$$

and autocovariance function

$$\gamma_{YY}(\tau) = \mathrm{Cov}[Y(t), Y(t+\tau)] = \sigma_\varepsilon^2 \int_0^\infty \psi(u)\psi(\tau+u)\, du. \tag{5.50}$$

As a special case, the variance becomes

$$\sigma_Y^2 = \gamma_{YY}(0) = \sigma_\varepsilon^2 \int_0^\infty \psi^2(\tau)\, d\tau. \tag{5.51}$$

The concept of a continuous white noise process as a generalized process with the autocovariance function in Definition 5.14 is an approach often used in signal processing—see Papoulis (1983). However, it turns that a more satisfactory introduction is obtained by using a process in the class of processes with independent increments. A classical reference to this discussion is Øksendal (1995), which also contains an excellent introduction to stochastic differential equations. More recently the use of stochastic differential equations for modeling has become very popular—see, e.g., Madsen, Holst, and Lindström (2007) for several applications, Nielsen, Vestergaard, and Madsen (2000) and Madsen et al. (2004) for the use in finance, Kristensen, Madsen, and Jørgensen (2004) for modeling chemical systems, Andersen, Madsen, and Hansen (2000) for describing the heat dynamics of a building, Jónsdóttir, Jacobsen, and Madsen (2001) for describing the oxygen contents in a small creek, and Vio et al. (2006) for the use in astronomy.

5.4 Stationary processes in the frequency domain

In this section we will consider (weakly) stationary processes, which in the time domain are denoted $\{Y_t\}$ (discrete time) or $\{Y(t)\}$ (continuous time). Such a process can be characterized by the mean value μ and autocovariance function $\gamma(\tau)$. In this section an alternative way of characterizing the process will be introduced, namely, by introducing the Fourier transform of the autocovariance function of the process.

It was shown (Theorem 5.3 on page 104) that the autocovariance function is non-negative definite. Following a theorem of Bochner such a non-negative definite function can be written as a Stieltjes integral

$$\gamma(\tau) = \int_{-\infty}^{\infty} e^{i\omega\tau}\, dF(\omega) \tag{5.52}$$

for a process in continuous time, or

$$\gamma(\tau) = \int_{-\pi}^{\pi} e^{i\omega\tau} \, dF(\omega) \tag{5.53}$$

for a process in discrete time, where the properties of $F(\omega)$ correspond to the properties of a distribution function. It holds that:

- $F(\omega)$ is non-decreasing.
- $F(-\infty \,/\, -\pi) = 0$ for processes in continuous/discrete time, respectively.
- $F(\infty \,/\, \pi) = \gamma(0)$ for processes in continuous/discrete time, respectively.

The existence of (5.52) is given by *Wiener-Khintchine's theorem*, while the existence of (5.53) is given by *Wold's theorem*, cf. Priestley (1981). $F(\omega)$ is referred to as the *spectral distribution* of the process.

We have seen that any stationary process can be formulated as a sum of a purely stochastic process and a purely deterministic process (Theorem 5.2). Similarly, the spectral distribution can be written

$$F(\omega) = F_S(\omega) + F_D(\omega), \tag{5.54}$$

where $F_S(\omega)$ is an even continuous function and $F_D(\omega)$ is a step function. Both functions are non-decreasing.

For a pure deterministic process

$$Y_t = \sum_{i=1}^{k} A_i \cos(\omega_i t + \phi_i), \tag{5.55}$$

F_S will become 0, and thus $F(\omega)$ will become a step function with steps at the frequencies $\pm\omega_i$, $i = 1, \ldots, k$. In this case F can be written as

$$F(\omega) = F_D(\omega) = \sum_{\omega_i \le \omega} f(\omega_i) \tag{5.56}$$

and $\{f(\omega_i); i = 1, \ldots, k\}$ is often called the *line spectrum*. An example of a line spectrum is shown in Figure 5.4.

Of greater interest is the case where the process is purely stochastic. In this case $F(\omega)$ will be a continuous function and if we disregard cases of purely theoretical interest, $F(\omega)$ will also be differentiable. Hereby we are able to introduce the *spectrum* (or the *spectral density*) for the stochastic process as

$$f(\omega) = \frac{dF(\omega)}{d\omega} \tag{5.57}$$

Since we assume that $F_D(\omega)$ is zero, we can substitute $dF(\omega) = f(\omega)d\omega$ into (5.52) and (5.53), thereby obtaining the following Fourier transformations.

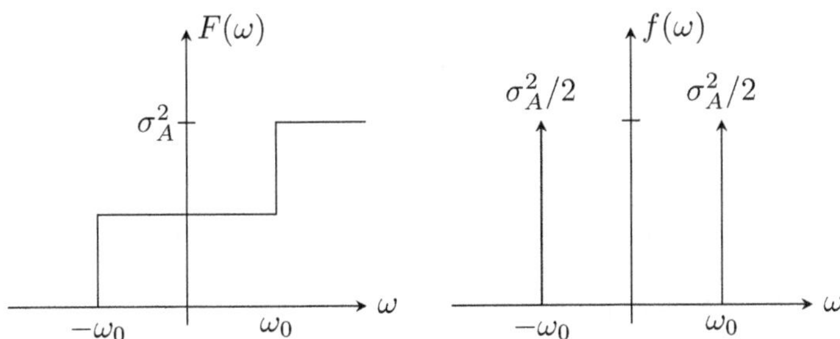

Figure 5.4: *Spectral distribution and density for the pure deterministic process given by* (5.61). *This also illustrates a very simple line spectrum.*

THEOREM 5.6 (RELATIONSHIP BETWEEN THE SPECTRUM AND THE AUTOCOVARIANCE)
For a purely stochastic process we have the following relations between the spectrum and the autocovariance function

$$\begin{aligned} f(\omega) &= \frac{1}{2\pi}\int_{-\infty}^{\infty} e^{-i\omega\tau}\gamma(\tau)\,d\tau \\ \gamma(\tau) &= \int_{-\infty}^{\infty} e^{i\omega\tau} f(\omega)\,d\omega \end{aligned} \qquad \textit{(continuous time)} \tag{5.58}$$

$$\begin{aligned} f(\omega) &= \frac{1}{2\pi}\sum_{k=-\infty}^{\infty} \gamma(k)e^{-i\omega k} \\ \gamma(k) &= \int_{-\pi}^{\pi} e^{ik\omega} f(\omega)\,d\omega \end{aligned} \qquad \textit{(discrete time)} \tag{5.59}$$

Proof Follows from the previous results. ■

▸ Remark 5.3

The above results can be exploited with (5.40) to calculate the variance of a process given its transfer function. According to (5.40) it holds that

$$f(\omega) = \frac{1}{2\pi}\,\Gamma(z)\big|_{z=e^{i\omega}} = \frac{\sigma_\varepsilon^2}{2\pi}\,\psi(z^{-1})\psi(z)\big|_{z=e^{i\omega}}\,. \tag{5.60}$$

Substituting this expression into (5.59) we have

$$\begin{aligned} \sigma^2 = \gamma(0) &= \int_{-\pi}^{\pi} e^{i\cdot 0\cdot\omega} f(\omega)d\omega \\ &= \frac{\sigma_\varepsilon^2}{2\pi i}\int_{-\pi}^{\pi} \psi(e^{i\omega})\psi(e^{-i\omega})e^{-i\omega}\,de^{i\omega} \\ &= \frac{\sigma_\varepsilon^2}{2\pi i}\oint_{|z|=1} \psi(z)\psi(z^{-1})\,dz/z \end{aligned}$$

where the variance of a process given by a transfer function ψ (see (5.34) on page 109) can be obtained by residue calculations. ◂

If we allow $f(\omega)$ to be completely or partly described by generalizing functions (delta functions), then (5.57) on page 114 and (5.58) on the previous page can be used for $F_D \neq 0$. This is illustrated in the following example.

Example 5.5 (The spectrum for a pure deterministic process)
Consider the pure deterministic process

$$Y(t) = A_1 \cos(\omega_0 t) + A_2 \sin(\omega_0 t) \tag{5.61}$$

where A_1 and A_2 are mutually uncorrelated random variables with mean values 0 and same variance σ_A^2. The autocovariance function becomes

$$\begin{aligned}\gamma(\tau) &= \sigma_A^2(\cos(\omega_0 t)\cos(\omega_0(t+\tau)) + \sin(\omega_0 t)\sin(\omega_0(t+\tau)))\\ &= \sigma_A^2 \cos(\omega_0 \tau)\end{aligned}$$

Applying (5.58) we get the spectrum

$$\begin{aligned}f(\omega) &= \frac{\sigma_A^2}{2\pi}\int_{-\infty}^{\infty} \mathrm{e}^{-i\omega\tau}\cos(\omega_0\tau)\,d\tau\\ &= \frac{\sigma_A^2}{2\pi}\int_{-\infty}^{\infty} \mathrm{e}^{-i\omega\tau}(\mathrm{e}^{i\omega_0\tau} + \mathrm{e}^{-i\omega_0\tau})/2\,d\tau\\ &= \frac{\sigma_A^2}{2}(\delta(\omega-\omega_0) + \delta(\omega+\omega_0))\end{aligned}$$

where the equal sign follows from $\int_{-\infty}^{\infty} \mathrm{e}^{-i(\omega_0-\omega)\tau}\,d\tau = 2\pi\delta(\omega-\omega_0)$. The spectral distribution and density will then become as sketched in Figure 5.4 on the preceding page. Please note that the spectrum in this case is a line spectrum.

Example 5.6 (The spectrum for white noise)
Since the autocovariance function for discrete white noise $\{\varepsilon_t\}$ is

$$\gamma(k) = \begin{cases}\sigma_\varepsilon^2 & \text{for } k = 0\\ 0 & \text{for } k = \pm 1, \pm 2, \dots\end{cases}$$

we get the spectral density for white noise

$$f(\omega) = \frac{\sigma_\varepsilon^2}{2\pi}, \quad \omega \in [-\pi, \pi]$$

i.e., uniformly distributed spectral density.

Often the spectrum can most easily be found using the transfer function for the linear process. Here we apply the following.

THEOREM 5.7 (THE SPECTRUM DETERMINED FROM A TRANSFER FUNCTION)
Consider the linear process $Y_t = \psi(\mathrm{B})\varepsilon_t$ *or* $\pi(\mathrm{B})Y_t = \varepsilon_t$ *where* $\{\varepsilon_t\}$ *is white noise. Then the spectrum for the process is given by* ($\omega \in [-\pi, \pi]$)

$$\begin{aligned} f(\omega) &= \frac{\sigma_\varepsilon^2}{2\pi}\left|\psi(\mathrm{e}^{-i\omega})\right| = \frac{\sigma_\varepsilon^2}{2\pi}\psi(\mathrm{e}^{-i\omega})\psi(\mathrm{e}^{i\omega}) \\ &= \frac{\sigma_\varepsilon^2}{2\pi}\pi^{-1}(\mathrm{e}^{-i\omega})\pi^{-1}(\mathrm{e}^{i\omega}). \end{aligned} \tag{5.62}$$

Proof Follows directly from (5.41) on page 110 and the definition on the autocovariance generating function (5.40) on page 109. ∎

It is seen that the integral of $f(\omega)$ equals the variance of the process, see (5.58) and (5.59) on page 115.

5.5 Commonly used linear processes

A very useful class of linear processes is the processes where $\psi(z)$ can be written as a rational function: $\psi(z) = \theta(z)/\phi(z)$, where $\theta(z)$ and $\phi(z)$ are polynomials.

5.5.1 The MA process

DEFINITION 5.16 (THE MA(q) PROCESS)
The process $\{Y_t\}$ given by

$$Y_t = \varepsilon_t + \theta_1\varepsilon_{t-1} + \cdots + \theta_q\varepsilon_{t-q}, \tag{5.63}$$

where $\{\varepsilon_t\}$ is white noise, is called a *Moving Average process* of order q. In short it is denoted an *MA(q) process.*

An MA(q) process is thus characterized by the fact that the ψ weights are zero from a given point. Using the backward shift operator B, the MA(q) process can also be written as

$$Y_t = (1 + \theta_1\mathrm{B} + \cdots + \theta_q\mathrm{B}^q)\varepsilon_t = \theta(\mathrm{B})\varepsilon_t. \tag{5.64}$$

The transfer operator $\theta(\mathrm{B})$ is thus given by (5.64). The corresponding transfer function is $\theta(z^{-1})$.

THEOREM 5.8 (MA PROCESS: STATIONARITY AND INVERTIBILITY)
An MA(q) process is

i) Always stationary.

ii) Invertible if the roots in $\theta(z^{-1}) = 0$, with respect to z, are all within the unit circle.

Proof

i) Follows from Theorem 5.5 on page 110.

ii) Follows from Definition 5.11 on page 111, where the π weights are the coefficients of the Taylor expansion of $\theta^{-1}(\mathrm{B})$. ∎

From (5.38) we find the *autocovariance function for the MA(q) process*

$$\gamma(k) = \begin{cases} \sigma_\varepsilon^2(\theta_k + \theta_1\theta_{k+1} + \cdots + \theta_{q-k}\theta_q) & |k| = 0, 1, \ldots, q \\ 0 & |k| = q+1, q+2, \ldots \end{cases} \tag{5.65}$$

In particular the variance

$$\sigma_Y^2 = \gamma(0) = \sigma_\varepsilon^2(1 + \theta_1^2 + \cdots + \theta_q^2). \tag{5.66}$$

Please note that the autocovariance function and, thus, the autocorrelation function are zero for lags greater than q. Furthermore, it should be noted that $\gamma(q) = \sigma_\varepsilon^2\theta_q$ will always be different from zero; otherwise, it would at most be an MA($q-1$) process.

From (5.62) it is seen that the *spectrum for the MA(q) process* is

$$f(\omega) = \frac{\sigma_\varepsilon^2}{2\pi}\theta(\mathrm{e}^{-i\omega})\theta(\mathrm{e}^{i\omega}) = \frac{\sigma_\varepsilon^2}{2\pi}|1 + \theta_1\mathrm{e}^{-i\omega} + \cdots + \theta_q\mathrm{e}^{-iq\omega}|^2, \tag{5.67}$$

where $\omega \in [-\pi, \pi]$.

If $\varepsilon_t \in \mathrm{N}(0, \sigma_\varepsilon^2)$, then $(Y_{t_1}, \ldots, Y_{t_n})^T$ follows an n-dimensional normal distribution and thus $\{Y_t\}$ is a normal process.

Example 5.7 (The MA(1) process)
Consider the MA(1) process

$$Y_t = (1 + \theta_1\,\mathrm{B})\varepsilon_t \tag{5.68}$$

Since $1 + \theta_1 z^{-1} = 0$ for $z = -\theta_1$, the process is invertible for $|\theta_1| < 1$.

The autocovariance is

$$\gamma(k) = \begin{cases} (1+\theta_1^2)\sigma_\varepsilon^2 & k = 0 \\ \theta_1\sigma_\varepsilon^2 & |k| = 1 \\ 0 & |k| = 2, 3, \ldots \end{cases} \tag{5.69}$$

and the autocorrelation becomes

$$\rho(k) = \frac{\gamma(k)}{\gamma(0)} = \begin{cases} 1 & k = 0 \\ \theta_1/(1+\theta_1^2) & |k| = 1 \\ 0 & |k| = 2, 3, \ldots \end{cases} \tag{5.70}$$

Using the autocovariance function $\gamma(k)$ the moment estimates for θ_1 and σ_ε^2 can be determined given that $\gamma(0) = \alpha$ and $\gamma(1) = \beta$ are known. This gives

$$\begin{aligned} \widetilde{\theta}_1 &= \frac{\alpha}{2\beta} \pm \sqrt{\left(\frac{\alpha}{2\beta}\right)^2 - 1} \\ \widetilde{\sigma}_\varepsilon^2 &= \beta/\widetilde{\theta}_1 \end{aligned} \tag{5.71}$$

where $\widetilde{\theta}_1$ is chosen as one of the two solutions that gives $|\widetilde{\theta}_1| < 1$ in order to ensure invertibility. More about this issue in Chapter 6.

Finally, the spectrum for the MA(1) process becomes

$$\begin{aligned} f(\omega) &= \frac{\sigma_\varepsilon^2}{2\pi}(1 + \theta_1 \mathrm{e}^{-i\omega})(1 + \theta_1 \mathrm{e}^{i\omega}) \\ &= \frac{\sigma_\varepsilon^2}{2\pi}(1 + \theta_1^2 + 2\theta_1 \cos(\omega)), \quad \omega \in [-\pi, \pi]. \end{aligned} \tag{5.72}$$

In Figure 5.5 on the following page the autocorrelation function and the spectrum for the MA(1) process are shown for $\theta_1 = 0.8$ and $\theta_1 = -0.8$.

For the MA(1) process it holds in general that when θ_1 is positive, $\rho(1)$ is positive and the spectrum is dominated by low frequencies. When θ_1 is negative the $\rho(1)$ is negative and the spectrum is dominated by high frequencies.

5.5.2 The AR process

DEFINITION 5.17 (THE AR(p) PROCESS)
The process $\{Y_t\}$ given by

$$Y_t + \phi_1 Y_{t-1} + \cdots + \phi_p Y_{t-p} = \varepsilon_t, \tag{5.73}$$

where $\{\varepsilon_t\}$ is white noise, is called an *autoregressive process* of order p (or an *AR(p) process*).

An AR(p) process is thus a linear process where the π weights are zero from point $p+1$. Alternatively, an AR(p) process can be written

$$(1 + \phi_1 \mathrm{B} + \cdots + \phi_p \mathrm{B}^p) Y_t = \varepsilon_t \tag{5.74}$$

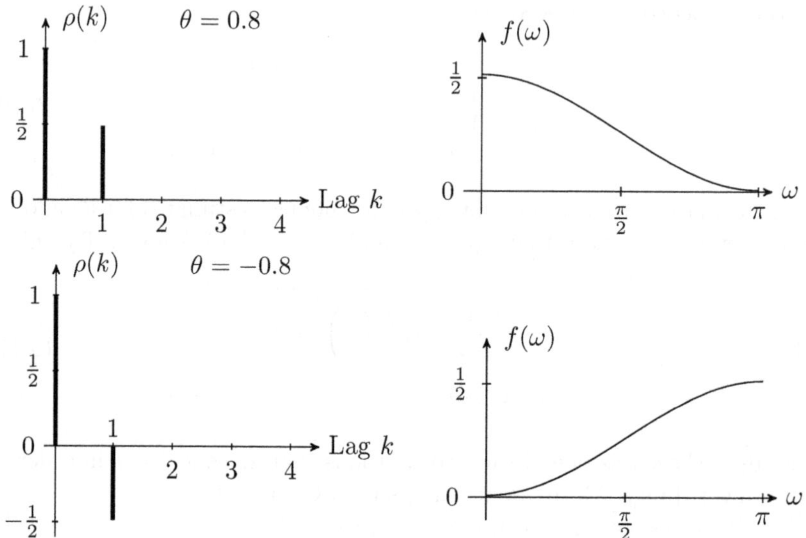

Figure 5.5: *The autocorrelation function (left) and spectrum (right) for the MA(1) process $Y_t = \varepsilon_t + \theta\varepsilon_{t-1}$ ($\sigma_\varepsilon^2 = 1$).*

or

$$\phi(\text{B})Y_t = \varepsilon_t, \tag{5.75}$$

where $\phi(\text{B})$ is a p'th order polynomial in B. The transfer operator is $\phi(\text{B})^{-1}$ and the transfer function is $1/\phi(z^{-1})$.

The name *autoregressive* is due to the fact that the value of Y_t can be seen as a regression on past values of the process, since

$$Y_t = -\phi_1 Y_{t-1} - \cdots - \phi_p Y_{t-p} + \varepsilon_t. \tag{5.76}$$

THEOREM 5.9 (AR PROCESS: STATIONARITY AND INVERTIBILITY)
An AR(p) process is

i) Always invertible.

ii) Stationary if the roots of $\phi(z^{-1}) = 0$, with respect to z, all lie within the unit circle.

Proof

i) Follows from the definition of invertibility.

ii) We can write $\phi(z^{-1}) = (1 - \lambda_1 z^{-1}) \cdots (1 - \lambda_p z^{-1})$. By decomposition of $\phi(z^{-1})^{-1}$, we get

$$Y(z) = \phi(z^{-1})^{-1}\varepsilon(z) = \sum_{i=1}^{p} \left[\frac{k_i}{1 - \lambda_i z^{-1}} \right] \varepsilon(z) = \psi(z)\varepsilon(z) \qquad (5.77)$$

According to Theorem 5.5 on page 110, the process is *stationary* if ψ_i converges for $|z| \geq 1$. This implies that in order to obtain stationarity we need $|\lambda_i| < 1$ for $i = 1, \ldots, p$. ■

Compare with the claim for stability for linear systems. The equation $\phi(z^{-1}) = 0$ is called the *characteristic equation* for the process. See, furthermore, Appendix A, "The solution to difference equations."

In the remaining part of this section only stationary processes will be considered, in this case the following holds.

THEOREM 5.10 (AUTOCOVARIANCE FUNCTION FOR AR PROCESSES)
The autocovariance function for an AR(p) process satisfies the linear difference equation

$$\gamma(k) + \phi_1\gamma(k-1) + \cdots + \phi_p\gamma(k-p) = 0\,, \quad k = 1, 2, \ldots \qquad (5.78)$$

with initial condition

$$\gamma(0) + \phi_1\gamma(-1) + \cdots + \phi_p\gamma(-p) = \sigma_\varepsilon^2 \qquad (5.79)$$

or, since $\gamma(k) = \gamma(-k)$,

$$\gamma(0) + \phi_1\gamma(1) + \cdots + \phi_p\gamma(p) = \sigma_\varepsilon^2. \qquad (5.80)$$

Proof By multiplying (5.73) by Y_{t-k} and finding expected values we get

$$\mathrm{E}[Y_{t-k}Y_t + \phi_1 Y_{t-k}Y_{t-1} + \cdots + \phi_p Y_{t-k}Y_{t-p}] = \mathrm{E}[Y_{t-k}\varepsilon_t]$$

Since

$$\mathrm{E}[Y_{t-k}\varepsilon_t] = \begin{cases} \sigma_\varepsilon^2 & k = 0 \\ 0 & k = 1, 2, \ldots \end{cases}$$

and

$$\begin{aligned} &\mathrm{E}[Y_{t-k}Y_t + \phi_1 Y_{t-k}Y_{t-1} + \cdots + \phi_p Y_{t-k}Y_{t-p}] \\ &\quad = \gamma(k) + \phi_1\gamma(k-1) + \cdots + \phi_p\gamma(k-p) \end{aligned}$$

the result follows. ■

If we write out (5.78) for $k = 1, 2, \ldots, p$, and divide by $\gamma(0)$, we get the *Yule-Walker equations*:

$$\begin{pmatrix} \rho(1) \\ \rho(2) \\ \vdots \\ \rho(p) \end{pmatrix} = \begin{pmatrix} 1 & \rho(1) & \cdots & \rho(p-1) \\ \rho(1) & 1 & \cdots & \rho(p-2) \\ \vdots & \vdots & \ddots & \vdots \\ \rho(p-1) & \rho(p-2) & \cdots & 1 \end{pmatrix} \begin{pmatrix} -\phi_1 \\ -\phi_2 \\ \vdots \\ -\phi_p \end{pmatrix} \tag{5.81}$$

and again by division by $\gamma(0)$ in (5.78), it is seen that also $\rho(k)$ satisfies the difference equation (5.78) for $k = 1, 2, \ldots$.

The difference equation can be solved explicitly (see Appendix A). If all the roots $\lambda_1, \ldots, \lambda_p$ in the characteristic equation, $\phi(z^{-1}) = 0$, are different, then the complete solution can be written

$$\gamma(k) = A_1 \lambda_1^k + \cdots + A_p \lambda_p^k, \quad k = p, p+1, \ldots, \tag{5.82}$$

where the roots are either real or complex. Especially, if it is assumed that we have p_1 real and positive roots, p_2 real and negative roots, and $2p_3$ complex conjugated roots, then the roots can be written

$$\begin{aligned} &\alpha_j, \quad j = 1, \ldots, p_1 \\ &-\beta_j, \quad j = 1, \ldots, p_2 \\ &\xi_j \mathrm{e}^{\pm i\theta_j}, \quad j = 1, \ldots, p_3, \end{aligned}$$

where α_j, β_j, and ξ_j all belong to the interval $[0, 1]$.

Now the complete solution can be written

$$\gamma(k) = \sum_{j=1}^{p_1} A_j \alpha_j^k + \sum_{j=1}^{p_2} \mathrm{B}_j (-\beta_j)^k + \sum_{j=1}^{p_3} C_j \xi_j^k \cos(\theta_j k + \phi_j) \tag{5.83}$$

for $k = p, p+1, \ldots$. It is seen that positive real roots imply an exponentially decreasing increment and negative real roots imply a shifting positive and negative exponentially decreasing increment, while two complex conjugated roots give an increment in terms of a damped wave. It is usually quite tedious to find an explicit solution.

An alternative (and often easier) procedure is to calculate the value of $\gamma(0), \ldots, \gamma(p-1)$ and then determine $\gamma(p), \gamma(p+1), \ldots$ by using (5.78) on the preceding page recursively. The same holds for the autocorrelation function, which is illustrated in the following example.

Example 5.8 (Autocorrelations calculated recursively)
Let us determine the autocorrelation function, for the AR(3) process

$$Y_t + \phi_1 Y_{t-1} + \phi_2 Y_{t-2} + \phi_3 Y_{t-3} = \varepsilon_t \tag{5.84}$$

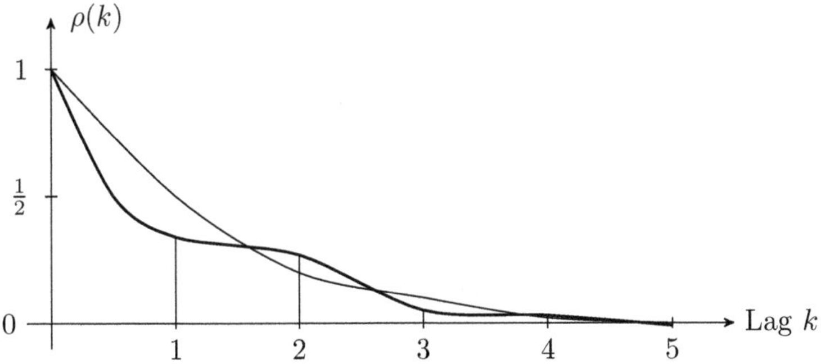

Figure 5.6: *The autocorrelation function for the AR(3) process* $Y_t - 0.3Y_{t-1} - 0.2Y_{t-2} + 0.1Y_{t-3} = \varepsilon_t$.

where $\phi_1 = -0.3$, $\phi_2 = -0.2$, and $\phi_3 = 0.1$.

Corresponding to the two first Yule-Walker equations we have

$$\begin{aligned} &\text{a)} \qquad \rho(1) + \phi_1 + \phi_2\rho(1) + \phi_3\rho(2) = 0 \\ &\text{b)} \qquad \rho(2) + \phi_1\rho(1) + \phi_2 + \phi_3\rho(1) = 0. \end{aligned}$$

From b) we get

$$\rho(2) = -\phi_2 - \rho(1)(\phi_1 + \phi_3),$$

which inserted in a) gives

$$\rho(1) = \frac{-\phi_1 + \phi_3\phi_2}{1 + \phi_2 - \phi_3(\phi_1 + \phi_3)}.$$

With the applied values we obtain $\rho(1) = 0.34$ and $\rho(2) = 0.27$. For the remaining values of k, we use the recursive formula

$$\rho(k) = -\phi_1\rho(k-1) - \phi_2\rho(k-2) - \phi_3\rho(k-3),$$

and find $\rho(3) = 0.05$, $\rho(4) = 0.03$, $\rho(5) = -0.01$, etc. The autocorrelation function is sketched in Figure 5.6.

We see that the autocorrelation function consists of an exponential decrease and an additive exponentially decreasing harmonic wave. The characteristic equation has one real root and two complex conjugated roots.

From (5.62) it is seen that *the spectrum for an AR(p) process* is given by

$$f(\omega) = \frac{\sigma_\varepsilon^2}{2\pi} \frac{1}{\phi(\mathrm{e}^{-i\omega})\phi(\mathrm{e}^{i\omega})} \tag{5.85}$$

$$= \frac{\sigma_\varepsilon^2}{2\pi} \frac{1}{|1 + \phi_1\mathrm{e}^{-i\omega} + \cdots + \phi_p\mathrm{e}^{-ip\omega}|^2} \tag{5.86}$$

where $\omega \in [-\pi, \pi]$.

5.5.2.1 The partial autocorrelation function

DEFINITION 5.18 (THE PARTIAL AUTOCORRELATION ϕ_{kk})
The partial autocorrelation ϕ_{kk} (PACF) is defined as

$$\phi_{kk} = \text{Cor}[Y_t, Y_{t+k} | Y_{t+1}, \ldots, Y_{t+k-1}], \tag{5.87}$$

i.e., ϕ_{kk} is the correlation coefficient between Y_t and Y_{t+k} given all the in-between observations.

It is shown in Appendix B that when the above definition is used, ϕ_{kk} is the k'th coefficient (with opposite sign) in an AR(k) process, which gives useful formulas for determining ϕ_{kk}. Let us consider the AR(k) process

$$Y_t = \phi_{k1} Y_{t-1} + \cdots + \phi_{kk} Y_{t-k} + \varepsilon_t, \tag{5.88}$$

where the partial autocorrelation thus is the coefficient for Y_{t-k}. Using this we can write the Yule-Walker equations

$$\begin{pmatrix} 1 & \rho_1 & \cdots & \rho_{k-1} \\ \rho_1 & 1 & \cdots & \rho_{k-2} \\ \vdots & \vdots & \ddots & \vdots \\ \rho_{k-1} & \rho_{k-2} & \cdots & 1 \end{pmatrix} \begin{pmatrix} \phi_{k1} \\ \phi_{k2} \\ \vdots \\ \phi_{kk} \end{pmatrix} = \begin{pmatrix} \rho_1 \\ \rho_2 \\ \vdots \\ \rho_k \end{pmatrix}, \tag{5.89}$$

where ϕ_{kk} is the partial autocorrelation at lag k. Equation (5.89) is written in short

$$\boldsymbol{P}_k \boldsymbol{\phi}_k = \boldsymbol{\rho}_k \tag{5.90}$$

THEOREM 5.11 (PARTIAL AUTOCORRELATION FOR AR PROCESSES)
For an AR(p) process it holds

$$\begin{aligned} \phi_{kk} &\neq 0, \quad k \leq p \\ \phi_{kk} &= 0, \quad k > p \end{aligned}$$

Proof Follows from the previous. ∎

On the other hand, the partial autocorrelation function for an MA(q) process will take values different from 0 for arbitrary large k.

The partial autocorrelation function is obtained by solving (5.89) repeatedly with respect to ϕ_{kk} for $k = 1, 2, 3, \ldots$. Hence, one can determine ϕ_{kk} as

$$\phi_{kk} = \frac{\det \boldsymbol{V}_k}{\det \boldsymbol{P}_k}, \tag{5.91}$$

where $\boldsymbol{V}_k$ is found by replacing the k'th (i.e., the last) column in $\boldsymbol{P}_k$ with $\boldsymbol{\rho}_k$; see Appendix B.

We hereby get

$$\phi_{11} = \rho_1$$

$$\phi_{22} = \frac{\begin{vmatrix} 1 & \rho_1 \\ \rho_1 & \rho_2 \end{vmatrix}}{\begin{vmatrix} 1 & \rho_1 \\ \rho_1 & 1 \end{vmatrix}} = \frac{\rho_2 - \rho_1^2}{1 - \rho_1^2}$$

$$\phi_{33} = \frac{\begin{vmatrix} 1 & \rho_1 & \rho_1 \\ \rho_1 & 1 & \rho_2 \\ \rho_2 & \rho_1 & \rho_3 \end{vmatrix}}{\begin{vmatrix} 1 & \rho_1 & \rho_2 \\ \rho_1 & 1 & \rho_1 \\ \rho_2 & \rho_1 & 1 \end{vmatrix}}$$

In Appendix B a recursive formula for determination of ϕ_{kk} is given. This formula is useful for large values of k.

5.5.3 The ARMA process

DEFINITION 5.19 (THE ARMA(p,q) PROCESS)
The process $\{Y_t\}$ given by

$$Y_t + \phi_1 Y_{t-1} + \cdots + \phi_p Y_{t-p} = \varepsilon_t + \theta_1 \varepsilon_{t-1} + \cdots + \theta_q \varepsilon_{t-q}, \tag{5.92}$$

where $\{\varepsilon_t\}$ is white noise is called an *ARMA(p,q) process.*

By introducing polynomials using the shift operator B, the ARMA(p,q) process can be written

$$\phi(\mathrm{B})Y_t = \theta(\mathrm{B})\varepsilon_t, \tag{5.93}$$

where $\phi(\mathrm{B})$ and $\theta(\mathrm{B})$ are polynomials of order p and q, respectively.

The transfer function from ε to Y becomes

$$H(z) = \frac{\theta(z^{-1})}{\phi(z^{-1})} \tag{5.94}$$

and the transfer operator is $\psi(\mathrm{B}) = \phi^{-1}(\mathrm{B})\theta(\mathrm{B})$.

THEOREM 5.12 (ARMA PROCESS: STATIONARITY AND INVERTIBILITY)
An ARMA(p,q) process is stationary if the roots of $\phi(z^{-1}) = 0$ lie within the unit circle and invertible if the roots of $\theta(z^{-1})$ lie within the unit circle.

Proof Omitted. There is a close relation between Theorem 5.12 and the corresponding theorems for AR and MA processes. ∎

In the remaining part of this section all processes are assumed to be stationary.

Below we show a general method to determine the *autocovariance function for an ARMA(p, q) process.*

Since Y_t can be written in the form

$$Y_t = \sum_{i=0}^{\infty} \psi_i \varepsilon_{t-i} \tag{5.95}$$

it is obvious that

$$\gamma_{\varepsilon Y}(k) = \text{Cov}[\varepsilon_t, Y_{t+k}] = \text{E}[Y_t \varepsilon_{t-k}] \begin{cases} = 0 & \text{for } k < 0 \\ \neq 0 & \text{for } k \geq 0 \end{cases} \tag{5.96}$$

By multiplying (5.92) on the preceding page by ε_{t-k} on both sides and taking expected values, we obtain the difference equation

$$\gamma_{\varepsilon Y}(k) + \phi_1 \gamma_{\varepsilon Y}(k-1) + \cdots + \phi_p \gamma_{\varepsilon Y}(k-p) = \theta_k \sigma_\varepsilon^2 , \quad k = 0, 1, \ldots \tag{5.97}$$

First we determine $\gamma_{\varepsilon Y}(k)$ using (5.96) and (5.97). For example,

$$\begin{aligned} k = 0: \quad & \gamma_{\varepsilon Y}(0) = \theta_0 \sigma_\varepsilon^2 \\ k = 1: \quad & \gamma_{\varepsilon Y}(1) + \phi_1 \gamma_{\varepsilon Y}(0) = \theta_1 \sigma_\varepsilon^2 \end{aligned}$$

From this it follows that

$$\gamma_{\varepsilon Y}(1) = (\theta_1 - \phi_1 \theta_0) \sigma_\varepsilon^2 \tag{5.98}$$

Subsequently by multiplying (5.92) by Y_{t-k} on both sides and taking expected values, we get that $\gamma(k)$ satisfies

$$\begin{aligned} &\gamma(k) + \phi_1 \gamma(k-1) + \cdots + \phi_p \gamma(k-p) \\ &\quad = \theta_k \gamma_{\varepsilon Y}(0) + \cdots + \theta_q \gamma_{\varepsilon Y}(q-k) , \quad \text{for } k = 0, 1, \ldots \end{aligned} \tag{5.99}$$

If $p > q$ we obtain

$$\gamma(k) + \phi_1 \gamma(k-1) + \cdots + \phi_p \gamma(k-p) = 0 , \quad k = p, p+1, \ldots \tag{5.100}$$

In this case the autocovariance—and, hence, the autocorrelation—function will consist of damped exponential and sine functions (cf. Appendix A on the solution of difference equations). The coefficients of the complete solution are found using (5.99) for $k = 0, 1, \ldots, p-1$.

If $p \leq q$, we obtain

$$\begin{aligned}
&\gamma(p)+\phi_1\gamma(p-1)+\cdots+\phi_p\gamma(0)\\
&\qquad=\theta_p\gamma_{\varepsilon Y}(0)+\cdots+\theta_q\gamma_{\varepsilon Y}(q-p)\\
&\gamma(p+1)+\phi_1\gamma(p)+\cdots+\phi_p\gamma(1)\\
&\qquad=\theta_{p+1}\gamma_{\varepsilon Y}(0)+\cdots+\theta_q\gamma_{\varepsilon Y}(q-p-1)\\
&\quad\vdots\\
&\gamma(q+1)+\phi_1\gamma(q)+\cdots+\phi_p\gamma(q+1-p)=0.
\end{aligned} \tag{5.101}$$

It is seen that only from lag $k = q + 1 - p$ will the autocovariance and, hence, the autocorrelation function consist of damped exponential and harmonic functions. The coefficients in the complete solution are determined using (5.99) for $k = q - p, \ldots, q$.

Example 5.9 (The autocorrelation for an ARMA(1, 1) process)
We consider the ARMA(1, 1) process:

$$Y_t + \phi_1 Y_{t-1} = \varepsilon_t + \theta_1 \varepsilon_{t-1}$$

We determine $\gamma_{\varepsilon Y}(k)$ (using (5.97)):

$$\begin{aligned}
\gamma_{\varepsilon Y}(0) &= \sigma_\varepsilon^2\\
\gamma_{\varepsilon Y}(1) &= (\theta_1 - \phi_1)\sigma_\varepsilon^2 && (5.102)\\
\gamma_{\varepsilon Y}(k) &= -\phi_1 \gamma_{\varepsilon Y}(k-1), \quad \text{for } k \geq 2 && (5.103)
\end{aligned}$$

Determination of $\gamma(k)$ (using. (5.101)):

$$\begin{aligned}
\gamma(0) + \phi_1\gamma(1) &= \gamma_{\varepsilon Y}(0) + \theta_1\gamma_{\varepsilon Y}(1)\\
\gamma(1) + \phi_1\gamma(0) &= \theta_1\gamma_{\varepsilon Y}(0) && (5.104)\\
\gamma(k) &= -\phi_1\gamma(k-1), \quad \text{for } k \geq 2 && (5.105)
\end{aligned}$$

Substituting $\gamma_{\varepsilon Y}(0)$ and $\gamma_{\varepsilon Y}(1)$ we obtain

$$\begin{aligned}
\gamma(0) + \phi_1\gamma(1) &= (1 + \theta_1(\theta_1 - \phi_1))\,\sigma_\varepsilon^2\\
\gamma(1) + \phi_1\gamma(0) &= \theta_1\sigma_\varepsilon^2 && (5.106)
\end{aligned}$$

or

$$\begin{aligned}
\gamma(0) + \phi_1\left(\theta_1\sigma_\varepsilon^2 - \phi_1\gamma(0)\right) &= (1 + \theta_1(\theta_1 - \phi_1))\,\sigma_\varepsilon^2\\
\gamma(1) &= -\phi_1\gamma(0) + \theta_1\sigma_\varepsilon^2 && (5.107)
\end{aligned}$$

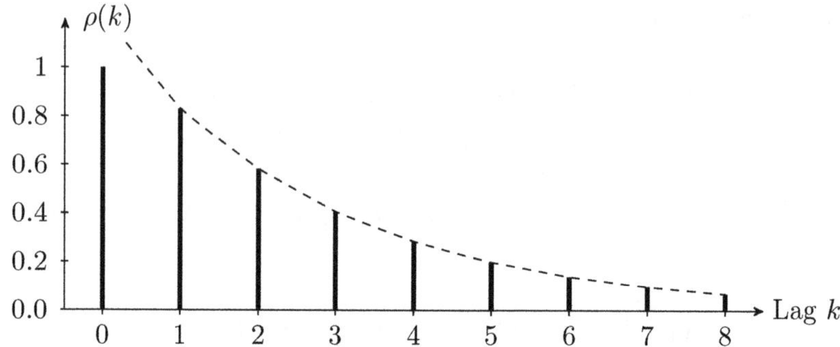

Figure 5.7: *The autocorrelation for the process* $Y_t - 0.7Y_{t-1} = \varepsilon_t + 0.5\varepsilon_{t-1}$.

or

$$\begin{aligned}
\gamma(0) &= \frac{1+\theta_1^2 - 2\phi_1\theta_1}{(1-\phi_1^2)}\sigma_\varepsilon^2, \\
\gamma(1) &= \frac{\theta_1(1-\phi_1^2) - \phi_1(1+\theta_1^2-2\phi_1\theta_1)}{(1-\phi_1^2)}\sigma_\varepsilon^2 \\
&= \frac{(1-\theta_1\phi_1)(\theta_1-\phi_1)}{(1-\phi_1^2)}\sigma_\varepsilon^2, \qquad (5.108)\\
\gamma(k) &= -\phi_1\gamma(k-1), \quad k \geq 2 \qquad (5.109)
\end{aligned}$$

The autocorrelation function becomes

$$\begin{aligned}
\rho(0) &= 1 \\
\rho(1) &= \frac{(1-\theta_1\phi_1)(\theta_1-\phi_1)}{1+\theta_1^2-2\phi_1\theta_1} \qquad (5.110)\\
\rho(k) &= (-\phi_1)^{k-1}\rho(1), \quad k \geq 2 \qquad (5.111)
\end{aligned}$$

If we consider for example an ARMA(1, 1) process where $\phi_1 = -0.7$ and $\theta_1 = 0.5$, we get $\rho(0) = 1$, $\rho(1) = 0.8308$, $\rho(2) = 0.5816$, $\rho(3) = 0.4071$, etc. The autocorrelation function is sketched in Figure 5.7.

As the figure indicates there is an exponential decrease from $\rho(1)$. This starting value depends on both ϕ_1 and θ_1.

▸ **Remark 5.4**

It is seen from (5.100) that for sufficiently large values of k, the autocovariance function fulfills a homogeneous difference equation determined by the autoregressive part of the model. ◂

THEOREM 5.13

The partial autocorrelation function for an ARMA(p, q) process can take values different from 0 for arbitrarily large values of k.

Proof Follows from the fact that the ARMA(p, q) process can be written as $\theta^{-1}(\mathrm{B})\phi(\mathrm{B})Y_t = \varepsilon_t$, i.e., as an infinite order autoregressive process. ∎

The spectrum for an ARMA(p, q) process is given by

$$\begin{aligned} f(\omega) &= \frac{\sigma_\varepsilon^2}{2\pi} \frac{\theta(\mathrm{e}^{-i\omega})\theta(\mathrm{e}^{i\omega})}{\phi(\mathrm{e}^{-i\omega})\phi(\mathrm{e}^{i\omega})} \\ &= \frac{\sigma_\varepsilon^2}{2\pi} \frac{|1 + \theta_1 \mathrm{e}^{-i\omega} + \cdots + \theta_q \mathrm{e}^{-iq\omega}|^2}{|1 + \phi_1 \mathrm{e}^{-i\omega} + \cdots + \phi_p \mathrm{e}^{-ip\omega}|^2} \end{aligned} \tag{5.112}$$

where $\omega \in [-\pi, \pi]$.

5.5.3.1 The inverse autocorrelation function

Consider the stationary ARMA(p, q) process, $\{Y_t\}$, given by

$$\phi(\mathrm{B})Y_t = \theta(\mathrm{B})\varepsilon_t \tag{5.113}$$

where $\{\varepsilon_t\}$ is white noise. If the process, $\{Y_t\}$, is invertible, then the process

$$\theta(\mathrm{B})Z_t = \phi(\mathrm{B})\varepsilon_t \tag{5.114}$$

is a stationary ARMA(q, p) process. This process is called the *dual process.*

DEFINITION 5.20 (THE INVERSE AUTOCORRELATION FUNCTION)
The inverse autocorrelation function (IACF) for the process, $\{Y_t\}$, given by (5.113), is found as the autocorrelation function for the dual process. The inverse autocorrelation at lag k is denoted $\rho i(k)$.

Please note that if the process is an AR(p) process, then the inverse autocorrelation corresponds to the autocorrelation function for an MA(p) process. Thus, we obtain the following.

THEOREM 5.14 (INVERSE AUTOCORRELATION FUNCTION FOR AR PROCESSES)
For an AR(p) process it holds that

$$\begin{aligned} \rho i(k) &\neq 0\,, \quad k \le p, \\ \rho i(k) &= 0\,, \quad k > p. \end{aligned} \tag{5.115}$$

Proof Follows from the previous paragraph. ∎

The inverse autocorrelation can be used as an alternative to the partial autocorrelation in the identification of pure AR processes. Compare Theorem 5.14 with Theorem 5.11 on page 124.

Since the inverse autocorrelation function can be found simply as the autocorrelation function for the dual process, the inverse autocorrelation for ARMA processes at sufficiently large lag-values consists of damped exponential and harmonic functions.

5.6 Non-stationary models

In practical applications, one often has to deal with data which cannot be described by a stationary process, e.g., due to long-term trends, periodic trends, or a more general time-varying behavior.

5.6.1 The ARIMA process

If a time series has a long-term trend, then this can be removed by a suitable differencing of the series. The differenced series will then often be adequately described by a stationary ARMA(p,q) process. Hereby, a simple model for non-stationary phenomena can be achieved.

DEFINITION 5.21 (THE ARIMA(p,d,q) PROCESS)
The process $\{Y_t\}$ is called an *integrated* (or summarized) *autoregressive moving average*, or in short an *ARIMA(p,d,q) process*, if it can be written in the form

$$\phi(\mathrm{B})\nabla^d Y_t = \theta(\mathrm{B})\varepsilon_t\,, \quad (d \in N), \tag{5.116}$$

where $\{\varepsilon_t\}$ is white noise, $\phi(z^{-1})$ is a polynomial of order p, and $\theta(z^{-1})$ is a polynomial of order q, both having all roots inside the unit circle.

It follows from the definition that the process

$$W_t = \nabla^d Y_t \tag{5.117}$$

is a stationary and invertible ARMA(p,q) process.

If we introduce a new polynomial $\varphi(\mathrm{B})$ by

$$\varphi(\mathrm{B}) = \phi(\mathrm{B})\nabla^d = \phi(\mathrm{B})(1-\mathrm{B})^d \tag{5.118}$$

the *transfer function* from the white noise input process to the output process, $\{Y_t\}$ for the ARIMA process can be written

$$H(z) = \frac{\theta(z^{-1})}{\varphi(z^{-1})} = \frac{\theta(z^{-1})}{\phi(z^{-1})(1-z^{-1})^d}, \tag{5.119}$$

where it follows that the transfer function has d poles in $z=1$ and p poles inside the unit circle.

The notation "integrated" in Definition 5.21 is due to the fact that $W_t = \nabla^d Y_t$ if, and only if, $Y_t = \mathrm{S}^d W_t$. This shows that Y_t is obtained by d summations of the stationary and invertible ARMA(p,q) process W_t.

As a special case of the ARIMA process, we have the *IMA*(d,q) *process* (Integrated Moving Average), which is a process in the form

$$\nabla^d Y_t = \theta(\mathrm{B})\varepsilon_t, \tag{5.120}$$

and the *ARI*(p,d) *process* (Integrated AutoRegressive) which is a process in the form

$$\phi(\mathrm{B})\nabla^d Y_t = \varepsilon_t. \tag{5.121}$$

It can be shown that the k'th order exponential smoothing (see Chapter 3) gives the optimal prediction if the underlying model follows an IMA(k,k) model

$$\nabla^k Y_t = (1 - \theta_1 \mathrm{B})^k \varepsilon_t. \tag{5.122}$$

For an ARIMA(p,d,q) process ($d \geq 1$), it holds that the sum of the π weights is zero, i.e.,

$$\pi(1) = \sum_{i=0}^{\infty} \pi_i = 1 + \sum_{i=1}^{\infty} \pi_i = 0, \tag{5.123}$$

which can be seen by

$$\pi(z^{-1}) = \frac{\phi(z^{-1})(1 - z^{-1})^d}{\theta(z^{-1})} = 0, \quad \text{for } z = 1.$$

Since an ARIMA(p,d,q) process can be written on the inverse form

$$\pi(\mathrm{B})Y_t = \varepsilon_t \tag{5.124}$$

we get from (5.123) that Y_t, for $d \geq 1$, can be written as

$$Y_t = \sum_{i=1}^{\infty} \pi_i Y_{t-i} + \varepsilon_t, \quad \sum_{i=1}^{\infty} \pi_i = -1 \tag{5.125}$$

as a weighted mean from past observations.

Finally it should be stressed, that the non-stationary process defined by

$$\nabla Y_t = \varepsilon_t \tag{5.126}$$

is called a *random walk*.

5.6.2 Seasonal models

Box and Jenkins (1970/1976), who have been the main contributors to making the ARIMA models popular, have proposed a class of models for describing a time series with periodical behavior. For a time series with the *seasonal period* s, the idea is to let Y_t depend on Y_{t-s}, Y_{t-2s}, etc.

DEFINITION 5.22 (MULTIPLICATIVE $(p,d,q)\times(P,D,Q)_s$ SEASONAL MODEL)
The process $\{Y_t\}$ is said to follow a *multiplicative* $(p,d,q)\times(P,D,Q)_s$ *seasonal model* if

$$\phi(\mathrm{B})\Phi(\mathrm{B}^s)\nabla^d\nabla_s^D Y_t = \theta(\mathrm{B})\Theta(\mathrm{B}^s)\varepsilon_t \tag{5.127}$$

where $\{\varepsilon_t\}$ is white noise, ϕ and θ are polynomials of order p and q, respectively, and Φ and Θ are polynomials of order P and Q, which have all the roots inside the unit circle.

The seasonal difference operator is denoted ∇_s, and Φ and Θ are polynomials in B^s,

$$\Phi(\mathrm{B}^s) = 1 + \Phi_1\,\mathrm{B}^s + \cdots + \Phi_P\,\mathrm{B}^{sP}, \tag{5.128}$$

$$\Theta(\mathrm{B}^s) = 1 + \Theta_1\,\mathrm{B}^s + \cdots + \Theta_Q\,\mathrm{B}^{sQ}. \tag{5.129}$$

When describing time series with a season of one year and monthly observations, we have $s = 12$.

From Definition 5.22, we get that ∇^d contributes with d poles in $z = 1$ whereas $\nabla_s^D = (1-\mathrm{B}^s)^D$ contributes with $D \times s$ poles on the unit circle that are determined by

$$z_k = \mathrm{e}^{ik2\pi/s}, \quad k = 0, 1, \ldots, s-1, \tag{5.130}$$

since they are the solution to $1 - z^s = 0$. The poles are, thus, uniformly distributed on the unit circle with the distance in angle of $2\pi/s$.

For $d = D = 0$ the process given by (5.127) is stationary, and it is easy to see that the autocorrelation function for a pure seasonal model $(0,0,0)\times(P,0,Q)_s$ follows the same pattern as the autocorrelation function for the ARIMA$(P,0,Q)$ model, if the lags in between $1, 2, \ldots, s-1, s+1, \ldots, 2s-1, 2s+1, \ldots$, etc., are removed. For a (mixed) seasonal model $(p,0,q)\times(P,0,Q)_s$, the autocorrelation function at lags that are not a multiple of s will be determined by the ordinary ARIMA part. This is illustrated in the following example.

Example 5.10 (Autocorrelation function for a seasonal model)
This example illustrates how the autocorrelation function for a seasonal model is determined. We will consider a multiplicative $(0,0,1)\times(1,0,0)_{12}$ seasonal model

$$(1 - \Phi\mathrm{B}^{12})Y_t = (1 + \theta\,\mathrm{B})\varepsilon_t \tag{5.131}$$

By multiplying (5.131) by Y_{t-k} $(k \geq 12)$ and calculating expected values

$$\mathrm{E}[Y_t Y_{t-k}] - \Phi\,\mathrm{E}[Y_{t-12} Y_{t-k}] = \mathrm{E}[\varepsilon_t Y_{t-k}] + \theta\,\mathrm{E}[\varepsilon_{t-1} Y_{t-k}]$$

we obtain

$$\gamma(k) = \Phi\gamma(k-12)\,, \quad (k \geq 12).$$

We will rewrite (5.131) in MA form

$$\begin{aligned} Y_t &= (1 - \Phi\,\mathrm{B}^{12})^{-1}(1 + \theta\,\mathrm{B})\varepsilon_t \\ &= \left((1 + \Phi\,\mathrm{B}^{12} + \Phi^2\,\mathrm{B}^{24} + \cdots) + (\theta\,\mathrm{B} + \theta\Phi\,\mathrm{B}^{13} + \theta\Phi^2\,\mathrm{B}^{25} + \cdots)\right)\varepsilon_t. \end{aligned}$$

We get

$$\begin{aligned} \gamma(k) &= \mathrm{Cov}[Y_t, Y_{t+k}] \\ &= \mathrm{Cov}\Big[\big[(\varepsilon_t + \Phi\varepsilon_{t-12} + \Phi^2\varepsilon_{t-24} + \cdots) \\ &\qquad + (\theta\varepsilon_{t-1} + \theta\Phi\varepsilon_{t-13} + \theta\Phi^2\varepsilon_{t-25} + \cdots)\big], \\ &\qquad \big[(\varepsilon_{t+k} + \Phi\varepsilon_{t-12+k} + \Phi^2\varepsilon_{t-24+k} + \cdots) \\ &\qquad + (\theta\varepsilon_{t-1+k} + \theta\Phi\varepsilon_{t-13+k} + \theta\Phi^2\varepsilon_{t-25+k} + \cdots)\big]\Big]. \end{aligned}$$

From this it is seen

$$\begin{aligned} \gamma(0) &= \{(1 + \Phi^2 + \Phi^4 + \dots) + (\theta^2 + \theta^2\Phi^2 + \theta^2\Phi^4 + \cdots)\}\sigma_\varepsilon^2 = \frac{1+\theta^2}{1-\Phi^2}\sigma_\varepsilon^2 \\ \gamma(1) &= \{\theta + \theta\Phi^2 + \theta\Phi^4 + \cdots\}\sigma_\varepsilon^2 = \frac{\theta}{1-\Phi^2}\sigma_\varepsilon^2 \\ \gamma(2) &= \gamma(3) = \cdots = \gamma(10) = 0 \\ \gamma(11) &= \{\theta\Phi + \theta\Phi^3 + \theta\Phi^5 + \cdots\}\sigma_\varepsilon^2 = \frac{\theta\Phi}{1-\Phi^2}\sigma_\varepsilon^2 \\ \gamma(k) &= \Phi\gamma(k-12)\,, \quad (k \geq 12). \end{aligned}$$

The autocorrelation function becomes

$$\rho(k) = \begin{cases} 1 & k = 0 \\ \theta/(1+\theta^2) & k = 1 \\ 0 & k = 2, 3, \dots, 10 \\ \theta\Phi/(1+\theta^2) & k = 11 \\ \Phi\rho(k-12) & k = 12, 13, 14, \dots \end{cases}$$

Applying $\Phi = 0.6$ and $\theta = 0.5$, we get the autocorrelation function shown in Figure 5.8 on the next page.

We observe an exponential decrease in the correlation pattern from an MA(1) process. The exponential decrease is due to the first order autoregressive seasonal component, $\mathrm{AR}(1)_{12}$.

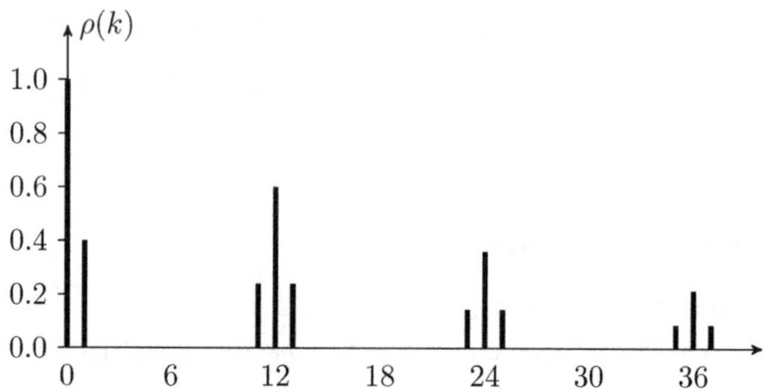

Figure 5.8: *The autocorrelation function for the process* $(1 - \Phi\mathrm{B}^{12})Y_t = (1 + \theta\mathrm{B})\varepsilon_t$.

5.6.3 Models with covariates

Some useful version of the models with time-varying means are models where, e.g., the general linear model from Chapter 3 is used to explain a static influence of some covariates or explanatory variables. The model structure is

$$\phi(\mathrm{B})(Y_t - \boldsymbol{X}_t^T \boldsymbol{\beta}) = \theta(\mathrm{B})\varepsilon_t$$

where $\boldsymbol{X}_t$ is a vector of p explanatory variables and $\boldsymbol{\beta}$ is the parameter vector of the general linear model $Y_t = \boldsymbol{X}_t^T \boldsymbol{\beta} + W_t$, where $\{W_t\}$ is an autocorrelated zero mean process.

5.6.4 Models with time-varying mean values

Sometimes it is desirable to describe a possible *variation of the mean* using, e.g., linear deterministic functions $f_1(t), \ldots, f_k(t)$. This can be achieved by estimating a model in the form

$$Y_t = \beta_1 f_1(t) + \cdots + \beta_k f_k(t) + W_t, \tag{5.132}$$

and subsequently modeling the variations of $\{W_t\}$ using an ARMA(p, q) process, $\phi(\mathrm{B})W_t = \theta(\mathrm{B})\varepsilon_t$. The total model becomes

$$\phi(\mathrm{B})[Y_t - \beta_1 f_1(t) - \cdots - \beta_k f_k(t)] = \theta(\mathrm{B})\varepsilon_t. \tag{5.133}$$

In the estimation of $\beta_1, \ldots, \beta_k$ the covariance structure of W_t has to be used. This can be done using a weighted least squares method, as described in Chapter 3, where $\boldsymbol{\Sigma}$ is chosen so that it is proportional to the covariance structure of W_t. For applications this can be done using the algorithm presented in Section 3.2.2 on page 40. A practical example is given in Madsen and Thyregod (1988).

Alternatively, the following model can be used

$$\phi(\mathrm{B})Y_t = \beta_1 f_1(t) + \cdots + \beta_k f_k(t) + \theta(\mathrm{B})\varepsilon_t. \tag{5.134}$$

In this model all the parameters are estimated simultaneously. However, it should be noted that (5.133) and (5.134) are not equivalent, because then $f_i(t-1), \ldots, f_i(t-p)$ should be included in (5.134).

5.6.5 Models with time-varying coefficients

For many phenomena it is observed that the *dynamical characteristics change over time.* In such cases it might be reasonable to model such a process using an ARMA model where the coefficients ϕ_i and θ_i are time-varying.

In principle, a model in the form

$$(1 + \phi(t)\,\mathrm{B})(Y_t - \mu(t)) = \varepsilon_t \tag{5.135}$$

where

$$\phi(t) = \phi_1 \sin(\omega_1 t + \alpha_1) \quad \text{and} \quad \mu(t) = \mu \sin(\omega_2 t + \alpha_2)$$

(often one will expect that $\omega_1 = \omega_2$) can be estimated by using an iterative method, e.g., in maximizing the corresponding likelihood function or minimizing the sum of squared residuals. However, a problem concerning this method is to provide methods for the identification of the model structure. We will consider this problem later when adaptive forecasting models are discussed in Chapter 11. Examples of identification of models with time-varying coefficients can be found in Madsen (1985) and Porsholt (1989).

5.7 Optimal prediction of stochastic processes

In this section we consider how to predict future values of the stochastic process $\{Y_t\}$ based on past observations from the same process. In a later chapter we consider the case where we wish to predict the stochastic process $\{Y_t\}$ based on observations of a correlated process $\{X_t\}$.

We assume that we have observations from the process up to time t, i.e., $\{Y_s; s \leq t\}$. We now want to predict the value of Y_{t+k}, and distinguish between

i) *Prediction*—make a forecast of a future observation; in other words, $k > 0$, and

ii) *Interpolation*—where Y_{t+k} in time are among the observations (a time series with missing observations); in other words, $k < 0$.

In this section we will consider only the issue of prediction and we assume that there are no missing values, i.e., we have the following observations $\{Y_s; s = t, t-1, \ldots\}$.

Based on the observations $Y_t, Y_{t-1}, \ldots$, we seek to predict Y_{t+k} $(k > 0)$, and thus introduce the predictor

$$\widehat{Y}_{t+k|t} = \mathrm{g}(Y_t, Y_{t-1}, \ldots). \tag{5.136}$$

As a criterion for the optimal predictor, we choose to *minimize the expected squared error.* This implies that $\mathrm{g}(Y_t, Y_{t-1}, \ldots)$ is chosen so that it minimizes the criterion function

$$\mathrm{E}\left[(Y_{t+k} - \mathrm{g}(Y_t, Y_{t-1}, \ldots))^2 | Y_t, Y_{t-1}, \ldots\right]. \tag{5.137}$$

From Theorem 3.9 on page 44, we get that (5.137) is minimized for

$$\mathrm{g}(Y_t, Y_{t-1}, \ldots) = \mathrm{g}_0(Y_t, Y_{t-1}, \ldots) = \mathrm{E}[Y_{t+k} | Y_t, Y_{t-1}, \ldots]. \tag{5.138}$$

The *optimal predictor* (or forecast) for Y_{t+k} becomes

$$\widehat{Y}_{t+k|t} = \mathrm{E}[Y_{t+k} | Y_t, Y_{t-1}, \ldots], \tag{5.139}$$

i.e., the optimal predictor is the conditional mean.

If we introduce the *prediction error*

$$e_{t+k|t} = Y_{t+k} - \widehat{Y}_{t+k|t} \tag{5.140}$$

we see from (5.137) and (5.138) that *the optimal predictor minimizes the variance of the prediction error.* The drawback of solving (5.139) is that the conditional mean is difficult to calculate because it is necessary to know the multidimensional distributions for the process. However, for normal processes it is easier.

THEOREM 5.15
For a normal process, $\{Y_t\}$, the optimal linear predictor is linear in the observations, i.e.,

$$\widehat{Y}_{t+k|t} = \sum_{j=0}^{n} \beta_j Y_{t-j} \tag{5.141}$$

or

$$\widehat{Y}_{t+k|t} = \lim_{n \to \infty} \sum_{j=0}^{n} \beta_j Y_{t-j}. \tag{5.142}$$

Proof Follows from (5.139) and a generalization of Theorem 2.8 on page 27. ∎

5.7.1 Prediction in the ARIMA process

We consider an ARIMA(p, d, q) process, $\{Y_t\}$, defined by

$$\phi(\mathrm{B})\nabla^d Y_t = \theta(\mathrm{B})\varepsilon_t, \tag{5.143}$$

where $\{\varepsilon_t\}$ is white noise with variance, σ^2.

The process can be formulated in MA-form

$$Y_t = \varepsilon_t + \psi_1\varepsilon_{t-1} + \psi_2\varepsilon_{t-2} + \cdots. \tag{5.144}$$

Similarly, we have

$$Y_{t+k} = \varepsilon_{t+k} + \psi_1\varepsilon_{t+k-1} + \cdots + \psi_k\varepsilon_t + \psi_{k+1}\varepsilon_{t-1} + \cdots. \tag{5.145}$$

By writing the process in inverse form (or AR-form),

$$\varepsilon_t = Y_t + \pi_1 Y_{t-1} + \pi_2 Y_{t-2} + \cdots, \tag{5.146}$$

we see that the values $\{\varepsilon_s\, , s = t, t-1, \dots\}$ are known when $\{Y_s\, , s = t, t-1, \dots\}$ are known. From (5.144) it is seen that the opposite also holds. Thus, if $\{\varepsilon_t\}$ is white noise we get

$$\mathrm{E}[\varepsilon_{t+k}|Y_t, Y_{t-1}, \dots] = \begin{cases} \varepsilon_{t+k} & \text{for } k \le 0 \\ 0 & \text{for } k > 0 \end{cases} \tag{5.147}$$

Since the optimal prediction is the conditional mean, we get from (5.145) the *predictor*.

$$\widehat{Y}_{t+k|t} = \mathrm{E}[Y_{t+k}|Y_t, Y_{t-1}, \dots] = \psi_k\varepsilon_t + \psi_{k+1}\varepsilon_{t-1} + \cdots. \tag{5.148}$$

By subtracting (5.148) from (5.145), we obtain the *prediction error*,

$$e_{t+k|t} = Y_{t+k} - \widehat{Y}_{t+k|t} = \varepsilon_{t+k} + \psi_1\varepsilon_{t+k-1} + \cdots + \psi_{k-1}\varepsilon_{t+1} \tag{5.149}$$

and the *variance of the prediction error* becomes

$$\sigma_k^2 = (1 + \psi_1^2 + \cdots + \psi_{k-1}^2)\sigma_\varepsilon^2. \tag{5.150}$$

If we assume that $\{\varepsilon_t\}$ is normally distributed, we get the following $(1-\alpha)$-*confidence interval for the value of* Y_{t+k}

$$\widehat{Y}_{t+k|t} \pm u_{\alpha/2}\sigma_k = \widehat{Y}_{t+k|t} \pm u_{\alpha/2}\sigma_\varepsilon\sqrt{1 + \psi_1^2 + \cdots + \psi_{k-1}^2}, \tag{5.151}$$

where $u_{\alpha/2}$ is the $\alpha/2$ quantile in the standard normal distribution. Usually the parameters of the ARIMA model will not be known exactly, but estimated based on observations. The prediction error will then be larger and (5.151) is thus an approximative confidence interval.

▸ **Remark 5.5**
In practice k is small, and hence, only a small number of the ψ weights are needed. The needed ψ weights are often most easily calculated by "sending a unit pulse through the system," as illustrated in Example 4.3 on page 72. ◂

We now assume that one time unit has passed and we get a new observation Y_{t+1}. We want to examine how the prediction can be updated (and improved). From (5.145) and (5.147) we have

$$\widehat{Y}_{t+k|t+1} = \psi_{k-1}\varepsilon_{t+1} + \psi_k\varepsilon_t + \psi_{k+1}\varepsilon_{t-1} + \cdots = \psi_{k-1}\varepsilon_{t+1} + \widehat{Y}_{t+k|t}. \quad (5.152)$$

This formula is used to *update the prediction.* (It is seen that when Y_{t+1} becomes available, then ε_{t+1} can be calculated.)

In practice the prediction (forecast) is most easily obtained by applying the conditional mean on the process written as a difference equation

$$Y_t + \varphi_1 Y_{t-1} + \cdots + \varphi_{p+d} Y_{t-p-d} = \varepsilon_t + \theta_1\varepsilon_{t-1} + \cdots + \theta_q\varepsilon_{t-q} \quad (5.153)$$

where the coefficients $\varphi_1, \dots, \varphi_{p+d}$ are given from the identity: $\varphi(\mathrm{B}) = \phi(\mathrm{B})\nabla^d$, see (5.118) on page 130.

If we want to calculate the k-step prediction, we write

$$\begin{aligned} Y_{t+k} &+ \varphi_1 Y_{t+k-1} + \cdots + \varphi_{p+d} Y_{t+k-p-d} \\ &= \varepsilon_{t+k} + \theta_1\varepsilon_{t+k-1} + \cdots + \theta_q\varepsilon_{t+k-q} \end{aligned} \quad (5.154)$$

The forecast equation becomes

$$\begin{aligned} \widehat{Y}_{t+k|t} = &-\varphi_1 \,\mathrm{E}[Y_{t+k-1}|Y_t, Y_{t-1}, \dots] - \cdots \\ &- \varphi_{p+d}\,\mathrm{E}[Y_{t+k-p-d}|Y_t, Y_{t-1}, \dots] \\ &+ \mathrm{E}[\varepsilon_{t+k}|Y_t, Y_{t-1}, \dots] + \theta_1\,\mathrm{E}[\varepsilon_{t+k-1}|Y_t, Y_{t-1}, \dots] + \cdots \\ &+ \theta_q\,\mathrm{E}[\varepsilon_{t+k-q}|Y_t, Y_{t-1}, \dots]. \end{aligned} \quad (5.155)$$

In the calculation of (5.155) we exploit that

$$\begin{aligned} \mathrm{E}[Y_{t-j}|Y_t, Y_{t-1}, \dots] &= Y_{t-j} & j &= 0, 1, 2, \dots \\ \mathrm{E}[Y_{t+j}|Y_t, Y_{t-1}, \dots] &= \widehat{Y}_{t+j|t} & j &= 1, 2, \dots \\ \mathrm{E}[\varepsilon_{t-j}|Y_t, Y_{t-1}, \dots] &= \varepsilon_{t-j} & j &= 0, 1, 2, \dots \\ \mathrm{E}[\varepsilon_{t+j}|Y_t, Y_{t-1}, \dots] &= 0 & j &= 1, 2, \dots \end{aligned}$$

which follows from the fact that $\{\varepsilon_s, s = t, t-1, \dots\}$ is known when $\{Y_s, s = t, t-1, \dots\}$ is known.

Example 5.11 (Prediction in an IMA(1, 3) process)
We have given an IMA(1, 3) process:

$$Y_t - Y_{t-1} = \varepsilon_t + \theta_1\varepsilon_{t-1} + \theta_2\varepsilon_{t-2} + \theta_3\varepsilon_{t-3}.$$

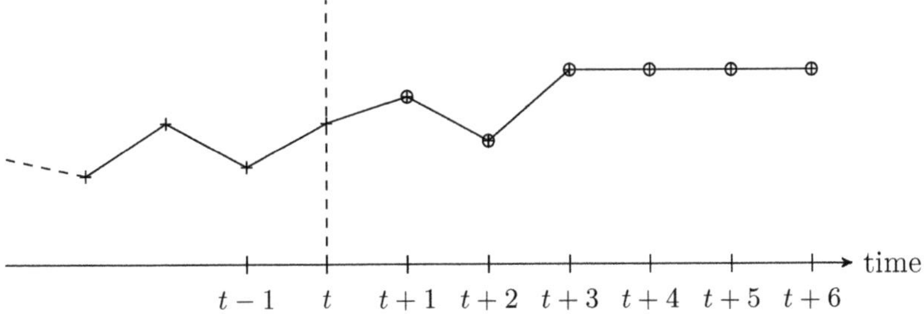

Figure 5.9: *The evaluation of the predictions of an IMA(3) process.*

From (5.155) we get

$$\begin{aligned}\widehat{Y}_{t+1|t} &= \mathrm{E}[Y_t|Y_t, Y_{t-1}, \dots] + \mathrm{E}[\varepsilon_{t+1}|Y_t, Y_{t-1}, \dots] \\ &\quad + \theta_1 \mathrm{E}[\varepsilon_t|Y_t, Y_{t-1}, \dots] + \theta_2 \mathrm{E}[\varepsilon_{t-1}|Y_t, Y_{t-1}, \dots] \\ &\quad + \theta_3 \mathrm{E}[\varepsilon_{t-2}|Y_t, Y_{t-1}, \dots] \\ &= Y_t + 0 + \theta_1\varepsilon_t + \theta_2\varepsilon_{t-1} + \theta_3\varepsilon_{t-2}\end{aligned}$$

Similarly we get

$$\begin{aligned}\widehat{Y}_{t+2|t} &= \widehat{Y}_{t+1|t} + \theta_2\varepsilon_t + \theta_3\varepsilon_{t-1} \\ \widehat{Y}_{t+3|t} &= \widehat{Y}_{t+2|t} + \theta_3\varepsilon_t \\ \widehat{Y}_{t+k|t} &= \widehat{Y}_{t+3|t}, \quad k \geq 4.\end{aligned}$$

The variance of the one-step predictions is $\sigma_1^2 = \sigma_\varepsilon^2$. The variance of the k-step prediction can easily be calculated after writing the process in MA-form:

$$\begin{aligned}Y_t &= \varepsilon_t + (1+\theta_1)\varepsilon_{t-1} + (1+\theta_1+\theta_2)\varepsilon_{t-2} \\ &\quad + (1+\theta_1+\theta_2+\theta_3)\varepsilon_{t-3} \\ &\quad + (1+\theta_1+\theta_2+\theta_3)\varepsilon_{t-4} + \cdots \\ &\quad + (1+\theta_1+\theta_2+\theta_3)\varepsilon_{t-i} + \cdots.\end{aligned}$$

From this we get

$$\begin{aligned}\sigma_2^2 &= (1 + (1+\theta_1)^2)\sigma_\varepsilon^2 \\ \sigma_3^2 &= (1 + (1+\theta_1)^2 + (1+\theta_1+\theta_2)^2)\sigma_\varepsilon^2 \\ \sigma_k^2 &= (1 + (1+\theta_1)^2 + (1+\theta_1+\theta_2)^2 \\ &\quad + (k-3)(1+\theta_1+\theta_2+\theta_3)^2)\sigma_\varepsilon^2, \quad \text{for } k \geq 3.\end{aligned}$$

We see that the variance of the prediction error goes to infinity when k goes to infinity.

Previously we have seen that the autocovariance function fulfills a homogeneous difference equation determined by the autogressive part of the model—see (5.99) on page 126. Exactly the same holds for the predictor $Y_{t+k|t}$.

THEOREM 5.16 (DIFFERENCE EQUATION FOR OPTIMAL PREDICTOR)
For the ARIMA(p, d, q) process, the optimal predictor fulfills the homogeneous difference equation

$$\varphi(B)\nabla^d \widehat{Y}_{t+k|t} = 0 \tag{5.156}$$

for $k > q$.

Proof Assume that we have observations until time t, write the ARIMA process for Y_{t+k} and take the expectations conditional on observations until time t. Doing this the MA part of the model will vanish for $k > q$. ∎

This is illustrated in the following example:

Example 5.12 (Prediction in the ARIMA$(0, d, q)$ process)
Consider the ARIMA$(0, d, q)$ process

$$\nabla^d Y_t = \theta(B)\varepsilon_t. \tag{5.157}$$

For $k > q$ we obtain the homogeneous difference equation for the predictor

$$\nabla^d \widehat{Y}_{t+k|t} = 0. \tag{5.158}$$

The characteristic equation for the difference equation has a d-double root in one. This means that the general solution is

$$\widehat{Y}_{t+k|t} = A_0^t + A_1^t k + A_2^t k^2 + \cdots + A_{d-1}^t k^{d-1} \tag{5.159}$$

where the superscipt t on the coefficients indicates that the particular solution is found using the information set $Y_t, Y_{t-1}, \ldots$ at time t. That is, the predictor is a polynomial of degree $d - 1$.

5.8 Problems

Exercise 5.1
Question 1 Show that the stochastic process $\{X_t\}$, defined by

$$X_t = \varepsilon_t + c(\varepsilon_{t-1} + \varepsilon_{t-2} + \cdots)$$

where c is a constant and $\{\varepsilon_t\}$ is white noise, is non-stationary.

Question 2 On basis of the process in Question 1, a new process $\{Y_t\}$ is introduced by

$$Y_t = \nabla X_t$$

Show that $\{Y_t\}$ is an MA(1) process. Is $\{Y_t\}$ stationary?

Question 3 Find the autocorrelation function for $\{Y_t\}$.

Exercise 5.2

Consider a two-state Markov chain model for a stochastic process $\{Y(t)\}$ where the states are denoted -1 and 1. The number of transitions is determined by a Poisson process with a mean event rate λ.

Question 1 Show that the process is stationary with the autocorrelation function

$$\rho(\tau) = \mathrm{e}^{-2\lambda|u|}$$

Question 2 Show that the spectral density is

$$f(\omega) = \frac{1}{2\pi}\frac{4\lambda}{4\lambda^2 + \omega^2}$$

Exercise 5.3

Let $\{X_t\}$ and $\{Y_t\}$ be mutually independent and stationary stochastic processes with the spectral densities $f_x(\omega)$ and $f_y(\omega)$.

Question 1 Show that the process $\{Z_t\}$ given by

$$Z_t = X_t + Y_t$$

is a stationary process with the spectral density

$$f_z(\omega) = f_x(\omega) + f_y(\omega)$$

Question 2 Consider now the process $\{Z_t\}$ given by

$$Z_t = X_t + Y_t$$

where $X_t = \alpha X_{t-1} + W_t$, $(|\alpha| < 1)$. $\{Y_t\}$ and $\{W_t\}$ are mutually independent white noise processes each with mean 0 and variance σ^2.

Question 3 Show that the spectral density of $\{Z_t\}$ is given by

$$f_z(\omega) = \frac{\sigma^2}{2\pi}\frac{2 + \alpha^2 - 2\alpha\cos(\omega)}{1 + \alpha^2 - 2\alpha\cos(\omega)}$$

Exercise 5.4

Let $\{\varepsilon_t\}$ be white noise with the mean value μ_ε and the variance σ_ε^2.

The process $\{X_t\}$ defined by

$$\nabla X_t = \varepsilon_t$$

is called a random walk. Set $X_1 = \varepsilon_1$, and consider the following set of time indices $\{t | t \geq 1\}$.

Question 1 Calculate the mean value $\mathrm{E}[X_t]$, the variance $\mathrm{Var}[X_t]$, and the covariance $\mathrm{Cov}[X_{t_1}, X_{t_2}]$.

Question 2 Is $\{X_t\}$ a stationary process?

Exercise 5.5

Question 1 Rewrite the process $\{X_t\}$ defined by

$$X_t - \phi X_{t-1} = \varepsilon_t$$

such that $\{X_t\}$ is exclusively described by past values of $\{\varepsilon_t\}$. If $\{\varepsilon_t\}$ is white noise, X_t is said to be expressed in an MA form.

Question 2 To which category of ARIMA(p, d, q) does each of the following processes belong? Are they stationary/invertible?

a) $X_t = \frac{5}{6}X_{t-1} - \frac{1}{6}X_{t-2} + \varepsilon_t - \frac{1}{4}\varepsilon_{t-1}$.

b) $X_t = \frac{4}{3}X_{t-1} - \frac{1}{3}X_{t-2} + \varepsilon_t - \frac{1}{4}\varepsilon_{t-1}$.

c) $X_t = 2X_{t-1} - X_{t-2} + \varepsilon_t - \frac{1}{4}\varepsilon_{t-1} - \frac{1}{4}\varepsilon_{t-2}$.

Exercise 5.6

Let the following ARMA(1, 1) process be given by

$$(1 - \phi\,\mathrm{B})X_t = (1 - \theta\,\mathrm{B})\varepsilon_t$$

with $\mathrm{E}[\varepsilon_t] = 0$ and $\mathrm{Var}[\varepsilon_t] = \sigma_\varepsilon^2$.

Question 1 Rewrite the process into an AR form.

Question 2 Rewrite the process into an MA form and find an expression for $\mathrm{Var}[X_t]$.

Exercise 5.7

Let $\{X_t\}$ be a stationary process with the autocovariance function $\gamma_X(k)$. Consider a new process $\{Y_t\}$ defined by

$$Y_t = \nabla_s X_t \tag{5.160}$$

Question 1 Determine the autocovariance function of $\{Y_t\}$ as a function of $\gamma_X(k)$.

Question 2 In the following, it is given that

$$X_t = \alpha X_{t-1} + \varepsilon_t\,, \quad |\alpha| < 1 \tag{5.161}$$

where $\{\varepsilon_t\}$ is white noise.

Determine the autocovariance function of $\{Y_t\}$.

Question 3 Use $s = 1$ in (5.160), i.e., $Y_t = \nabla X_t$, and determine for which values of α the variance of Y_t will be less than the variance of X_t.

Exercise 5.8

Consider the following ARMA(2, 1) process

$$(1 - 1.27\,\mathrm{B} + 0.81\,\mathrm{B}^2)X_t = (1 - 0.3\,\mathrm{B})\varepsilon_t$$

where $\{\varepsilon_t\}$ is white noise.

Question 1 Check if the process is stationary.

Question 2 Sketch the impulse response function for the transfer from ε_t to X_t (only for $k \leq 10$).

Question 3 Determine the corresponding frequency response function, and sketch the amplitude function.

Exercise 5.9

Question 1 Consider the seasonal difference

$$Y_t = \nabla_s X_t = X_t - X_{t-s}$$

Put $s = 4$ and find the impulse-, step-, and frequency-response functions.
Find and sketch the amplitude function.

Question 2 Consider the summation

$$Y_t = SX_t = X_t + X_{t-1} + X_{t-2} + \cdots$$

Find the impulse, step and frequency-response functions.
Find $G^2(\omega)$.

Question 3 Exponential smoothing is given by

$$\begin{aligned} Y_t &= \alpha X_t + \alpha(1-\alpha)X_{t-1} + \alpha(1-\alpha)^2 X_{t-2} + \ldots \\ &= (1-\alpha)Y_{t-1} + \alpha X_t \end{aligned}$$

Determine the amplitude function and sketch it for $\alpha = 0.1$

Question 4 Based on Y_t a new time series Z_t is defined as

$$Z_t = X_t - Y_t$$

Find the amplitude function related to the transfer from X_t to Z_t and sketch it for $\alpha = 0.1$.

Exercise 5.10

Consider a process $\{X_t\}$ defined by

$$(1 - \mathrm{B} + 0.5\,\mathrm{B}^2)X_t = (1 + 0.5\,\mathrm{B})\varepsilon_t$$

where $\{\varepsilon_t\}$ is white noise with $\mathrm{E}[\varepsilon_t] = 0$ and $\mathrm{Var}[\varepsilon_t] = 1$.

Question 1 Examine whether $\{X_t\}$ is stationary/invertible. Characterize the process.

Question 2 Determine the autocovariance and autocorrelation functions for $\{X_t\}$. Sketch the autocorrelation function for lags ≤ 6.

Question 3 Sketch the partial autocorrelation function for lags ≤ 4.

Exercise 5.11

Consider the class of multiplicative $(p, d, q) \times (P, D, Q)_s$ seasonal models:

$$\phi(\mathrm{B})\Phi(\mathrm{B}^s)\nabla^d\nabla_s^D X_t = \theta(\mathrm{B})\Theta(\mathrm{B}^s)\varepsilon_t$$

where $\{\varepsilon_t\}$ is white noise.

Question 1 Sketch the autocorrelation function for the seasonal model:

$$X_t = (1 - 0.6\,\mathrm{B}^{12})\varepsilon_t$$

Question 2 Sketch the autocorrelation function for the seasonal model:

$$(1 - 0.6\,\mathrm{B}^{12})X_t = \varepsilon_t$$

Question 3 Sketch the autocorrelation function for the seasonal model:

$$(1 - 0.6\,\mathrm{B}^{12})X_t = (1 + 0.5\,\mathrm{B})\varepsilon_t$$

CHAPTER 6

Identification, estimation, and model checking

6.1 Introduction

In Chapter 5, we introduced a variety of stochastic models which are often used to describe stochastic phenomena of dynamic systems, (water levels, mail loads, stock indices, air temperatures, etc.). In this chapter, methods for determining suitable models for a given phenomena based on an observed time series will be described. The main part of these methods—especially the methods for estimation—are directly related to linear stochastic processes.

The *model building* procedure can be divided into the three stages as sketched in Figure 6.1. The model identification is based on data from one or more time series that represent observations of the stochastic process which

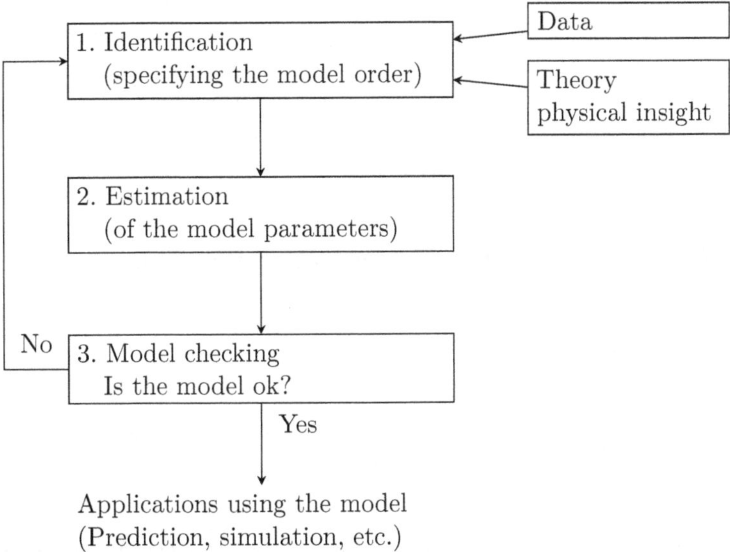

Figure 6.1: *Principles for model building.*

is being modelled. The primary tool for identification of the process is the estimated covariance or correlation functions. For many processes—especially for the processes containing periodical components—the estimated spectrum is an important supplementary tool. The method for estimation of spectra is discussed later in a chapter on spectral analysis. Estimation of the covariance and correlation is treated in Section 6.2. This section also describes how considerations of the correlation functions can be used to determine reasonable starting values for p, d, and q in an ARIMA(p, d, q) process. As indicated in Figure 6.1, the identification process is usually an iterative process, and often the procedure has to be repeated several times before an adequate model is found. Several tools for identification are therefore closely related to model checking, and these tools are thus treated in the section on model checking. When the model structure is determined in Stage 1, the model parameters can be estimated. The method of estimating the parameters in the ARMA model is treated in Section 6.4. Finally, some tools for model selection and model checking are described in Sections 6.5 and 6.6, respectively.

6.2 Estimation of covariance and correlation functions

Estimation of autocovariance and autocorrelation function, the partial autocorrelation function, and the cross-covariance and cross-correlation functions is described.

We will assume that we have observed a realization of a stochastic process $\{Y_t\}$, i.e., that we have a time series $Y_1, Y_2, \ldots, Y_N$. Furthermore, we will assume that the underlying process is ergodic (and hereby can be assumed stationary), so that the ensemble mean and the mean over time are equivalent, cf. Section 5.2 on page 97. This is the assumption that ensures that inference of the stochastic process can be based on a single time series, which is the only realization available of the stochastic process.

6.2.1 Autocovariance and autocorrelation functions

First we consider the autocovariance function. As an *estimator for the autocovariance function*, we apply

$$\mathrm{C}_{YY}(k) = \mathrm{C}(k) = \frac{1}{N} \sum_{t=1}^{N-|k|} (Y_t - \overline{Y})(Y_{t+|k|} - \overline{Y}) \tag{6.1}$$

for $|k| = 0, 1, \ldots, N-1$. Furthermore, $\overline{Y} = (\sum_{t=1}^{N} Y_t)/N$. In Priestley (1981), a detailed treatment of the properties of the above estimator for the autocovariance function is given. If we neglect that μ is estimated by $\overline{Y}$ in

(6.1), we obtain

$$\mathrm{E}[\mathrm{C}(k)] = \frac{1}{N} \sum_{t=1}^{N-|k|} \gamma(k) = \left(1 - \frac{|k|}{N}\right) \gamma(k), \tag{6.2}$$

This shows that the estimator (6.1) is a biased estimator for $\gamma(k)$, but it is also seen that the estimator for a fixed k is *consistent* (since $\mathrm{E}[\mathrm{C}(k)] \to \gamma(k)$ for $N \to \infty$).

Instead, it might be tempting to apply the estimator

$$\mathrm{C}^*(k) = \frac{1}{N - |k|} \sum_{t=1}^{N-|k|} (Y_t - \overline{Y})(Y_{t+|k|} - \overline{Y}), \tag{6.3}$$

which is unbiased when we again (as a first order approximation) neglect the influence from μ being estimated by $\overline{Y}$. However, $\mathrm{Var}[\mathrm{C}(k)] = O(1\,/\,N)$ (cf. Priestley 1981, pp. 150 and 324), whereas $\mathrm{Var}[\mathrm{C}^*(k)] = O(1\,/\,(N - |k|))$, and also the estimator $\mathrm{C}(k)$, in contrast to $\mathrm{C}^*(k)$, is a non-negative definite function (see Theorem 5.3 on page 104) just as the theoretical autocovariance function.

The estimates of the autocovariance function are correlated within themselves. It holds that

$$\begin{aligned} &\mathrm{Cov}[\mathrm{C}(r), \mathrm{C}(r+v)] \\ &\simeq \frac{1}{N} \sum_{m=-\infty}^{\infty} \big(\gamma(m)\gamma(m+v) + \gamma(m+r+v)\gamma(m-r)\big). \end{aligned} \tag{6.4}$$

This autocorrelation of the estimates implies that there can occur systematic behavior in an estimated autocovariance function, which, especially at large lags, does not necessarily imply that the observations corresponding to these lags are correlated. Often one will see a pattern at large lags which is also seen in the low lags, and this is only due to the fact that the estimates of the autocovariance are correlated within themselves. This fact is sketched in Figure 6.2 on the following page, which shows the theoretical and estimated autocovariance functions for an AR(1) process and an ARMA(2, 2) process, respectively, where $N = 500$.

From (6.4) we have as a special case *the variance of the estimator for the variance*:

$$\mathrm{Var}[\mathrm{C}(0)] \simeq \frac{2}{N} \sum_{k=-\infty}^{\infty} \gamma^2(k) = \frac{2\sigma^4}{N} \sum_{k=-\infty}^{\infty} \rho^2(k), \tag{6.5}$$

where $\sigma^2 = \gamma(0)$ and $\rho(k)$ is autocorrelation function for the process.

For an AR(1) process with the coefficient ϕ we have $\rho(k) = \phi^{|k|}$. In this case, we get from (6.5) that

$$\mathrm{Var}[\mathrm{C}(0)] \simeq \frac{2\sigma^4}{N} \left[\frac{1 + \phi^2}{1 - \phi^2}\right]. \tag{6.6}$$

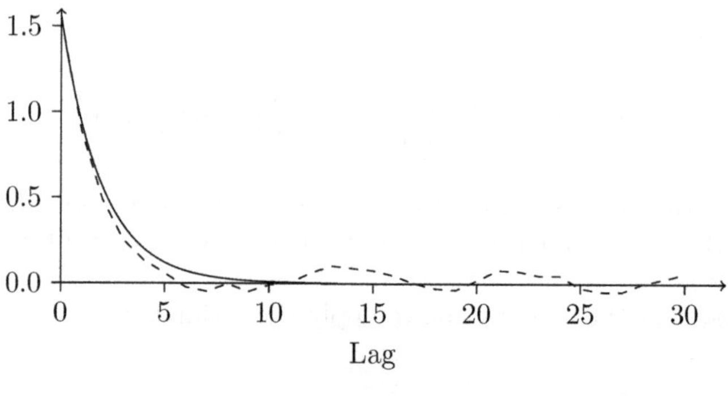

(a) *An AR(1) process*

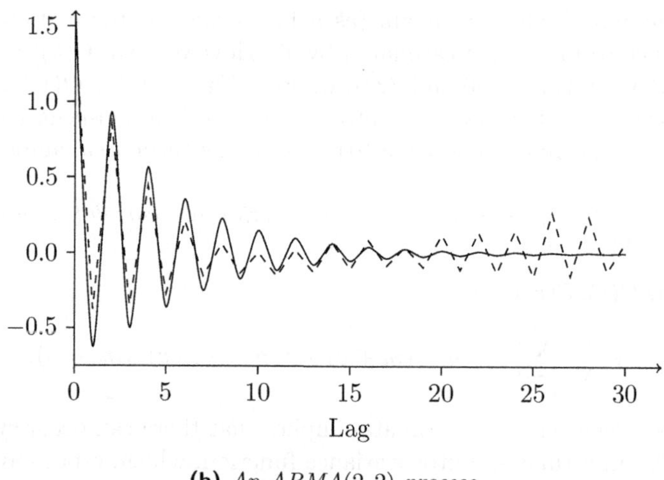

(b) *An ARMA(2,2) process*

Figure 6.2: *Theoretical (solid line) and estimated (dashed line) autocovariance functions for an AR(1) process (a) and an ARMA(2,2) process (b).*

This result can be compared with the variance of the estimated variance given N independent normally distributed observations, where

$$\text{Var}[S^2] \simeq \frac{2\sigma^4}{N'}. \tag{6.7}$$

If we compare this with (6.6) we see that *the equivalent number of independent observations* for the AR(1) process is

$$N' = N\left[\frac{1-\phi^2}{1+\phi^2}\right]. \tag{6.8}$$

We also see that if the observations in the AR(1) process are strongly correlated in time (ϕ close to 1), then the variance of the process is not determined well, even for relatively large values of N.

If a time series contain non-stationarities or very slow variations compared with the sampling frequency (for an AR(1) this corresponds to ϕ close to 1), it is recommendable to remove these trends by applying a suitable filter, e.g.,

$$W_t = Y_t - aY_{t-1}, \tag{6.9}$$

where $a \in [-1, 1]$. We have used $a = 1$ (Box and Jenkins 1970/1976), which corresponds to *a single differencing.*

Based on the *estimated autocovariance function* we find the *estimated autocorrelation function* as

$$\widehat{\rho}(k) = r_k = \frac{\mathrm{C}(k)}{\mathrm{C}(0)}, \tag{6.10}$$

for $k = 0, 1, \ldots, N-1$. In practical situations one will at most calculate the autocorrelation up to lag $k = N/4$. Both notations used in (6.10) for the estimate of autocorrelation function at lag k are often seen in the literature.

The properties for $\widehat{\rho}(k)$ are largely the same as we have discussed for the properties of $\mathrm{C}(k)$, and especially there can occur a *high correlation between neighbouring values of the estimated autocorrelation function.* The remarks on the estimated autocovariance function related to Figure 6.2 can, therefore, be used directly for the autocorrelation function.

Asymptotically (for $N \to \infty$) both the estimates of the autocovariance and the autocorrelation will be normally distributed; but the mean value and the variance are in general dependent on the theoretical autocovariance function and thus in practice impossible to calculate. Albeit, if $\{Y_t\}$ is white noise we have the following.

Theorem 6.1 (Properties in case of white noise)
For white noise the mean value and variance are

$$\mathrm{E}[\widehat{\rho}(k)] \simeq 0\,, \quad k \neq 0, \tag{6.11}$$

$$\mathrm{Var}[\widehat{\rho}(k)] \simeq \frac{1}{N}, \quad k \neq 0. \tag{6.12}$$

Furthermore, $\widehat{\rho}(k)$ is asymptotically normally distributed.

Proof Omitted. ∎

Hence, based on the asymptotical normality and under the hypothesis of white noise, one can make *an (approximative)* 95% *confidence interval* as $\pm 2\sqrt{1/N}$. When judging several lag values one has to bear in mind that even if the process is white noise, about 5% of the values will be outside this confidence interval.

The estimates in the partial autocorrelation function are determined based on the estimated autocorrelation function in the exact similar manner as for the theoretical functions; see Section 5.5.2 on page 119 or Appendix B.

6.2.2 Cross-covariance and cross-correlation functions

We assume that we observe two time series $Y_1, \dots, Y_N$ and $X_1, \dots, X_N$. As *estimator for the cross-covariance function* we use

$$\mathrm{C}_{XY}(k) = \frac{1}{N}\sum_{t=1}^{N-k}(X_t - \overline{X})(Y_{t+k} - \overline{Y}) \tag{6.13a}$$

$$\mathrm{C}_{XY}(-k) = \frac{1}{N}\sum_{t=1}^{N-k}(X_{t+k} - \overline{X})(Y_t - \overline{Y}) \tag{6.13b}$$

for $k = 0, 1, \dots, N-1$, and where $\overline{X} = (\sum_{t=1}^{N} X_t)/N$ and $\overline{Y} = (\sum_{t=1}^{N} Y_t)/N$. The estimator (6.13) is biased for fixed values of N but consistent.

As for the autocovariance function, neighboring values in an estimated cross-covariance function may be strongly correlated. Furthermore, *the autocorrelation in each of the two time series, $\{X_t\}$ and $\{Y_t\}$, may introduce a pattern in the estimated cross-covariance function*, even though the theoretical cross-covariance function is 0 for all lags. For two such mutually independent processes, it holds that

$$\mathrm{Cov}[\mathrm{C}_{XY}(k), \mathrm{C}_{XY}(\ell)] \simeq \frac{1}{N}\sum_{m=-\infty}^{\infty}\gamma_{XX}(m)\gamma_{YY}(m+\ell-k). \tag{6.14}$$

If $\{X(t)\}$ and $\{Y(t)\}$ are AR(1) processes with coefficients α and β, i.e.,

$$\gamma_{XX}(k) = \sigma_X^2\alpha^{|k|}, \quad \gamma_{YY}(k) = \sigma_Y^2\beta^{|k|}, \tag{6.15}$$

we get from (6.14) for $k = \ell$

$$\mathrm{Var}[\mathrm{C}_{XY}(k)] \simeq \frac{\sigma_X^2\sigma_Y^2}{N}\left[\frac{1+\alpha\beta}{1-\alpha\beta}\right]. \tag{6.16}$$

This implies that an autocorrelation in each of the two time series can generate a large cross-covariance between the two time series which are not mutually correlated.

Based on the estimated cross-covariance function, the *cross-correlation function* is estimated by

$$\widehat{\rho}_{XY}(k) = \frac{\mathrm{C}_{XY}(k)}{\sqrt{\mathrm{C}_{XX}(0)\,\mathrm{C}_{YY}(0)}}, \tag{6.17}$$

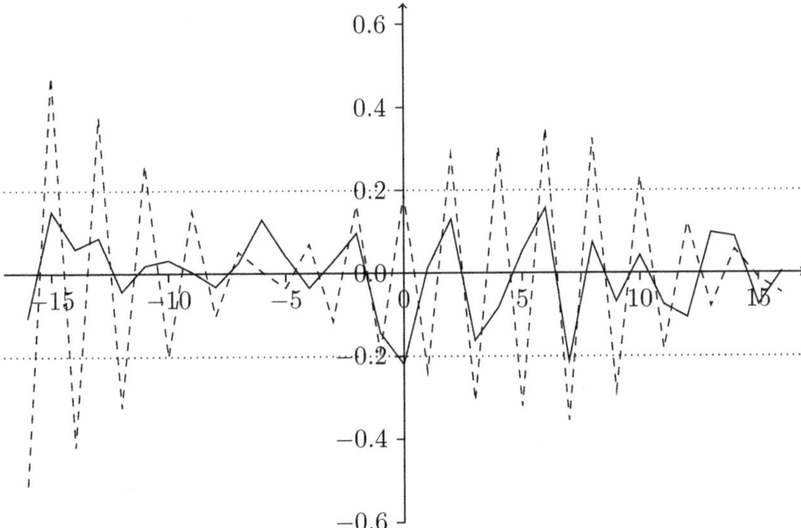

Figure 6.3: *Estimated cross-correlation between two mutually uncorrelated* $AR(1)$ *processes. Before (dashed line) and after filtering (solid line).*

where $\mathrm{C}_{XX}(0)$ and $C_{YY}(0)$ are the estimated variances of X_t and Y_t, respectively. As a test for zero cross-correlation, we have the following theorem.

THEOREM 6.2
The estimated cross-correlation between two mutually uncorrelated time series is asymptotically normally distributed having

$$\mathrm{E}[\widehat{\rho}_{XY}(k)] \simeq 0 \tag{6.18}$$

$$\mathrm{Var}[\widehat{\rho}_{XY}(k)] \simeq \frac{1}{N}, \tag{6.19}$$

conditioned that at least one of the processes is white noise. If both the processes are white noise then we have

$$\mathrm{Cov}[\widehat{\rho}_{XY}(k), \widehat{\rho}_{XY}(\ell)] = 0\,, \quad k \neq \ell. \tag{6.20}$$

Proof Omitted. Follows from (6.14). ∎

The problem with autocorrelated time series is also seen in the cross-correlation function. This is illustrated in Figure 6.3. Based on two mutually uncorrelated (no cross-correlation) AR(1) processes with $\alpha = \beta = -0.9$ and $N = 100$, the estimated cross-correlation function is illustrated by the dashed line in Figure 6.3. We observe that the shifting positive and negative values of the autocorrelation functions (see (6.15)) are also observed in the estimated cross-correlation function.

If one wishes to *test if two time series are cross-correlated*, one should use the following procedure.

Test for cross-correlation

i) Determine a model $\phi(\text{B})X_t = \theta(\text{B})\varepsilon_t$ for the $\{X_t\}$ time series so that $\{\varepsilon_t\}$ is white noise.

ii) Based on the model perform a "pre-whitening" on $\{X_t\}$ given by the sequence of residuals $\varepsilon_t = \theta^{-1}(\text{B})\phi(\text{B})X_t$.

iii) Perform the similar "pre-whitening" on $\{Y_t\}$ given by $w_t = \theta^{-1}(\text{B})\phi(\text{B})Y_t$.

iv) Test whether the cross-correlation $\rho_{\varepsilon w}(k) = 0$.

If we compare the proposed procedure with the variance (6.16) for the AR(1) model, it is seen that the procedure implies that $\alpha\beta$ becomes 0, and hereby we obtain an undisturbed estimate of the variance.

The solid line in Figure 6.3 shows the estimated cross-correlation between the residuals of both time series, and we observe an estimated cross-correlation function that corresponds more to the true one, namely, that $\{X_t\}$ and $\{Y_t\}$ are mutually uncorrelated.

6.3 Identification

The purpose of the identification of an ARIMA(p, d, q) model is to determine the model order, i.e., to determine values for p, d, and q. As indicated in Figure 6.1 on page 145, the identification is based on information from data in conjunction with considerations about the physics. However, for some phenomena it may be impossible to apply any information from the physics. As examples of such phenomena we have stock prices, mail loads, and traffic. In such cases the identification is based solely on the information from data. In cases where it is possible to apply physical knowledge, one should most certainly not disregard such knowledge but combine it with information from data.

A preliminary identification of an ARIMA(p, d, q) model is based mainly on the estimated autocorrelation function, where one compares the estimated autocorrelations with knowledge of the corresponding theoretical autocorrelation functions for stationary processes ($d = 0$). The preliminary identification and especially the choice of differencing based on the autocorrelation functions is widely used, which is especially due to Box and Jenkins (1970/1976).

6.3.1 Identification of the degree of differencing

In Section 5.5.3 on page 125 we considered ARMA(p, q) processes, written as $\phi(\mathrm{B})Y_t = \theta(\mathrm{B})\varepsilon_t$. If the process is stationary, the corresponding theoretical autocorrelation function satisfies the following difference equation (see (5.100) on page 126 and (5.101) on page 127):

$$\phi(\mathrm{B})\rho(k) = 0\,, \quad k = q+1, q+2, \ldots\,. \tag{6.21}$$

If all the roots for the characteristic equation are different, the solution to (6.21) can be written in the form

$$\rho(k) = A_1\lambda_1^k + \cdots + A_p\lambda_p^k\,, \quad k = q-p+1, q-p+2, \ldots \tag{6.22}$$

where the claim for stationarity ensures that $\lambda_1, \ldots, \lambda_p$ all lie within the unit circle. It should also be noted that if one of the roots is close to the unit circle, then the autocorrelation will only very slowly decrease toward zero. Similarly, as we have seen in the last section, a root close to the unit circle implies that the variance of the process for relatively large values of N is badly determined.

If a time series $\{Y_t, t = 1, 2, \ldots, N\}$ contains non-stationarities or very slow variations compared with the sampling frequency, the estimated autocorrelation will decrease very slowly to zero. In Box and Jenkins (1970/1976) it is recommended that such an apparently non-stationarity should be removed using a suitable order of differencing

$$W_t = \nabla^d Y_t \tag{6.23}$$

and subsequently describing $\{W_t\}$ by an ARMA(p, q) model. Hence, we model $\{Y_t\}$ using an ARIMA(p, d, q) model.

Based on the above considerations, we select d as the lowest order of differencing for which the estimated autocorrelation decreases sufficiently rapid toward 0. In practical situations this corresponds to d equal to 0, 1 or 2.

In certain situations it will not be adequate to perform a differencing. This would be the case if, e.g., from physical considerations, one can assume that the process is stationary and the model is intended for simulation. Instead one has to ensure that there are enough observations available to estimate all the parameters of the process. Furthermore, one can apply a method which is robust against mal-recorded observations (see, e.g., Huber (1964) or Martin and Yohai (1985)), since a model with roots close to the unit circle is very sensitive against such erroneous observations.

6.3.2 Identification of the ARMA part

We now assume that we have chosen a suitable d. The next step in the preliminary identification is to identify values of p and q in the ARMA part, based on the estimated autocorrelation functions for $W_t = \nabla^d Y_t$, and evaluate which combination of p and q can yield theoretical autocorrelations that best approximate the estimated autocorrelation functions. For this evaluation, the former derived characteristics for the theoretical autocorrelations functions should be applied. These characteristics are summarized in Table 6.1.

As a supplement to the evaluation of the estimated autocorrelation functions, one can exploit the following theorem for testing the hypothesis of whether a time series can be described by a pure AR(p) process or by a pure MA(p) process.

THEOREM 6.3 (PROPERTY FOR AR PROCESSES)
For an AR(p) process *it holds*

$$\left.\begin{aligned} \mathrm{E}[\widehat{\phi}_{kk}] &\simeq 0 \\ \mathrm{Var}[\widehat{\phi}_{kk}] &\simeq \frac{1}{N} \end{aligned}\right\}, \quad k = p+1, p+2, \dots \tag{6.24}$$

where N is the number of observations in the stationary time series.

Proof Omitted—see Daniells (1956). ∎

THEOREM 6.4 (PROPERTY FOR MA PROCESSES)
For an MA(q) process *it holds*

$$\left.\begin{aligned} \mathrm{E}[\widehat{\rho}(k)] &\simeq 0 \\ \mathrm{Var}[\widehat{\rho}(k)] &\simeq \frac{1}{N}\left[1 + 2\left(\widehat{\rho}^2(1) + \cdots + \widehat{\rho}^2(q)\right)\right] \end{aligned}\right\}, \quad k = q+1, q+2, \dots$$

where N is the number of observations in the stationary time series.

Proof Omitted—follows from some approximate results due to Bartlett (1946). ∎

Furthermore, it can be shown that the estimates of $\widehat{\rho}(k)\,, (k > q)$ and $\widehat{\phi}_{kk}\,, (k > p)$ are approximately normally distributed, and thus, tests for pure processes are based on the normal distribution.

It is much more difficult to determine the model order for mixed processes. In that case one has to make a "good guess" for a model order, and often a model check reveals that a re-identification is needed.

It is important to bear in mind that when evaluating estimates of autocorrelation functions, the variance of the estimates may be large and the

Table 6.1: *Characteristics for the autocorrelation functions of ARMA processes.*

	ACF $\rho(k)$	PACF ϕ_{kk}	IACF $\rho i(k)$
AR(p)	Damped exponential and/or sine functions	$\phi_{kk} = 0$ for $k > p$	$\rho i(k) = 0$ for $k > p$
MA(q)	$\rho(k) = 0$ for $k > q$	Dominated by damped exponential and/or sine functions	Damped exponential and/or sine functions
ARMA(p, q)	Damped exponential and/or sine functions after lag $q - p$	Dominated by damped exponential and/or sine functions after lag $p - q$	Damped exponential and/or sine functions after lag $p - q$

estimates may be mutually correlated. Especially for large lags, there can be a systematic pattern in the estimated autocorrelation after the theoretical autocorrelation has faded out, cf. the discussion in the previous section.

Example 6.1 (Sample autocorrelation functions for cloud cover)
A model for simulations of cloud cover is sought. In order to identify the order of the model the autocorrelation and partial autocorrelation are estimated.

The measurements of cloud cover express the percentage of the sky which is covered by clouds. In the present study hourly observations of total cloud cover in octas are used. Actually, the cloud cover is measured as an integer between 0 and 9, where 0 corresponds to completely clear sky, 8 to completely overcast sky, and the observation 9 indicates that the cloud cover is unobservable which usually is the case in foggy weather. The observations are based on a subjective evaluation.

Hourly values from January 1, 1959, to December 31, 1973, observed at Værløse near Copenhagen are used in estimating the sample autocorrelation function shown in Figure 6.4 on the next page, and the partial autocorrelation function in Figure 6.5. This gives a total of 131,496 hourly values.

With this huge amount of data, testing should be used with care. It is seen, however, that only ϕ_{11} seems to be significantly different from zero. According to Theorem 5.11 it is concluded that in the class of ARMA(p, q) processes, the variation of cloud cover must be described by the AR(1) process.

The AR(1) process is, as mentioned in Example 5.1 on page 101, a first order Markov process. Since the sample space of the cloud cover observations is discrete, a more appropriate model would be to use a first order Markov

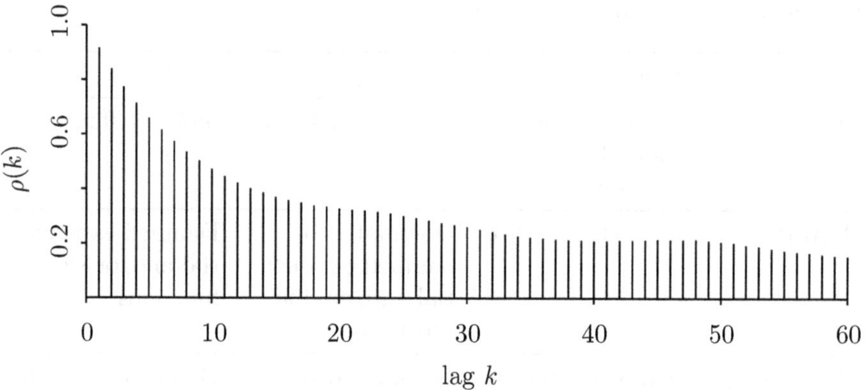

Figure 6.4: *Sample autocorrelation function for cloud cover near Copenhagen. Notice the exponential decay.*

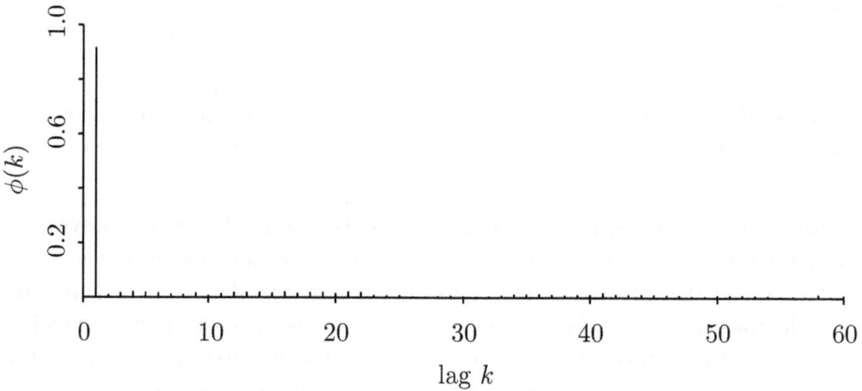

Figure 6.5: *Sample partial autocorrelation function for cloud cover near Copenhagen. Notice that* $\widehat{\phi}_{kk} \approx 0$ *for* $k = 2, 3, \ldots$.

chain, see, e.g., Grimmit and Stirzaker (1992). Actually this model is successfully used to model the variation in cloud cover, see Madsen, Spliid, and Thyregod (1985).

6.3.3 Cointegration

Cointegration is a property of time series variables often used in econometrics.

DEFINITION 6.1 (COINTEGRATION)
If two or more series are themselves non-stationary, but a linear combination of them is stationary, then the series are said to be cointegrated.

In econometrics, like for stock market index values, the variables often move through time roughly following a random walk. Testing for relation between such index values could most appropriately be done by testing for a cointegration.

Let us consider a simple example of cointegration more closely. If a time series X_t shows a linear trend and $Z_t = (1 - B)X_t = \nabla X_t$ is stationary, then X_t is said to be an integrated process of order one, which we write $X_t \in I(1)$. If a time series X_t shows a quadratic trend and $Z_t = \nabla^2 X_t$ is stationary, then X_t is said to be integrated of order two, which similarly is written as $X_t \in I(2)$.

Additional variables are often introduced to model and eventually predict the variations in the process X_t. If X_t is an $I(1)$ process, an additional variable Y_t is also an $I(1)$ process and $X_t - \alpha Y_t$ is stationary. The process $(X_t - \alpha Y_t)$ is then $I(0)$, and X_t and Y_t are said to be *cointegrated.* If we now want to build a model for these two variables, instead of first differencing both of the variables and then building a model, the constraint implied by the stationary linear combination $X_t - \alpha Y_t$ needs to be incorporated in the model.

Cointegration is a means for correctly testing hypotheses concerning the relationship between non-stationary processes having *unit roots*, i.e., integrated of order one. See, e.g., Johansen (2001) and Engle and You (1985) for a further introduction to cointegration analysis.

Example 6.2 (Cointegration)
Consider the data in Figure 6.6 on the next page. It is clearly seen that the money series should be differenced once to obtain stationarity. This also pertains to the bond rate, although it is less clear. By comparing the plots to the left, it is seen that the money demand decreases when the bond rate increases and vice versa. Thus, it is expected that these series are cointegrated.

6.4 Estimation of parameters in standard models

In this section some methods for estimation of parameters in AR, MA, and ARMA models are described. The methods may be interpreted as a generalization to dynamic linear models of the methods described in Chapter 3 concerning static linear models.

6.4.1 Moment estimates

The so-called *moment estimates* are found by substituting estimated values of the autocorrelation or autocovariance function into the theoretical relations between parameters and the values of the autocorrelation or autocovariance function and, subsequently, solving these equations with respect to the parameters.

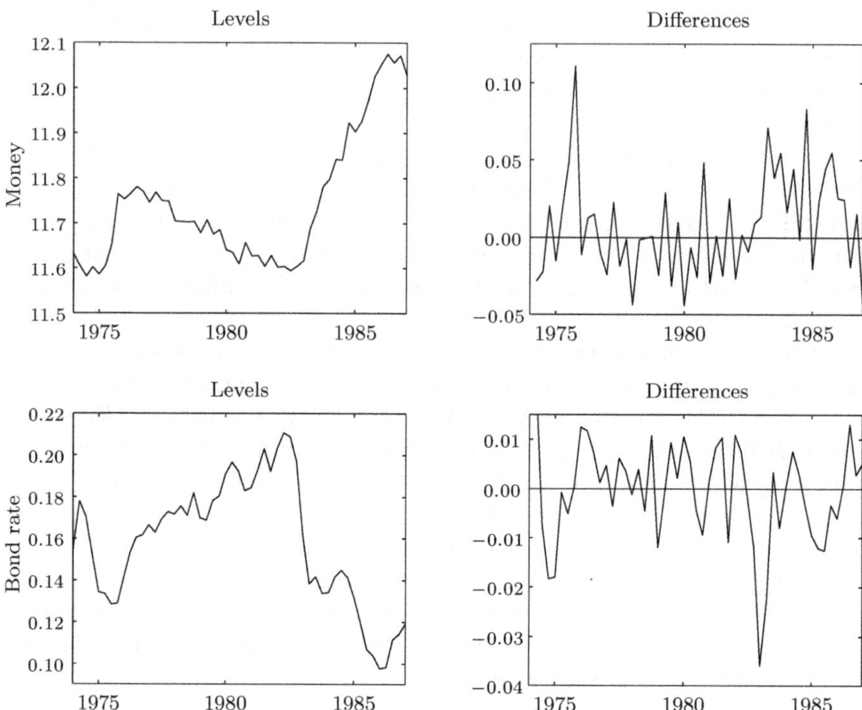

Figure 6.6: *The upper-left plot shows observations from 1974:1 to 1987:3 of log real money (m2), where the log transformation has been applied to stabilize the variance. The upper-right plot shows the differenced series. The lower-left plot shows observations from the same time period of the bond rate, and the lower-right plot shows the differenced series.*

The method is very ad hoc, and it is difficult to say anything general about the properties of the estimator. Furthermore, it is possible to find several relations between parameters and values of the autocorrelation function; in this case one should look for relations with the lowest possible lag values of the autocorrelation function. Moment estimates can serve as good starting values for other estimation methods. We have for the *AR(p) processes* the *Yule-Walker estimates* by solving the linear equations

$$\begin{pmatrix} \widehat{\rho}(1) \\ \widehat{\rho}(2) \\ \vdots \\ \widehat{\rho}(p) \end{pmatrix} = \begin{pmatrix} 1 & \widehat{\rho}(1) & \cdots & \widehat{\rho}(p-1) \\ \widehat{\rho}(1) & 1 & \cdots & \widehat{\rho}(p-2) \\ \vdots & \vdots & \ddots & \vdots \\ \widehat{\rho}(p-1) & \widehat{\rho}(p-2) & \cdots & 1 \end{pmatrix} \begin{pmatrix} -\phi_1 \\ -\phi_2 \\ \vdots \\ -\phi_p \end{pmatrix} \tag{6.25}$$

with respect to $\phi_1, \phi_2, \ldots, \phi_p$.

For the *MA(q) and ARMA(p, q) processes* the equations become a set of non-linear equations. Let us illustrate the method of moments in an example.

Example 6.3 (Moment estimates: MA(2) process)
For an MA(2) process we observe the following:

$$\widehat{\rho}(1) = r_1 = 0.6 \quad \text{and} \quad \widehat{\rho}(2) = r_2 = 0.3. \tag{6.26}$$

For the MA(2) process

$$Y_t = (1 + \theta_1 \mathrm{B} + \theta_2 \mathrm{B}^2)\varepsilon_t$$

we have the theoretical autocorrelation function (see Equation (5.65) on page 118)

$$\begin{aligned}
\rho(0) &= 1 \\
\rho(1) &= \frac{\theta_1 + \theta_1\theta_2}{(1 + \theta_1^2 + \theta_2^2)} \\
\rho(2) &= \frac{\theta_2}{(1 + \theta_1^2 + \theta_2^2)} \\
\rho(k) &= 0\,, \quad \text{for } k \geq 3.
\end{aligned} \tag{6.27}$$

The moment estimates for (θ_1, θ_2) are obtained by solving the equations

$$\text{a)} \qquad \frac{\theta_1 + \theta_1\theta_2}{1 + \theta_1^2 + \theta_2^2} = r_1$$

$$\text{b)} \qquad \frac{\theta_2}{1 + \theta_1^2 + \theta_2^2} = r_2$$

with respect to θ_1, θ_2. When a) is divided by b), we obtain

$$\theta_1 = \frac{r_1\theta_2}{r_2(1 + \theta_2)}.$$

Inserting this into b) we get the 4'th order equation

$$r_2^3\theta_2^4 + (2r_2^3 - r_2^2)\theta_2^3 + (2r_2^3 + r_2r_1^2 - 2r_2^2)\theta_2^2 + (2r_2^3 - r_2^2)\theta_2 + r_2^3 = 0.$$

With the observed values of r_1 and r_2 we accept the solution that yields an invertible MA(2) process for $\widehat{\theta}_2 = 0.5247$ and, hereby, $\widehat{\theta}_1 = 0.6883$.

In general the moment estimates are often doubtful for non-AR(p) processes.

6.4.2 The LS estimator for linear dynamic models

First, let us consider the AR(p) model

$$Y_t + \phi_1 Y_{t-1} + \cdots + \phi_p Y_{t-p} = \varepsilon_t\,, \quad t = p+1, p+2, \ldots \tag{6.28}$$

where $\{\varepsilon_t\}$ is white noise. We assume, that we have the observations $Y_1, \ldots, Y_N$.

If we introduce

$$\boldsymbol{X}_t^T = (-Y_{t-1}, \ldots, -Y_{t-p}), \quad t = p+1, p+2, \ldots \tag{6.29}$$

$$\boldsymbol{\theta}^T = (\phi_1, \ldots, \phi_p) \tag{6.30}$$

then (6.28) can be written in the *linear* form

$$Y_t = \boldsymbol{X}_t^T \boldsymbol{\theta} + \varepsilon_t, \quad t = p+1, p+2, \ldots \tag{6.31}$$

Similarly to the approach described in Chapter 3, we will write the model (6.31) for all the observations. We introduce

$$\boldsymbol{Y}^T = (Y_{p+1}, \ldots, Y_N), \tag{6.32}$$

$$\boldsymbol{X}^T = (\boldsymbol{X}_{p+1}, \ldots, \boldsymbol{X}_N), \tag{6.33}$$

$$\boldsymbol{\varepsilon}^T = (\varepsilon_{p+1}, \ldots, \varepsilon_N). \tag{6.34}$$

Hereby the model (6.28) can be written in a compact linear form for all the observations:

$$\boldsymbol{Y} = \boldsymbol{X}\boldsymbol{\theta} + \boldsymbol{\varepsilon}. \tag{6.35}$$

Compare with (3.33) on page 36.

The LS estimator is found as in Chapter 3 by minimizing

$$S(\boldsymbol{\theta}) = (\boldsymbol{Y} - \boldsymbol{X}\boldsymbol{\theta})^T (\boldsymbol{Y} - \boldsymbol{X}\boldsymbol{\theta}), \tag{6.36}$$

and the solution is given in the following theorem.

THEOREM 6.5 (THE LS ESTIMATOR FOR LINEAR DYNAMIC MODELS)
The LS estimate for $\boldsymbol{\theta}$ in the linear model (6.35) *is found by solving the* normal equation

$$(\boldsymbol{X}^T \boldsymbol{X})\widehat{\boldsymbol{\theta}} = \boldsymbol{X}^T \boldsymbol{Y}, \tag{6.37}$$

and as an estimate of $\sigma_\varepsilon^2 = \mathrm{Var}[\varepsilon_t]$ we apply

$$\widehat{\sigma}_\varepsilon^2 = \frac{S(\widehat{\boldsymbol{\theta}})}{N-p} = \frac{\sum_{t=p+1}^N \varepsilon_t^2(\widehat{\boldsymbol{\theta}})}{N-p}. \tag{6.38}$$

Proof As in Chapter 3. ■

Example 6.4 (LS estimates: AR(2) process)
Let us consider the AR(2) model

$$Y_t = -\phi_1 Y_{t-1} - \phi_2 Y_{t-2} + \varepsilon_t. \tag{6.39}$$

Having the observations $Y_1, Y_2, \ldots, Y_N$, we obtain

$$Y^T = (Y_3, Y_4, \ldots, Y_N),$$

$$X = \begin{pmatrix} X_3^T \\ \vdots \\ X_N^T \end{pmatrix} = \begin{pmatrix} -Y_2 & -Y_1 \\ \vdots & \vdots \\ -Y_{N-1} & -Y_{N-2} \end{pmatrix}$$

which implies that

$$X^T X = \begin{pmatrix} \sum_{t=2}^{N-1} Y_t^2 & \sum_{t=2}^{N-1} Y_t Y_{t-1} \\ \sum_{t=2}^{N-1} Y_t Y_{t-1} & \sum_{t=1}^{N-2} Y_t^2 \end{pmatrix}$$

$$X^T Y = \begin{pmatrix} -\sum_{t=2}^{N-1} Y_t Y_{t+1} \\ -\sum_{t=1}^{N-2} Y_t Y_{t+2} \end{pmatrix}$$

Now, the LS estimates of $\theta = (\theta_1, \theta_2)^T$ can be found by solving the normal equation

$$\begin{pmatrix} \sum_{t=2}^{N-1} Y_t^2 & \sum_{t=1}^{N-2} Y_{t+1} Y_t \\ \sum_{t=1}^{N-2} Y_{t+1} Y_t & \sum_{t=1}^{N-2} Y_t^2 \end{pmatrix} \begin{pmatrix} \phi_1 \\ \phi_2 \end{pmatrix} = \begin{pmatrix} -\sum_{t=2}^{N-1} Y_t Y_{t+1} \\ -\sum_{t=1}^{N-2} Y_t Y_{t+2} \end{pmatrix} \tag{6.40}$$

It is seen that

$$\widehat{\theta} = (\widehat{\phi}_1, \widehat{\phi}_2)^T = (X^T X)^{-1} X^T Y \tag{6.41}$$

is not, as in Chapter 3, linear in Y_t since the X matrix contains Y variables.

Since

$$\mathrm{C}(k) = \frac{1}{N} \sum_{t=1}^{N-k} Y_t Y_{t+k} \quad \text{(mean value} = 0\text{)},$$

we get from (6.25) on page 158 that the *Yule-Walker estimates* are determined by

$$\begin{pmatrix} \sum_{t=1}^{N} Y_t^2 & \sum_{t=1}^{N-1} Y_t Y_{t+1} \\ \sum_{t=1}^{N-1} Y_t Y_{t+1} & \sum_{t=1}^{N} Y_t^2 \end{pmatrix} \begin{pmatrix} -\phi_1 \\ -\phi_2 \end{pmatrix} = \begin{pmatrix} \sum_{t=1}^{N-1} Y_t Y_{t+1} \\ \sum_{t=1}^{N-2} Y_t Y_{t+2} \end{pmatrix} \tag{6.42}$$

It is seen that the *LS estimates* (determined by (6.40)) and the *Yule-Walker estimates*, (6.42), are *asymptotically identical.*

When comparing *dynamic models* with *static models*, it is important to take note of the following properties:

i) The LS estimator $\widehat{\theta}$ is *not* a linear function of the observations since the X matrix contains Y variables.

ii) If $\{\varepsilon_t\}$ is not white noise, then $\widehat{\theta}$ will often be non-consistent ($\widehat{\theta}$ will not converge to the true value of θ for $N \to \infty$).

THEOREM 6.6 (PROPERTIES FOR THE LS ESTIMATOR FOR DYNAMIC MODELS)
Assume $\{\varepsilon_t\}$ is normally distributed white noise, then

i) $\widehat{\boldsymbol{\theta}}$ *is consistent.*

ii) $\widehat{\boldsymbol{\theta}}$ *is asymptotically efficient.*

iii) $\widehat{\boldsymbol{\theta}}$ *is asymptotically normally distributed with*

$$\mathrm{Var}[\widehat{\boldsymbol{\theta}}] \approx \sigma_\varepsilon^2(\boldsymbol{X}^T\boldsymbol{X})^{-1}. \tag{6.43}$$

If $\{\varepsilon_t\}$ is not normally distributed then the estimator is not asymptotically efficient.

Proof Omitted. ■

An estimate for σ_ε^2 is given by (6.38). This estimate can be used in (6.43) by calculating the variance matrix belonging to the estimate.

It should be noted that it is not possible to extend the description to include MA models and, at the same time, use the normal equations (6.37) to estimate the parameters. This is due to the fact that ε_t is unknown and the residuals, which should otherwise be included in an extended $\boldsymbol{X}$ matrix, are a function of $\boldsymbol{\theta}$. In the following some alternative methods will be illustrated.

In contrast, the LS method can be applied to the *extended linear model class*

$$\begin{aligned}\phi(\mathrm{B})Y_t &= \omega_1(\mathrm{B})u_{1,t} + \cdots + \omega_k(\mathrm{B})u_{k,t}\\ &\quad + d_1 f_1(t) + \cdots + d_\ell f_\ell(t) + \varepsilon_t,\end{aligned} \tag{6.44}$$

where $\{\varepsilon_t\}$ is white noise and uncorrelated with $\{u_{i,t}\}$. Here $\{u_{i,t}\}$ and $f_i(t)$ are deterministic functions and, hence, (6.44) can be written in the linear form (6.35).

Finally, it should be noted that there are two ways to treat the mean value of the process $\{Y_t\}$, $\{u_{1,t}\}, \ldots, \{u_{k,t}\}$: either by subtracting the estimated means before the estimation so that the model (6.44) describes the variations around the mean or by subtracting the stationary value, as illustrated in the following example of a model in the extended class:

$$\phi(\mathrm{B})(Y_t - \mu_Y) = \omega(\mathrm{B})(u_t - \mu_u) + \varepsilon_t. \tag{6.45}$$

Alternatively, we can consider the reformulation

$$\phi(\mathrm{B})Y_t = \omega(\mathrm{B})u_t + (\phi(1)\mu_Y - \omega(1)\mu_u) + \varepsilon_t, \tag{6.46}$$

which can be written in the form

$$\phi(\mathrm{B})Y_t = \omega(\mathrm{B})u_t + d + \varepsilon_t, \tag{6.47}$$

which is a model in the class (6.44) with ($f(t) = 1$). Equation (6.47) is different from (6.45) in that it is linear in all the parameters.

6.4.3 The prediction error method

The LS estimation of static models in Chapter 3 is obtained by minimizing the sum of squared residuals. The residuals can be written as

$$\varepsilon_t = Y_t - \mathrm{E}[Y_t|X] \tag{6.48}$$

where $\mathrm{E}[Y_t|X]$ is the conditional mean or the prediction of Y_t given X.

By the *LS estimation of dynamic models*, as exemplified in the AR(p) model, the residuals can be written

$$\varepsilon_t = Y_t - \mathrm{E}[Y_t|Y_{t-1}, \ldots, Y_{t-p}], \tag{6.49}$$

and the estimate for the parameters is again found by minimizing the sum of squared residuals.

In the prediction error method, this principle is generalized to (among others) ARMA(p, q) processes, i.e., models in the form

$$Y_t + \phi_1 Y_{t-1} + \cdots + \phi_p Y_{t-p} = \varepsilon_t + \theta_1 \varepsilon_{t-1} + \cdots + \theta_q \varepsilon_{t-q}, \tag{6.50}$$

where $\{\varepsilon_t\}$ is white noise.

For a given set of parameters,

$$\boldsymbol{\theta}^T = (\phi_1, \ldots, \phi_p, \theta_1, \ldots, \theta_q), \tag{6.51}$$

we can *recursively* calculate the *one-step prediction*:

$$\mathrm{E}[Y_t|\boldsymbol{\theta}, \mathbf{Y}_{t-1}] = \widehat{Y}_{t|t-1}(\boldsymbol{\theta}) = -\sum_{i=1}^{p} \phi_i Y_{t-i} + \sum_{i=1}^{q} \theta_i \varepsilon_{t-i}(\boldsymbol{\theta}), \tag{6.52}$$

where

$$\mathbf{Y}_t = (Y_t, Y_{t-1}, \ldots)^T. \tag{6.53}$$

In principle, the calculations assume that at time t there are infinitely many past observations. In practical applications, one will often use either conditioned or unconditioned estimates as described in the following.

6.4.3.1 Conditioned estimation

This procedure is sometimes called the *conditioned least squares method* (CLS) (see, e.g., Abraham and Ledolter (1983)).

In this method we simply put

$$\varepsilon_p = \varepsilon_{p-1} = \cdots = \varepsilon_{p+1-q} = 0, \tag{6.54}$$

and hereafter, $\widehat{Y}_{t|t-1}$ can be calculated for $t = p+1, p+2, \ldots$ since the prediction errors are calculated recursively as

$$\varepsilon_t(\boldsymbol{\theta}) = Y_t - \widehat{Y}_{t|t-1}(\boldsymbol{\theta}). \tag{6.55}$$

Analogous to the LS method in Section 6.4.2, we find the *prediction error estimate* as

$$\widehat{\boldsymbol{\theta}} = \arg\min_{\theta} \left\{ S(\boldsymbol{\theta}) = \sum_{t=p+1}^{N} \varepsilon_t^2(\boldsymbol{\theta}) \right\}. \tag{6.56}$$

In general the criteria is a non-quadratic function of $\boldsymbol{\theta}$.

An estimate for σ_ε^2 is

$$\widehat{\sigma}_\varepsilon^2 = \frac{S(\widehat{\boldsymbol{\theta}})}{N-p}. \tag{6.57}$$

An alternative estimate is obtained as $S(\widehat{\boldsymbol{\theta}})$ divided by the number of observations utilized in (6.56), $N-p$, minus the number of estimated parameters, $p+q$ or $p+q+1$.

6.4.3.2 Unconditioned estimation

This method is sometimes called the *unconditioned least squares method.* For unconditioned estimation, the estimates are found by

$$\widehat{\boldsymbol{\theta}} = \arg\min_{\theta} \left\{ S_u(\boldsymbol{\theta}) = \sum_{t=-\infty}^{N} \varepsilon_t^2(\boldsymbol{\theta}) \right\}, \tag{6.58}$$

where

$$\varepsilon_t(\boldsymbol{\theta}) = \mathrm{E}[\varepsilon_t | \boldsymbol{\theta}, Y_1, \ldots, Y_N]. \tag{6.59}$$

This conditioned expected value can be calculated using *back-forecasting*, which will be described briefly in the following.

Assume that $\{Y_t\}$ is generated by the stationary model

$$\phi(\mathrm{B})Y_t = \theta(\mathrm{B})\varepsilon_t. \tag{6.60}$$

It is also possible to consider $\{Y_t\}$ as being generated by the *backward shifted model*

$$\phi(F)Y_t = \theta(F)e_t. \tag{6.61}$$

This backward shifted model can be applied in the calculation of the backward shifted predictions

$$\mathrm{E}[Y_t | \boldsymbol{\theta}, Y_1, \ldots, Y_N], \quad t = 0, -1, -2, \ldots \tag{6.62}$$

since $\mathrm{E}[e_t | \boldsymbol{\theta}, Y_1, \ldots, Y_N] = 0$ for $t \leq 0$. Because the model (6.60) is stationary, the predictions will be very close to 0 after a given step, $t = -Q$. Then the usual prediction procedure using the forward shifted equations (6.60) can be used to calculate $\varepsilon_t(\boldsymbol{\theta})$, which is then used in (6.58). For a more detailed description of the procedure, see Box and Jenkins (1970/1976).

For unconditioned estimation, we have

$$\widehat{\sigma}_\varepsilon^2 = \frac{S_u(\widehat{\boldsymbol{\theta}})}{N}. \tag{6.63}$$

As for (6.57), one can alternatively divide by N minus the number of estimated parameters.

For minimizing (6.56), as well as (6.58), one may apply a Newton-Raphson method for iterative minimization, see (3.31).

THEOREM 6.7 (PROPERTIES FOR THE PREDICTION ERROR METHOD)
Assume $\{\varepsilon_t\}$ is white noise, then

i) $\widehat{\boldsymbol{\theta}}$ *is consistent.*

ii) $\widehat{\boldsymbol{\theta}}$ *is asymptotical normally distributed with* $\mathrm{E}[\widehat{\boldsymbol{\theta}}] = \boldsymbol{\theta}$ *and the variance*

$$\mathrm{Var}[\widehat{\boldsymbol{\theta}}] \simeq 2\sigma_\varepsilon^2 \boldsymbol{H}^{-1}, \tag{6.64}$$

where the Hessian matrix, $\boldsymbol{H}$, *is determined by*

$$\{h_{lk}\} = \left. \frac{\partial^2 S(\boldsymbol{\theta})}{\partial \theta_l \partial \theta_k} \right|_{\boldsymbol{\theta}=\widehat{\boldsymbol{\theta}}}, \tag{6.65}$$

i.e., $\boldsymbol{H}$ *is the curvature of* $S(\boldsymbol{\theta})$ *in the point* $\boldsymbol{\theta} = \widehat{\boldsymbol{\theta}}$.

Proof Omitted. ∎

Example 6.5 (Influence of the number of observations)
We have simulated 1000 values of Y_t defined by the ARMA(1, 1) process

$$(1 + 0.7B)Y_t = (1 - 0.4B)\varepsilon_t.$$

Figure 6.7 on the next page shows the contour curves $S(\boldsymbol{\theta}) = S(\phi_1, \theta_1)$. On the left figure, a subset of 100 values of Y_t was used while on the right figure, the whole series was used in the minimization. It can be seen that the minimum of the objective function lies at the true value of the parameters, marked by an ✖, when using the whole series. However, when using only 100 observations the minimum does not lie near the true value of the parameters.

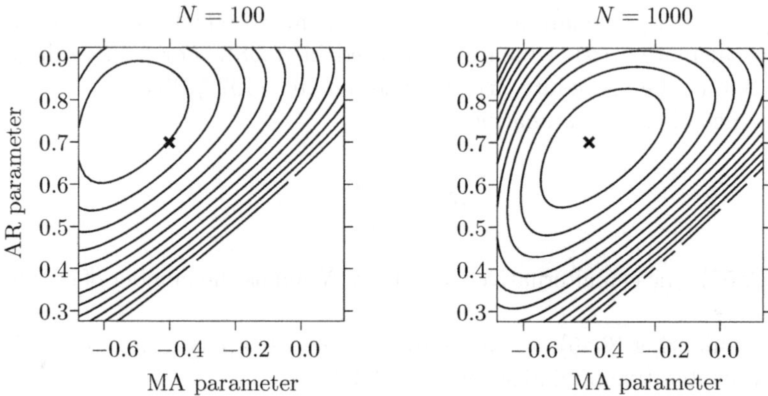

Figure 6.7: *Contour curves for $S(\boldsymbol{\theta}) = S(\phi_1, \theta_1)$ determined from 100 and 1000 simulated values of the ARMA(1,1) process:* $(1 + 0.7\mathrm{B})Y_t = (1 - 0.4\mathrm{B})\varepsilon_t$.

6.4.4 The ML method for dynamic models

Contrary to the methods mentioned so far, the ML method (Maximum Likelihood) requires an assumption of the distribution. In general, we will assume that $\{\varepsilon_t\}$ *is normally distributed* white noise with the variance $\mathrm{Var}[\varepsilon_t] = \sigma_\varepsilon^2$.

We consider the ARMA(p, q) process

$$Y_t + \phi_1 Y_{t-1} + \cdots + \phi_p Y_{t-p} = \varepsilon_t + \theta_1 \varepsilon_{t-1} + \cdots + \theta_q \varepsilon_{t-q} \tag{6.66}$$

Furthermore, we introduce

$$\boldsymbol{\theta}^T = (\phi_1, \ldots, \phi_p, \theta_1, \ldots, \theta_q), \tag{6.67}$$

$$\boldsymbol{Y}_t^T = (Y_t, Y_{t-1}, \ldots, Y_1), \tag{6.68}$$

i.e., $\boldsymbol{Y}_t$ contains all observations up to time t.

The likelihood function is the joint probability distribution function for all observations for given values of $\boldsymbol{\theta}$ and σ_ε^2, i.e.,

$$\begin{aligned} L(\boldsymbol{Y}_N; \boldsymbol{\theta}, \sigma_\varepsilon^2) &= f(\boldsymbol{Y}_N | \boldsymbol{\theta}, \sigma_\varepsilon^2) \\ &= f(Y_N | \boldsymbol{Y}_{N-1}, \boldsymbol{\theta}, \sigma_\varepsilon^2) f(\boldsymbol{Y}_{N-1} | \boldsymbol{\theta}, \sigma_\varepsilon^2) \\ &= \left(\prod_{t=p+1}^{N} f(Y_t | \boldsymbol{Y}_{t-1}, \boldsymbol{\theta}, \sigma_\varepsilon^2) \right) f(\boldsymbol{Y}_p | \boldsymbol{\theta}, \sigma_\varepsilon^2). \end{aligned} \tag{6.69}$$

This is the general formula for the likelihood function for time series data. If we assume that there are infinitely many past Y values, as for the prediction error method, we can calculate

$$\widehat{Y}_{t|t-1}(\boldsymbol{\theta}) = \mathrm{E}[Y_t | \boldsymbol{Y}_{t-1}, \boldsymbol{\theta}] = -\sum_{i=1}^{p} \phi_i Y_{t-i} + \sum_{i=1}^{q} \theta_i \varepsilon_{t-i}(\boldsymbol{\theta}) \tag{6.70}$$

which is the *mean value* for $(Y_t|\boldsymbol{Y}_{t-1}, \boldsymbol{\theta}, \sigma_\varepsilon^2)$.

The prediction errors become

$$\varepsilon_t(\boldsymbol{\theta}) = Y_t - \widehat{Y}_{t|t-1}(\boldsymbol{\theta}) \tag{6.71}$$

with *variance* σ_ε^2. Since ε_t is normally distributed, we have

$$\begin{aligned} f(Y_t|\boldsymbol{Y}_{t-1}, \boldsymbol{\theta}, \sigma_\varepsilon^2) &= \frac{1}{\sigma_\varepsilon\sqrt{2\pi}} \exp\big(-(Y_t - \mathrm{E}[Y_t|\boldsymbol{Y}_{t-1}, \boldsymbol{\theta}])^2/2\sigma_\varepsilon^2\big) \\ &= \frac{1}{\sigma_\varepsilon\sqrt{2\pi}} \exp\big(-\varepsilon_t^2(\boldsymbol{\theta})/2\sigma_\varepsilon^2\big) \end{aligned} \tag{6.72}$$

We can now write the *conditional likelihood function* (conditioned on $\boldsymbol{Y}_p$)

$$\begin{aligned} L(\boldsymbol{Y}_N; \boldsymbol{\theta}, \sigma_\varepsilon^2) &= \prod_{t=p+1}^{N} f(Y_t|\boldsymbol{Y}_{t-1}, \boldsymbol{\theta}, \sigma_\varepsilon^2) \\ &= (\sigma_\varepsilon^2 2\pi)^{-\frac{N-p}{2}} \exp\left(-\frac{1}{2\sigma_\varepsilon^2}\sum_{t=p+1}^{N} \varepsilon_t^2(\boldsymbol{\theta})\right). \end{aligned} \tag{6.73}$$

Taking the logarithm yields

$$\log L(\boldsymbol{Y}_N; \boldsymbol{\theta}, \sigma_\varepsilon^2) = -\frac{N-p}{2}\log(\sigma_\varepsilon^2) - \frac{1}{2\sigma_\varepsilon^2}\sum_{t=p+1}^{N} \varepsilon_t^2(\boldsymbol{\theta}) + c \tag{6.74}$$

where c is a constant. Differentiating (6.74) with respect to σ_ε^2, and setting this equal to zero gives

$$\frac{\partial \log L}{\partial \sigma_\varepsilon^2} = -\frac{(N-p)}{2}\frac{1}{\sigma_\varepsilon^2} + \frac{1}{2\sigma_\varepsilon^4}\sum_{t=p+1}^{N} \varepsilon_t^2(\boldsymbol{\theta}) = 0.$$

The minimum is obtained for

$$\widehat{\sigma}_\varepsilon^2 = \left(\sum_{t=p+1}^{N} \varepsilon_t^2(\boldsymbol{\theta})\right)/(N-p). \tag{6.75}$$

Inserting into (6.74) yields

$$\log L(\boldsymbol{Y}_N; \boldsymbol{\theta}, \widehat{\sigma}_\varepsilon^2) = -\frac{(N-p)}{2}\log\left(\sum_{t=p+1}^{N} \varepsilon_t^2(\boldsymbol{\theta})\right) + c. \tag{6.76}$$

This shows that *the ML estimate* $\boldsymbol{\theta}$ is obtained by minimizing

$$S(\boldsymbol{\theta}) = \sum_{t=p+1}^{N} \varepsilon_t^2(\boldsymbol{\theta}), \tag{6.77}$$

where the *ML estimate for* σ_ε^2 is obtained by

$$\widehat{\sigma}_\varepsilon^2 = \frac{S(\widehat{\boldsymbol{\theta}})}{N-p}. \tag{6.78}$$

Compared with the prediction error method, it is seen that if the assumption of normality does not hold, then the estimates (6.77) and (6.78) can be interpreted as prediction error estimates; see (6.56) and (6.57) on page 164.

Since the above mentioned estimates relate to the conditional likelihood function (6.73), they are called conditioned ML estimates. A method for calculating the *exact likelihood function* is based on the back-forecasting method, which was described in the previous section.

THEOREM 6.8 (PROPERTIES OF THE ML ESTIMATOR)

i) $\widehat{\boldsymbol{\theta}}$ *is consistent.*

ii) $\widehat{\boldsymbol{\theta}}$ *is asymptotically normally distributed with* $\mathrm{E}[\widehat{\boldsymbol{\theta}}] = \boldsymbol{\theta}$ *and variance*

$$\mathrm{Var}[\widehat{\boldsymbol{\theta}}] \simeq 2\sigma_\varepsilon^2 \boldsymbol{H}^{-1}, \tag{6.79}$$

where the matrix $\boldsymbol{H}$ *is determined by*

$$\{h_{lk}\} = \left.\frac{\partial^2 S(\boldsymbol{\theta})}{\partial\theta_l \partial\theta_k}\right|_{\boldsymbol{\theta}=\widehat{\boldsymbol{\theta}}}. \tag{6.80}$$

iii) $\widehat{\boldsymbol{\theta}}$ *is asymptotically efficient.*

Proof Omitted. Follows from the general properties of the ML estimator (see Pawitan (2001), Olsson (2002), or Thyregod and Madsen (2006)). ■

The variance estimate in (6.79) is also based on the assumption that for ML estimators, it holds asymptotically that

$$\mathrm{Var}[\widehat{\boldsymbol{\theta}}] = \boldsymbol{I}^{-1}(\boldsymbol{\theta}) \tag{6.81}$$

where $\boldsymbol{I}(\boldsymbol{\theta})$ is the *Fisher information matrix*

$$\boldsymbol{I}(\boldsymbol{\theta}) = -\mathrm{E}\left[\frac{\partial^2 \log L}{\partial \boldsymbol{\theta}^2}\right]. \tag{6.82}$$

Since (see (6.76))

$$\log L = -\frac{N-p}{2} \log S(\boldsymbol{\theta}) + c$$

we obtain

$$\frac{\partial \log L}{\partial \boldsymbol{\theta}} = -\frac{N-p}{2} S^{-1}(\boldsymbol{\theta}) \frac{\partial S(\boldsymbol{\theta})}{\partial \boldsymbol{\theta}}$$

$$\frac{\partial^2 \log L}{\partial \boldsymbol{\theta}^2} = \frac{N-p}{2} \left(S^{-2}(\boldsymbol{\theta}) \frac{\partial S(\boldsymbol{\theta})}{\partial \boldsymbol{\theta}} - S^{-1}(\boldsymbol{\theta}) \frac{\partial^2 S(\boldsymbol{\theta})}{\partial \boldsymbol{\theta}^2} \right).$$

For $\boldsymbol{\theta} = \widehat{\boldsymbol{\theta}}$, we have $\frac{\partial S(\boldsymbol{\theta})}{\partial \boldsymbol{\theta}} = 0$, and hereby

$$\left. \frac{\partial^2 \log L}{\partial \boldsymbol{\theta}^2} \right|_{\boldsymbol{\theta}=\widehat{\boldsymbol{\theta}}} = -\frac{N-p}{2} \left(S^{-1}(\boldsymbol{\theta}) \frac{\partial^2 S(\boldsymbol{\theta})}{\partial \boldsymbol{\theta}^2} \right) \bigg|_{\boldsymbol{\theta}=\widehat{\boldsymbol{\theta}}}.$$

If we apply the observed curvature of the logarithm of the likelihood function in $\boldsymbol{\theta} = \widehat{\boldsymbol{\theta}}$ as an approximation to the Fisher information matrix, we obtain

$$\operatorname{Var}[\widehat{\boldsymbol{\theta}}] \simeq 2 \frac{S(\boldsymbol{\theta})}{N-p} \boldsymbol{H}^{-1}$$

where $\boldsymbol{H}$ is given by (6.80). Since $S(\boldsymbol{\theta})/(N-p) = \sigma_\varepsilon^2$, the result follows.

6.4.4.1 Extended model classes

Both the prediction error method and the ML method can be used directly for extended model classes. For the considered methods the idea is that given a model with the parameters $\boldsymbol{\theta}$, one can calculate

$$\varepsilon_t(\boldsymbol{\theta}) = Y_t - \widehat{Y}_{t|t-1}(\boldsymbol{\theta}) \tag{6.83}$$

where

$$\widehat{Y}_{t|t-1}(\boldsymbol{\theta}) = \operatorname{E}[Y_t | \boldsymbol{Y}_{t-1}, \boldsymbol{\theta}]. \tag{6.84}$$

The problem is in any case to calculate the conditional mean $\operatorname{E}[Y_t | \boldsymbol{Y}_{t-1}, \boldsymbol{\theta}]$.

Example 6.6 (Model with deterministic input)

As an example of an extended model and the principles, consider the *tranfer function* model

$$Y_t = H_1(\mathrm{B}) u_t + H_2(\mathrm{B}) \varepsilon_t \tag{6.85}$$

where $\{u_t\}$ is deterministic. Here $H_1(\mathrm{B})$ and $H_2(\mathrm{B})$ are rational transfer operators. Furthermore, we will demand that $H_2(0) = 1$. (For example we get for $H_1(\mathrm{B}) = 0$ and $H_2(\mathrm{B}) = \phi(\mathrm{B})^{-1}\theta(\mathrm{B})$ the previously considered ARMA models).

For the model (6.85) we get the conditional expected value

$$\begin{aligned} \widehat{Y}_{t|t-1} &= H_1(\mathrm{B}) u_t + [H_2(\mathrm{B}) - 1] \varepsilon_t \\ &= H_1(\mathrm{B}) u_t + [1 - H_2(\mathrm{B})^{-1}][Y_t - H_1(\mathrm{B}) u_t] \\ &= H_2(\mathrm{B})^{-1} H_1(\mathrm{B}) u_t + [1 - H_2(\mathrm{B})^{-1}] Y_t \end{aligned} \tag{6.86}$$

Since $H_2(0) = 1$, the conditional expected value will depend only on $Y_{t-1}, Y_{t-2}, \ldots$ as well as the known values of $\{u_t\}$. With given initial conditions, we can subsequently calculate the prediction error by using (6.83). And hereby, as we have seen earlier, apply either the prediction error method or the ML method for estimating the unknown parameters, $\boldsymbol{\theta}$, in $H_1(\mathrm{B})$ and $H_2(\mathrm{B})$.

For the ARMAX model (described in Chapter 8)

$$\phi(\mathrm{B})Y_t = \omega(\mathrm{B})u_t + \theta(\mathrm{B})\varepsilon_t \tag{6.87}$$

we have $H_1(\mathrm{B}) = \omega(\mathrm{B})/\phi(\mathrm{B})$ and $H_2(\mathrm{B}) = \theta(\mathrm{B})/\phi(\mathrm{B})$. Using (6.86) we get

$$\widehat{Y}_{t|t-1} = \frac{\omega(\mathrm{B})}{\theta(\mathrm{B})}u_t + \left[1 - \frac{\phi(\mathrm{B})}{\theta(\mathrm{B})}\right] Y_t \tag{6.88}$$

or

$$\theta(\mathrm{B})\widehat{Y}_{t|t-1} = \omega(\mathrm{B})u_t + [\theta(\mathrm{B}) - \phi(\mathrm{B})]Y_t. \tag{6.89}$$

With given initial conditions (most often $Y_{t|t-1} = 0$, $u_t = 0$, and $Y_t = 0$ for $t \leq 0$), one may proceed as described previously and calculate the prediction error from (6.83).

Please note that models in the form

$$\phi(\mathrm{B})Y_t = d_1 f_1(t) + \cdots + d_\ell f_\ell(t) + \theta(\mathrm{B})\varepsilon_t \tag{6.90}$$

where $f_1(t)$ up to $f_\ell(t)$ are known functions and d_1 up to d_ℓ are unknown parameters, can also be estimated by the prediction error method or the ML method.

6.5 Selection of the model order

The model building is often an iterative procedure, since a model checking of an estimated model with a given structure will commonly indicate alternative models. The model building first stops when all likely models have been tested and the most adequate model is found.

A very important problem is to determine the number of parameters, which with the chosen model structure, is the most suitable to describe the observed variation. The number of parameters is called the *model order*. If one seeks the most adequate model among the class of the ARMA processes, the process consists of determining (p, q).

In the following some tools and tests for determining the model order will be described.

6.5.1 The autocorrelation functions

The autocorrelation function and the partial autocorrelation function are usually applied to determine reasonable guesses for (p, q), which are described in Section 6.3.2 on page 154.

Furthermore the autocorrelation functions for the residuals serve as good guesses for how the estimated model should be extended. This is illustrated in the following.

Assume that at a given stage in the *iterative model building procedure*, we have estimated the following model:

$$\phi(\mathrm{B})Y_t = \theta(\mathrm{B})W_t \tag{6.91}$$

and that the estimated autocorrelation function for the residuals, $\{W_t\}$, indicates that the residuals can be described by the following:

$$\phi^*(\mathrm{B})W_t = \theta^*(\mathrm{B})\varepsilon_t. \tag{6.92}$$

Inserting $W_t = \phi^{*-1}(\mathrm{B})\theta^*(\mathrm{B})\varepsilon_t$ in (6.91), we see that the multiplicative model

$$\phi^*(\mathrm{B})\phi(\mathrm{B})Y_t = \theta(\mathrm{B})\theta^*(\mathrm{B})\varepsilon_t \tag{6.93}$$

would be an obvious model to examine in the next step. Alternatively to (6.93) one could obviously select the corresponding ordinary (non-multiplicative) model. The described procedure is especially useful for seasonal models.

6.5.2 Testing the model

This section contains a number of tools for testing the model. An example is a simple test for significance of the estimated parameters.

6.5.2.1 Likelihood based inference

In time series analysis the maximum likelihood method, as described previously, is often used. The likelihood principle is not just a method for obtaining an estimate of parameters; it is a method for an objective reasoning with data. It is the entire likelihood function, not just its maximizer, that captures all the information in the data about a certain parameter. The likelihood principle also serve the basis for a rich family of methods for selecting the most appropriate model.

In this section we describe methods for testing hypotheses using the likelihood function. The basic idea is to determine the maximum likelihood estimates under both the null and alternative hypothesis. For a more rigious presentation we refer to, e.g., Pawitan (2001), Thyregod and Madsen (2006), and Harvey (1981).

Consider the hypothesis $H_0 : \theta \in \Omega_0$ against the alternative $H_0^a : \theta \in \Omega \backslash \Omega_0$, where $\dim(\Omega_0) = r$ and $\dim(\Omega) = m$.

The *likelihood ratio* is defined as

$$\lambda(Y) = \frac{\sup_{\theta \in \Omega_0} L(\theta; Y)}{\sup_{\theta \in \Omega} L(\theta; Y)} \tag{6.94}$$

Clearly if λ is small, then the data are seen to be more plausible under the alternative hypothesis than under the null hypothesis. Hence the hypothesis (H_0) is rejected for small values of λ. It is sometimes possible to transform the likelihood ratio into a statistic, the exact distribution of which is known under H_0. This is for instance the case for the General Linear Model for Gaussian data.

In most cases, however, we must use the following important result regarding the asymptotic behavior.

THEOREM 6.9 (WILK'S LIKELIHOOD RATIO TEST)
For $\lambda(Y)$ defined by (6.94), then under the null hypothesis H_0 (as above), the random variable $-2\log\lambda(Y)$ converges in law to a χ^2 random variable with $(m - r)$ degrees of freedom, i.e.,

$$-2\log\lambda(Y) \to \chi^2(m - r) \tag{6.95}$$

under H_0.

Note, that the model must be estimated under both H_0 and the alternative hypothesis. A variant which does not require an evaluation in both cases is the Wald test; see, e.g., Harvey (1981).

6.5.2.2 Test whether a parameter θ_i is zero

In Sections 6.4.2 through 6.4.4, it is shown how LS, ML, and prediction error estimates of the model parameters can be found. In all cases, formulas for determining the (asymptotic) variance of the estimates for each of the methods have been given. Since the estimators can be assumed to be asymptotically normally distributed, we have a tool for testing whether a parameter θ_i is zero.

Given an estimator $\widehat{\theta}_i$ with the (asymptotic) variance $\widehat{\sigma}^2_{\widehat{\theta}_i}$, the variance is found as the diagonal element of the covariance matrix $\text{Var}[\widehat{\boldsymbol{\theta}}]$, see Theorems 6.6, 6.7, and 6.8.

We want to *test whether θ_i is zero*, i.e., we want to test the hypothesis

$$H_0\colon \theta_i = 0 \quad \text{against} \quad H_1\colon \theta_i \neq 0. \tag{6.96}$$

Based on the asymptotic normality we introduce the test statistic

$$T = \frac{\widehat{\theta}_i}{\widehat{\sigma}_{\widehat{\theta}_i}}, \tag{6.97}$$

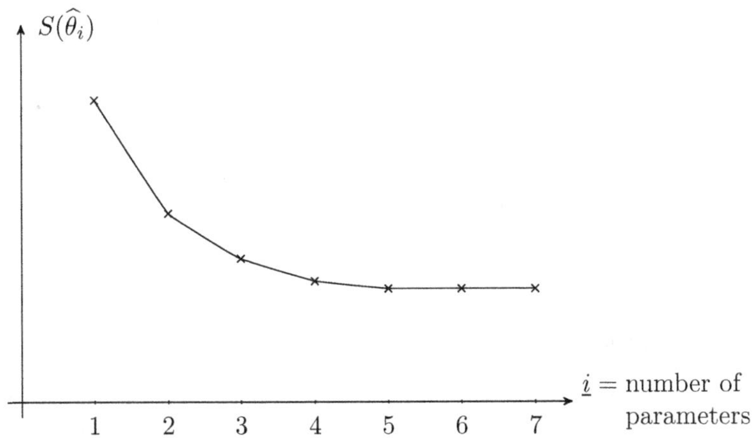

Figure 6.8: *The loss functions dependency on the number of parameters.*

which under H_0 follows a t-distribution with $f = N - p - q$ degrees of freedom, where N is the number of observations used in the estimation of $\theta_{\underline{i}}$. If we have estimated a mean value, then the degrees of freedom become $f = N - 1 - p - q$.

Alternatively, we can form the test statistic

$$F = T^2 = \frac{\widehat{\theta}_i}{\widehat{\sigma}_{\widehat{\theta}_i}}, \tag{6.98}$$

which under H_0 is $F(1, f)$-distributed (see Section 2.7 on page 23).

6.5.2.3 Test for lower model order

In the previous section we introduced the *sum of squared residuals* or *the loss function* for a given set of parameters θ_i by

$$S(\theta_{\underline{i}}) = \sum^{N} \varepsilon_t^2(\theta_{\underline{i}}), \tag{6.99}$$

where the index $\underline{i}$ is the number of estimated parameters. It will also hold, that if the above model is extended with one extra parameter, i.e., to $\underline{i}+1$ parameters, then $S(\theta_{\underline{i}+1}) < S(\theta_{\underline{i}})$. But it is also obvious that the "gain" from including one more parameter becomes less as the number of parameters increases. This is sketched in Figure 6.8.

Suppose that we have N observations of a stochastic process which can be described by n_0 parameters. Based on these observations, models containing n_1 and n_2 parameters have been estimated, where $n_0 \leq n_1 < n_2$. Furthermore, the n_0 parameters are a subset of the n_1 parameters, and the n_1 parameters are a subset of the n_2 parameters. The corresponding loss functions are denoted S_1 and S_2, and it holds asymptotically that

$$\frac{(S_1 - S_2)/(n_2 - n_1)}{S_2/(N - n_2)} \in \mathrm{F}(n_2 - n_1, N - n_2). \tag{6.100}$$

The test statistics in (6.100) can be used to test in the general linear model as described in Chapter 3. However, it can be shown that (6.100) can also be used for AR models, and in general, the test is regarded as being adequate for ARMA models.

6.5.3 Information criteria

The literature proposes various different methods and procedures for the determination of the most "adequate" model order. Probably the most commonly used is based on *Akaike's Information Criteria* (AIC), which is defined as

$$\text{AIC} = -2\log(\text{max. likelihood}) + 2n_i, \tag{6.101}$$

where n_i is the number of estimated model parameters. In the case of the ARMA(p, q) model we get (including an additive constant)

$$\text{AIC} = N\log\widehat{\sigma}_\varepsilon^2 + 2(p+q), \tag{6.102}$$

where $\widehat{\sigma}_\varepsilon^2 = S(\widehat{\boldsymbol{\theta}})/N$. The method basically consists of *choosing* (p, q) *such that AIC* (p, q) *is minimized.* Akaike's information criteria has a tendency, especially for large N, to accept too many parameters (Lütkepohl 1991).

An alternative criterion is *BIC (Bayesian Information Criterion)*, which, including an additive constant, is

$$\text{BIC} = N\log\widehat{\sigma}_\varepsilon^2 + (p+q)\log N. \tag{6.103}$$

Contrary to the AIC, the BIC yields a consistent estimate of the model order. The information criterion (6.103) is sometimes called *Schwartz's Bayesian Criterion (SBC)*.

6.6 Model checking

After having determined the model structure, determined the model order, and estimated the model parameters, the next step should be checking the model. By checking the model one can examine whether the observations can be described by the estimated model adequately. Often a model check will reveal that a model is not satisfactory, but at the same time indicate alternative model orders. It is thus often seen that model checking is applied on all the obvious model choices and that the outcome from checking the model is used in finding the final and most satisfactory model.

In the following sections, some of the most commonly used methods for checking the model will be discussed.

6.6.1 Cross-validation

The properties of the model are examined on a dataset which has not been used for estimation. Specifically it is examined if the variance of the prediction errors, $\widehat{\sigma}_\varepsilon^2$, is of the same order of magnitude as for the observations used for estimation.

6.6.2 Residual analysis

The general assumption regarding the ARMA process is that $\{\varepsilon_t\}$ is white noise. Thus, if a model is adequate, we have to require that $\{\varepsilon_t(\widehat{\boldsymbol{\theta}})\}$ is white noise. The *residual analysis* consists of *testing whether* $\{\varepsilon_t(\widehat{\boldsymbol{\theta}})\}$ *can be assumed to be white noise.*

a) Plot of $\{\varepsilon_t(\widehat{\boldsymbol{\theta}})\}$. This plot can reveal non-stationarities and outliers (e.g., from erroneous measurements).

b) Test for change in signs. If $\{\varepsilon_t\}$ is white noise (with mean value 0), one will expect ε_t on average will change its sign every second time, i.e.,

$$\mathrm{P}\{\text{changes in sign from } t-1 \text{ to } t\} = \tfrac{1}{2}.$$

Since $\{\varepsilon_t\}$ is white noise, each of the changes will be independent and we have

$$\text{Number of changes in sign} \in \mathrm{B}(N-1, \tfrac{1}{2}), \tag{6.104}$$

where N is the number of residuals. For large values of N, the binomial distribution can be approximated with the normal distribution because we have $\mathrm{B}(N-1, 1/2) \simeq \mathrm{N}\left(\frac{N-1}{2}, \frac{N-1}{4}\right)$ for large values of N.

c) Test in the autocorrelation function. If $\{\varepsilon_t\}$ is white noise it holds (see Section 6.2)

$$\widehat{\rho}_\varepsilon(k) \in_{\text{approx.}} \mathrm{N}\Big(0, \frac{1}{N}\Big). \tag{6.105}$$

Based on this, one can plot, e.g., 2σ limits (corresponding to an approximate 95% confidence interval), and thereby test if the individual values of $\widehat{\rho}_\varepsilon(1), \widehat{\rho}_\varepsilon(2), \dots$ are significantly different from zero.

However, we cannot observe $\{\varepsilon_t(\boldsymbol{\theta})\}$, but only $\{\varepsilon_t(\widehat{\boldsymbol{\theta}})\}$. It can be shown that $1/N$ is the upper limit for the variance of $\widehat{\rho}_{\varepsilon_t(\widehat{\boldsymbol{\theta}})}(k)$. Especially for small values of k, the variance can be significantly smaller. Improved estimates of the variance of $\widehat{\rho}_{\varepsilon_t(\widehat{\boldsymbol{\theta}})}(k)$ can be found in McLeod and Li (1983).

d) Portmanteau lack-of-fit test. We define

$$Q^{*2} = \left(\sqrt{N}\widehat{\rho}_\varepsilon(1)\right)^2 + \left(\sqrt{N}\widehat{\rho}_\varepsilon(2)\right)^2 + \cdots + \left(\sqrt{N}\widehat{\rho}_\varepsilon(m)\right)^2, \tag{6.106}$$

which according to (6.105) approximates a sum of squared independent N(0,1)-distributed stochastic variables, i.e., $Q^* \in \chi^2(m)$.

Since we base the test on $\varepsilon(\widehat{\boldsymbol{\theta}})$, we obtain

$$Q^2 = N(\widehat{\rho}_{\varepsilon(\widehat{\theta})}(1)^2 + \widehat{\rho}_{\varepsilon(\widehat{\theta})}(2)^2 + \cdots + \widehat{\rho}_{\varepsilon(\widehat{\theta})}(m)^2) \in_{\text{approx.}} \chi^2(m-n) \quad (6.107)$$

where n is the number of estimated parameters in the model. A reasonable choice of m is in most cases $15 \leq m \leq 25$.

e) Test in the cumulated periodogram. For the frequencies $\nu_i = i/N$, $i = 0, 1, \ldots, \lfloor N/2 \rfloor$ we calculate the *periodogram* for the residuals as

$$\widehat{I}(\nu_i) = \frac{1}{N}\left[\left(\sum_{t=1}^{N} \varepsilon_t \cos(2\pi\nu_i t)\right)^2 + \left(\sum_{t=1}^{N} \varepsilon_t \sin(2\pi\nu_i t)\right)^2\right], \quad (6.108)$$

which is a description of the variations of $\{\varepsilon_t\}$ in the frequency domain where $I(\nu_i)$ denotes the amount of variation of $\{\varepsilon_t\}$ that is related to the frequency ν_i. The periodogram will be discussed more closely in Section 7.1 on page 187.

The *scaled cumulated periodogram* is

$$\widehat{C}(\nu_j) = \frac{\left[\sum_{i=1}^{j} \widehat{I}(\nu_i)\right]}{\left[\sum_{i=1}^{\lfloor N/2 \rfloor} \widehat{I}(\nu_i)\right]}, \quad (6.109)$$

which is a non-decreasing function defined for the frequencies $\nu_i = i/N\,, i = 0, 1, \ldots, \lfloor N/2 \rfloor$.

For white noise, $\{\varepsilon_t\}$, we get, as discussed in Chapter 5, that the variation is uniformly distributed on all frequencies. Thus, for white noise the theoretical periodogram is constant. Furthermore, the total variation from N observations is $N\sigma_\varepsilon^2$, and thus the theoretical periodogram for white noise is

$$I(\nu_i) = 2\sigma_\varepsilon^2. \quad (6.110)$$

The theoretical accumulated periodogram $C(\nu_i)$ for white noise is thus a straight line from the point $(0,0)$ to $(0.5,1)$. If $\{\varepsilon_t(\widehat{\boldsymbol{\theta}})\}$ is white noise one will thus expect that $\widehat{C}(\nu_i)$ is located around this straight line. A confidence interval for the straight line can be obtained in a similar manner to the Kolmogorov-Smirnov test for distribution, see Chapter 4 of Conradsen (1984) or Hald (1952). The confidence limits in the Kolmogorov-Smirnov test for distribution are determined so that they include all the $\widehat{C}(\nu_i)$ values at a given probability $(1-\alpha)$, conditioned that $\{\varepsilon_t(\widehat{\boldsymbol{\theta}})\}$ is white noise, as illustrated in Figure 6.9.

The lines are drawn at a distance $\pm K_\alpha/\sqrt{q}$ above and below the theoretical line, where $q = (N-2)/2$ for N even and $q = (N-1)/2$ for N odd. Approximate values for K_α are given in Table 6.2.

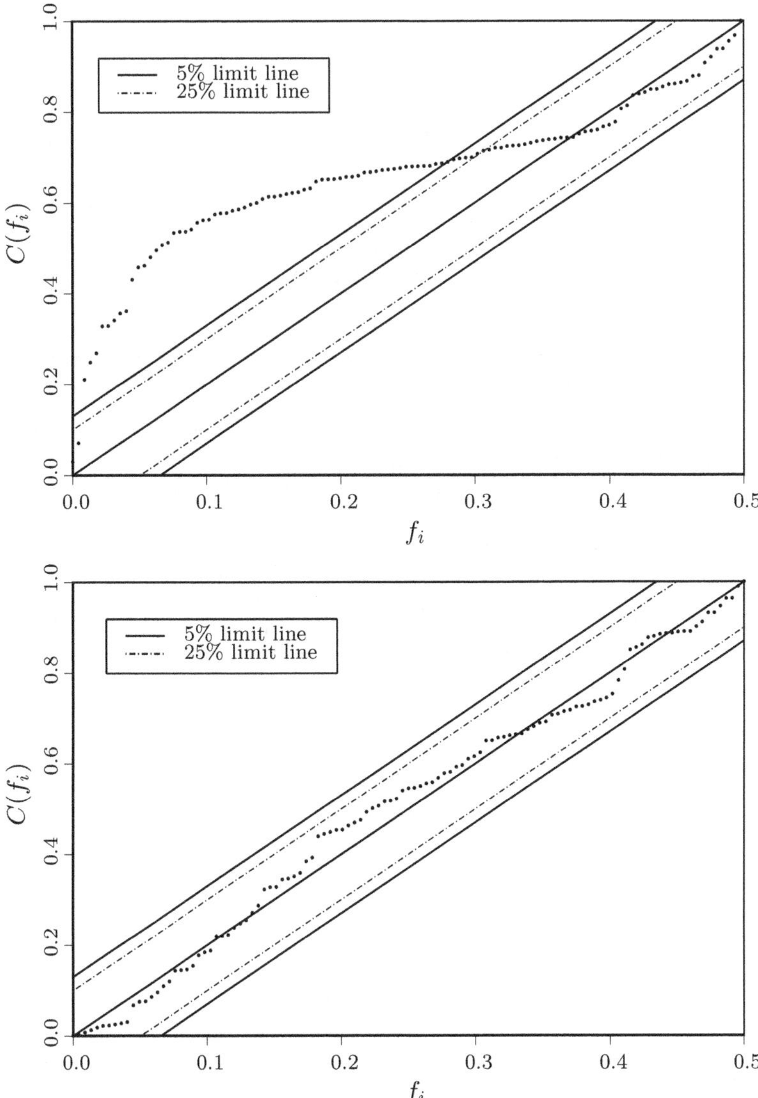

Figure 6.9: *The cumulated periodogram for an inadequate model (top) and an adequate model (bottom).*

Table 6.2: *Coefficients involved in calculating probability limits for test in the cumulative periodogram.*

α	0.01	0.05	0.10	0.25
K_α	1.63	1.36	1.22	1.02

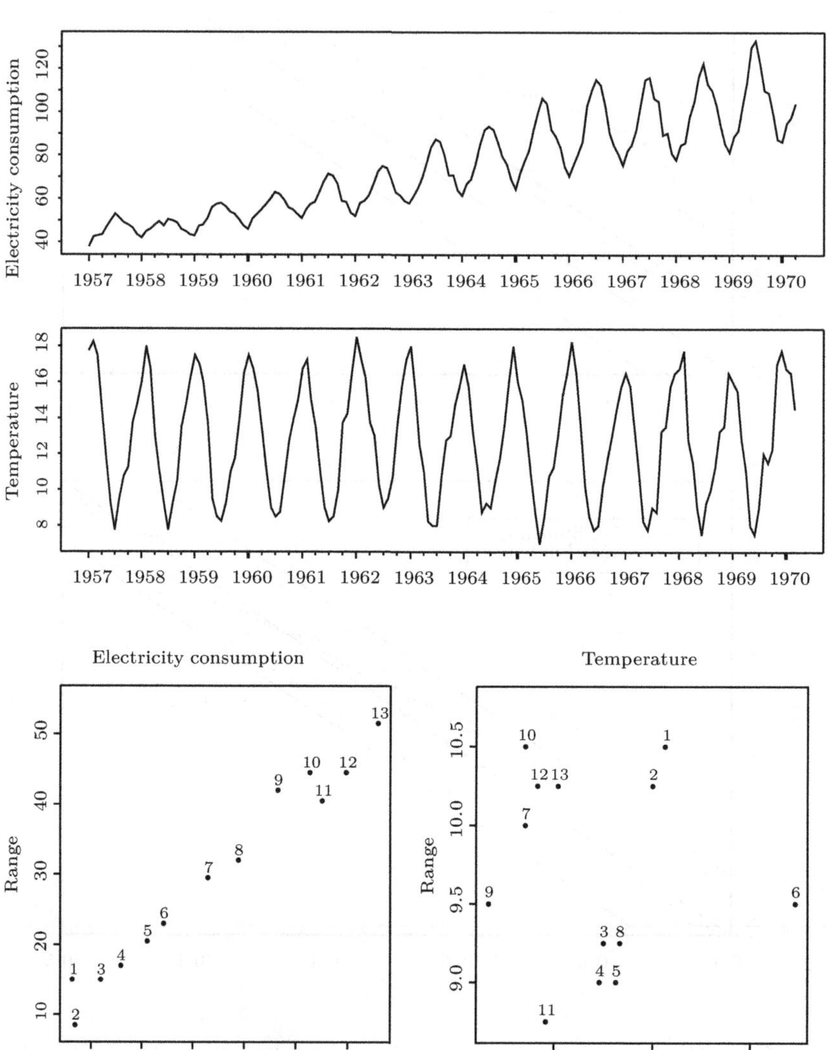

Figure 6.10: *Plot of the electricity consumption and temperature and the corresponding range-mean plots (sub-series of size 12 have been applied in the range-mean plot).*

6.7 Case study: Electricity consumption

We are considering a "real life" example taken from Jenkins (1979). In this example we are looking for a model for electricity consumption with the purpose of prediction. In a later example in Chapter 8, the model will be extended to an input-output model, which exploits the monthly mean ambient temperature in order to improve the model. The upper half of Figure 6.10 shows a plot of the monthly energy use, Y_t, and an input series, X_t, consisting of the corresponding monthly mean temperatures. The lower half of the plot shows the corresponding *range-mean plot*.

A range-mean plot is used to examine whether a transformation in the form

$$Y_t^{(\lambda)} = \begin{cases} (Y_t^\lambda - 1)/\lambda & \lambda \neq 0 \\ \ln Y_t & \lambda = 0 \end{cases} \tag{6.111}$$

is needed to yield a constant variance. This class of transformations is called *Box-Cox transformations*, or variance stabilizing transformations—see Box and Cox (1964).

Method: The time series is divided sequentially into groups consisting of n (e.g., 4 to 12) observations (depending on the length of the time series, or if seasonal variations equal the length of the season). Subsequently the range $Y_{(n)} - Y_{(1)}$ and mean value in each group are calculated, and corresponding values are plotted in the range-mean plot.

If the range is independent of the mean values, no transformation is needed ($\lambda = 1$), but any dependencies on the mean value indicate that a transformation should be considered. If there is a (positive) linear dependence, a logarithmic transformation ($\lambda = 0$) should be applied, and if there is an upwards or downwards curvature, then $\lambda = -1/2$ and $\lambda = 1/2$ should be applied, respectively. Figure 6.11 on the following page indicates Box-Cox transformations for various range-mean relations

The linear relationship in Figure 6.10 between the range and the mean value of the electricity consumption indicates that a logarithmic transformation should be considered. Regarding the temperature series, there is no need for transformation.

In Figure 6.12 on the next page, a sample of the series consisting of the logarithm of the electricity consumption and the estimated autocorrelation function for the entire series is shown. The autocorrelation function is characterized by a period of 12 months, and the values decrease to 0 slowly, which indicates that the series is non-stationary and that a differencing is necessary.

Figure 6.12(b) shows a sample of the series $\nabla \ln Y_t$ and the corresponding autocorrelation function. It is seen that the differencing has removed the non-stationarity in the original series, but a seasonal differencing is needed in order to remove the slowly decreasing autocorrelations at lag $12, 24, 36, \ldots$.

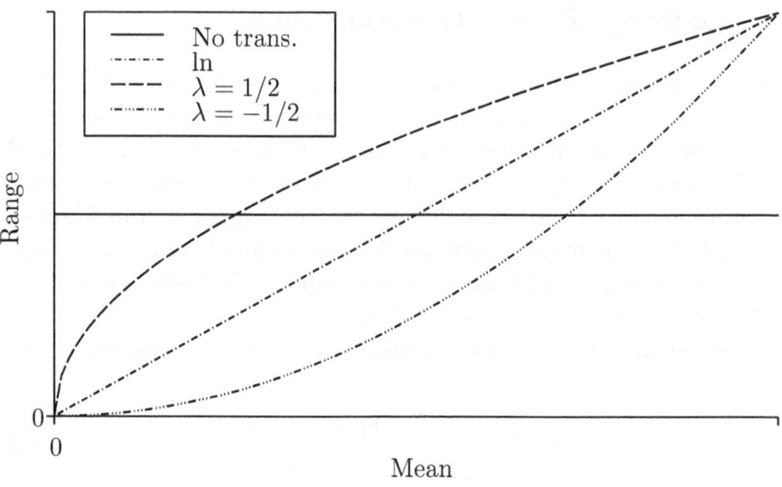

Figure 6.11: *Box-Cox transformations.*

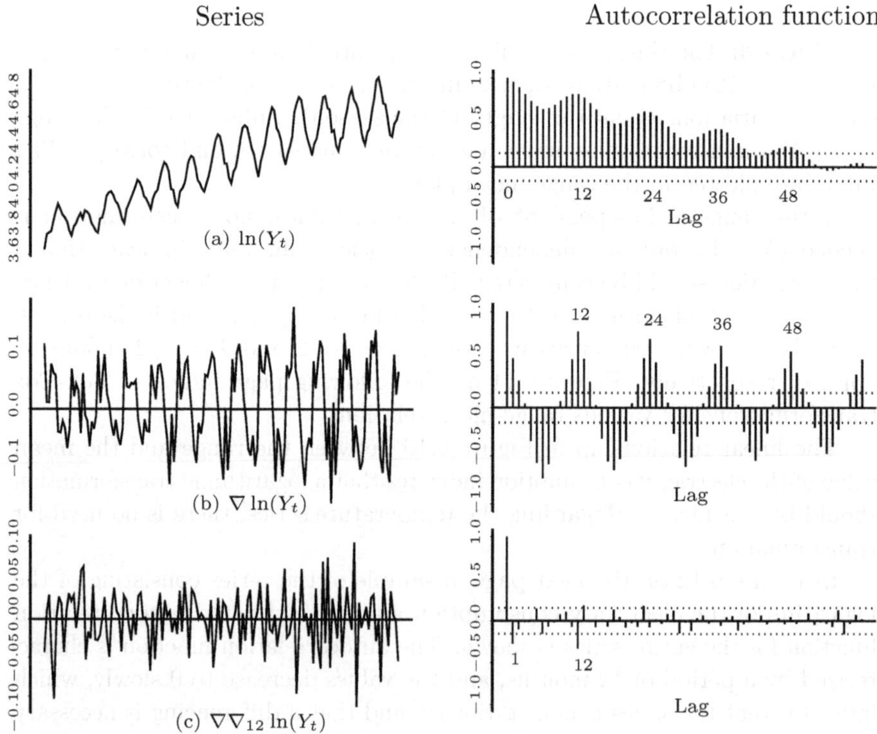

Figure 6.12: *Different orders of differencing of the series for the electricity consumption and the corresponding autocorrelation functions.*

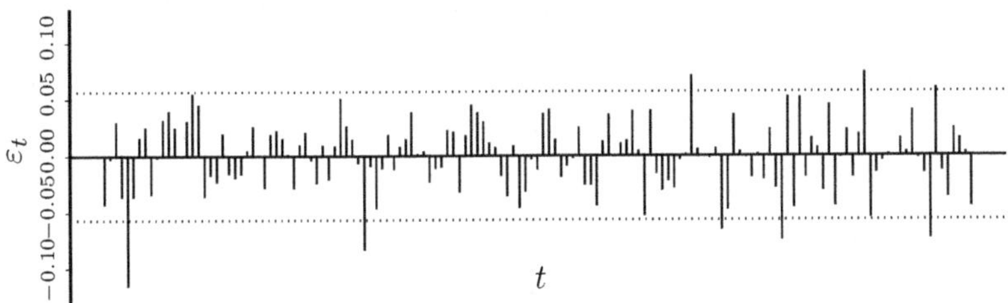

Residual autocorrelation function

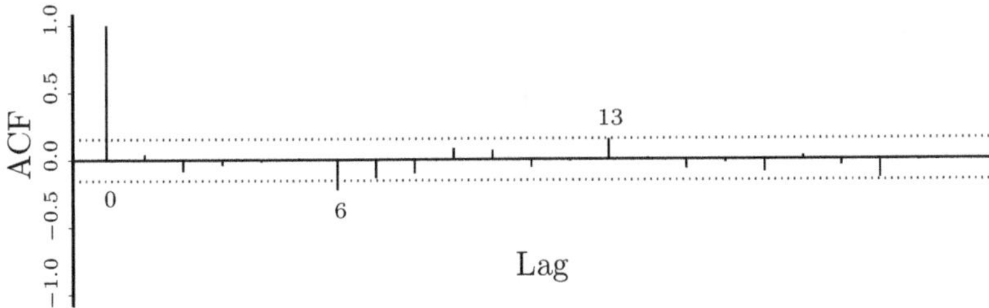

Figure 6.13: *Residuals, ε_t, and corresponding autocorrelation function for the model* (6.113) *based on the whole series.*

In Figure 6.12(c), the results from applying a seasonal differencing ∇_{12} are shown. Apparently, there are no need for further differencing the series, i.e., the series $\nabla\nabla_{12} \ln Y_t$ can be assumed stationary. The largest values of the autocorrelation function for $\nabla\nabla_{12} \ln Y_t$ are seen to be at lag 1 and 12, thus initially the following structure is applied:

$$\nabla\nabla_{12} \ln Y_t = (1 + \theta \mathrm{B})(1 + \Theta \mathrm{B}^{12})\varepsilon_t. \tag{6.112}$$

Based on the autocorrelation function at lag 1, $r_1 = -0.27$, for $\nabla\nabla_{12} \ln Y_t$, we select $\widehat{\theta} = -0.30$ as a starting value for θ. Similarly we use the autocorrelation at lag 12, $r_{12} = -0.33$, as a starting value $\widehat{\Theta} = -0.35$ for Θ. In both cases the starting values correspond closely to the moment estimate of the pure MA(1) and $\text{MA}(1)_{12}$ process, respectively.

The starting values are now applied for parameter estimation by the prediction error method. Based on the entire series (160 observations) we find the following model

$$\begin{aligned} \nabla\nabla_{12} \ln Y_t &= (1 - 0.30^{\pm 0.08}\mathrm{B})(1 - 0.55^{\pm 0.02}\mathrm{B}^{12})\varepsilon_t \\ \widehat{\sigma}_\varepsilon^2 &= 0.0283^2. \end{aligned} \tag{6.113}$$

In Figure 6.13 the series of residuals and corresponding autocorrelation function are shown. It is seen, that the autocorrelation is larger than two standard deviations at lag 6 and 13. The number of values outside 2 standard deviations is considerably larger than it should be, and it is concluded that the model is not adequate. Further improvements of the model have not been successful.

In order to examine the reason for the inadequacy of the model the time series has been divided into two sub-series of 64 and 96 respectively, and a model has been fitted to each of these series. The results are:

First half

$$(N=64) \qquad \begin{aligned} \nabla\nabla_{12} \ln Y_t &= (1+0.14^{\pm 0.12}\mathrm{B})(1-0.83^{\pm 0.04}\mathrm{B}^{12})\varepsilon_t \\ \widehat{\sigma}_\varepsilon^2 &= 0.0237^2. \end{aligned}$$

Second half

$$(N=96) \qquad \begin{aligned} \nabla\nabla_{12} \ln Y_t &= (1-0.73^{\pm 0.08}\mathrm{B})(1-0.83^{\pm 0.05}\mathrm{B}^{12})\varepsilon_t \\ \widehat{\sigma}_\varepsilon^2 &= 0.0255^2. \end{aligned}$$

It is seen that the estimate of the seasonal component is relatively constant comparing the two sub-series. On the other hand, the coefficients for the MA parts are very different. The explanation is most likely that this parameter is varying in time. It is recommended, depending on the application, to apply either an adaptive estimation procedure or a direct modeling of the time-varying coefficient. The methods to do so are discussed in Chapter 11. Another possible explanation could be that the input series behaves differently in each series, but this does not seem to be an obvious feature. In Example 8.1 on page 231 the information from the input series will be exploited in an input-output model. An analysis of the estimated autocorrelation function for each of the sub-series has shown that the models for each part are adequate.

Concerning the model checking, one should also apply other tests for the residuals, e.g., a Portmanteau lack-of-fit test and test in the cumulated periodogram. Thus, the previous example is not complete concerning the model checking. However, the example illustrates the possible need for transformation and differencing when building models of real life phenomena.

6.8 Problems

Exercise 6.1

The following time series $\{y_t\}$ is supposed to be the Danish krone (DKK) to the US dollar (USD) exchange rate (DKK/100\$) over a period of 20 weeks. Each number in Table 6.3 shows the average exchange rate of the corresponding week.

Question 1 Sketch the time series. Is the time series stationary?

Question 2 Find a transformation that gives an acceptable stationary time series and sketch the generated time series x_t.

Table 6.3: *Exchange rate: DKK to USD.*

Week no.	1	2	3	4	5	6	7	8	9	10
DKK/100$	1084	1052	1028	1035	1013	1029	998	980	976	967
Week No.	11	12	13	14	15	16	17	18	19	20
DKK/100$	955	934	922	949	956	940	942	919	906	885

Question 3 Estimate the autocorrelation function (lag ≤ 5) for $\{X_t\}$ and sketch the result.

Question 4 Based on the estimated autocorrelation function, investigate whether $\{X_t\}$ can be assumed to be white noise.

Question 5 Establish a model for the variations in the exchange rate. Use this model to predict the average exchange rate in week 21.

Exercise 6.2
As suggested in this chapter the autocovariance at lag k is estimated based on a time-series $x_1, x_2, \ldots, x_N$ using

$$c_k = \frac{1}{N} \sum_{t=1}^{N-k} (x_t - \overline{x})(x_{t+k} - \overline{x}).$$

Is the estimator

$$C_k = \frac{1}{N} \sum_{t=1}^{N-k} (X_t - \overline{X})(X_{t+k} - \overline{X})$$

an unbiased estimator? Comment.

Exercise 6.3
Consider the following AR(2) process:

$$X_t + \phi_1 X_{t-1} + \phi_2 X_{t-2} = \varepsilon_t$$

where ε_t is white noise.

Question 1 Sketch the allowed area for the parameters, i.e., the set of (ϕ_1, ϕ_2) for which the AR(2) process is stationary.

Question 2 Sketch that part of the allowed area of the parameters where the autocorrelation function shows damped harmonic oscillations.

Question 3 For a process the following autocorrelations have been estimated:

$$r_1 = 0.6, \qquad r_2 = -0.1, \qquad r_3 = -0.4, \qquad r_4 = -0.2$$

Provide a moment estimate for (ϕ_1, ϕ_2) based on the observed values of the autocorrelation function, assuming that the process can be approximated by an AR(2) model.

Table 6.4: *Sample partial autocorrelation.*

k	$\widehat{\phi}_{kk}$	k	$\widehat{\phi}_{kk}$	k	$\widehat{\phi}_{kk}$
1	−0.4	6	−0.15	11	0.18
2	0.19	7	0.05	12	−0.05
3	0.01	8	0.00	13	0.09
4	−0.07	9	−0.10	14	0.18
5	−0.07	10	0.05	15	0.01

Exercise 6.4
Based on observed values from an MA(2) process, we have found that $r_1 = -0.6222$ and $r_2 = 0.2222$.

Estimate the parameters of the process.

Exercise 6.5
For a process the sample autocorrelations $r_1 = 0.57$ and $r_2 = 0.5$ have been found. It is assumed that the process can be described by

$$(1 + \phi_1 \mathrm{B})X_t = (1 + \theta_1 \mathrm{B})\varepsilon_t$$

Estimate ϕ_1 and θ_1.

Exercise 6.6
The estimates of the partial autocorrelation function of a process are provided in Table 6.4. The total number of measurements behind the estimates is $N = 70$.

Find a suitable model (including parameter values) in the class of AR(p) models.

Exercise 6.7
Consider the following ARMA(1, 1) process

$$X_t - 0.9X_{t-1} = \varepsilon_t + 0.8\varepsilon_{t-1}$$

where $\{\varepsilon_t\}$ is white noise with $\mathrm{Var}[\varepsilon_t] = \sigma^2$.

Question 1 Write the one-step prediction $\widehat{X}_{t+1|t}$ as a function of the past values of X_t. Determine the variance of the one-step prediction error.

Question 2 Write the k-step prediction $\widehat{X}_{t+k|t}$ and determine the variance of the prediction error.

Exercise 6.8
A company has analyzed the daily price of shares Z_t in a consecutive period of 300 days. The average rate was 198.

Table 6.5: *Estimated variance and sample autocorrelation functions.*

		Sample autocorrelation in lag k ($\widehat{\rho}_k$)							
Process	Var.	1	2	3	4	5	6	7	8
Z_t	70.7	0.99	0.99	0.98	0.97	0.96	0.96	0.95	0.94
∇Z_t	52.5	0.14	−0.01	−0.04	−0.02	−0.02	0.06	0.07	0.04
$\nabla^2 Z_t$	96.3	−0.45	−0.02	−0.04	0.00	−0.07	0.06	−0.01	0.04

The last 10 days' prices (day 1 through day 10) were:

$$233, 219, 205, 174, 182, 193, 202, 206, 195, 179$$

The estimated autocorrelations for the observations, and their 1st and 2nd order differences, are given in Table 6.5.

Question 1 Formulate a suitable model for the process and estimate the parameters of the model.

Question 2 Predict the rate 3 days ahead (day 13).

Question 3 The observed rates of day 11 and day 12 are 184 and 196, respectively. Carry out an update of the prediction for day 13.

Question 4 Compare the calculated predictions with the actual rate of day 13, which became 202, using suitable confidence bands.

Exercise 6.9
In a gas-oven, the CO_2 concentration of the flue gas air has been registered every 10th second. The last 5 registrations were (in %):

$$78, 74, 73, 76, 79$$

The desired value is 75%. Based on 100 registrations the following estimates of the first four autocovariances for the deviation from 75% have been found:

$$C(0) = 2.25, \qquad C(1) = 1.58, \qquad C(2) = 1.13, \qquad C(3) = 0.90$$

Question 1 Formulate a model for registrations and use suitable tests to verify if the model provides an adequate (with respect to the available information) description.

Question 2 Estimate the unknown parameters of the identified process.

Question 3 Predict the next 3 values and determine a 95% confidence band for the predictions.

Exercise 6.10

In order to establish models for improving the predictions of the water level, it is found that deviations Y_t between the hourly calculated water level and the observed water level can be described by a linear process of the form

$$(1 - 0.8\mathrm{B})(1 - 0.2\mathrm{B}^6)(1 - \mathrm{B})Y_t = \varepsilon_t$$

where ε_t is white noise with the variance σ_ε^2. Based on about 1500 observations, it is found that

$$\widehat{\sigma}_\varepsilon^2 = 0.31\,\mathrm{dm}^2$$

Question 1 Based on the observations of Y_t shown below, use the given model in order to provide a prediction of the value of Y_t corresponding to $t = 12$.

t (in hours)	1	2	3	4	5	6	7	8	9	10
Y_t (in dm)	2	1	−1	−2	−3	1	4	4	0	−3

Table 6.6: *Observed water levels.*

Question 2 Provide a 95% confidence interval for the prediction.

CHAPTER 7

Spectral analysis

This chapter concerns the estimation of the spectrum for one or more time series, also called the sample spectrum. In Section 5.4 on page 113, it was demonstrated that the spectrum for a stationary stochastic process is a *frequency domain* description of the autocovariance function. Correspondingly, the sample spectrum is a frequency domain description of the estimated autocovariance function. The *spectrum* contains a description of the variations in the *frequency domain*, whereas the corresponding description in the *time domain* is the *autocovariance function*. Spectral analysis is thus applied when the frequency properties of a phenomena are investigated, e.g., when a time series contains periodicities. If more than one time series, or signal, is available, the spectral analysis can be applied to obtain a non-parametric description of, e.g., the amplitude and the phase. This will be used as a possible approach to identify transfer function models with multiple input in Section 8.5.

Sections 7.1 and 7.2 concern the estimation of the spectrum for one time series, while Section 7.3 concerns the estimation of the cross-spectrum between two time series. Mainly *non-parametric* methods will be considered. The characteristics of the *parametric methods* (also called time domain methods) are that the spectrum is described by parameters in an estimated model (e.g., by the estimated coefficient in an AR(p)-model), and the spectrum is found as the corresponding theoretical spectrum (e.g., (5.112) on page 129) for the estimated model.

The spectrum is closely linked to linear systems and processes since it is simply a description of the autocovariance function in the frequency domain. Among the many classical references to spectral analysis let us mention Jenkins and Watts (1968), Brillinger (1981), Priestley (1981), and Papoulis (1984). Higher order spectra such as the bispectrum are described in, e.g., Subba Rao and Gabr (1984).

7.1 The periodogram

It was shown in Section 5.4 (Theorem 5.6 on page 115) that the spectrum for a stationary process can be determined by Fourier transforming the autocovariance function. Hence, it seems obvious to apply the following estimate for the spectrum

$$I_N(\omega) = \frac{1}{2\pi} \sum_{k=-(N-1)}^{N-1} \mathrm{C}(k) \mathrm{e}^{-i\omega k}, \quad |\omega| \le \pi, \tag{7.1}$$

where $\mathrm{C}(k)$ is the estimate (6.1) on page 146 of the autocovariance function based on N observations: $Y_1, \dots, Y_N$.

If we assume that $\{Y_t\}$ has the mean 0, then we can write $I_N(\omega)$ as

$$\begin{aligned} I_N(\omega) &= \frac{1}{2\pi} \sum_{k=-(N-1)}^{N-1} \mathrm{C}(k) \mathrm{e}^{-i\omega k}, \quad |\omega| \le \pi \\ &= \frac{1}{2\pi} \sum_{k=-(N-1)}^{N-1} \left(\frac{1}{N} \sum_{t=1}^{N-|k|} Y_t Y_{t+|k|} \right) \mathrm{e}^{-i\omega k} \\ &= \frac{1}{2\pi N} \left(\sum_{t=1}^{N} Y_t \mathrm{e}^{-i\omega t} \right) \left(\sum_{t=1}^{N} Y_t \mathrm{e}^{i\omega t} \right) \\ &= \frac{1}{2\pi N} \left| \sum_{t=1}^{N} Y_t \mathrm{e}^{-i\omega t} \right|^2, \end{aligned} \tag{7.2}$$

which we can formulate as

$$I_N(\omega) = \frac{1}{2\pi N} \left[\left(\sum_{t=1}^{N} Y_t \cos(\omega t) \right)^2 + \left(\sum_{t=1}^{N} Y_t \sin(\omega t) \right)^2 \right]. \tag{7.3}$$

$I_N(\omega)$ is called the *periodogram* or *sample spectrum*, and can be determined by either (7.1), (7.2), or (7.3). The periodogram is defined for all ω in $[-\pi, \pi]$, but in order to achieve independence between $I_N(\omega)$ at different values of ω (more about this later), it is advisable to calculate the periodogram only at the so-called *fundamental frequencies*,

$$\omega_p = \frac{2\pi p}{N}, \quad p = 0, 1, \dots, \lfloor N/2 \rfloor. \tag{7.4}$$

It is seen from (7.2) that the sample spectrum is proportional to the squared amplitude of the Fourier transformed time series: $Y_1, \dots, Y_N$.

There also exists an inversion formula. Corresponding to (7.1) we have

$$\mathrm{C}(k) = \int_{-\pi}^{\pi} I_N(\omega) \mathrm{e}^{ik\omega} \, \mathrm{d}\omega, \quad |k| \le N - 1. \tag{7.5}$$

For $k = 0$ it is seen that $I_N(\omega)$ *shows how the empirical variance is distributed on the different frequencies.* It should be noted that alternative placements of the factor 2 in the Fourier transform are often seen in the literature for (7.1) through (7.5); cf. the discussion in Chapter 4.

7.1.1 Harmonic analysis

For the fundamental frequencies in (7.4), we have the following *orthogonal relations*:

$$\sum_{t=1}^{N} \cos(\omega_p t)\cos(\omega_q t) = \begin{cases} 0 & p \neq q \\ N/2 & 0 < p = q < N/2 \\ N & p = q = 0\,, \quad \text{(or } N/2 \text{ for } N \text{ even)} \end{cases}$$

$$\sum_{t=1}^{N} \sin(\omega_p t)\cos(\omega_q t) = 0\,, \quad \text{for all } p, q \tag{7.6}$$

$$\sum_{t=1}^{N} \sin(\omega_p t)\sin(\omega_q t) = \begin{cases} 0 & p \neq q \\ N/2 & 0 < p = q < N/2 \\ 0 & p = q = 0\,, \quad \text{(or } N/2 \text{ for } N \text{ even)} \end{cases}$$

Based on these orthogonal relations we see that for N even, we can write

$$Y_t = a_0 + \sum_{p=1}^{N/2-1} \big(a_p \cos(\omega_p t) + b_p \sin(\omega_p t)\big) + a_{N/2}\cos(\pi t) \tag{7.7}$$

for $t = 1, 2, \ldots, N$, and the coefficients are given by

$$\begin{aligned} a_0 &= \frac{1}{N}\sum_{t=1}^{N} Y_t = \overline{Y} \\ a_p &= \frac{2}{N}\sum_{t=1}^{N} Y_t \cos(\omega_p t)\,, \quad p = 1, \ldots, N/2-1 \\ b_p &= \frac{2}{N}\sum_{t=1}^{N} Y_t \sin(\omega_p t)\,, \quad p = 1, \ldots, N/2-1 \\ a_{N/2} &= \frac{1}{N}\sum_{t=1}^{N} Y_t \cos(\pi t) = \frac{1}{N}\sum_{t=1}^{N}(-1)^t Y_t. \end{aligned} \tag{7.8}$$

For example, the coefficient, a_q, can be found by multiplying $\cos(\omega_q t)$ on both sides of (7.7) and subsequently summing over t. Equation (7.7) is a finite *Fourier series representation* of the observations $Y_1, \ldots, Y_N$. It should be noted that the N observations correspond exactly to N coefficients in the Fourier representation.

If we introduce the amplitude of the p'th harmonic as ($p \neq N/2$)

$$R_p = \sqrt{a_p^2 + b_p^2}, \tag{7.9}$$

it can be shown that (again the orthogonal relations are applied)

$$\frac{1}{N}\sum_{t=1}^{N}(Y_t - \overline{Y})^2 = \sum_{p=1}^{N/2-1} R_p^2/2 + a_{N/2}^2, \tag{7.10}$$

which is known as the *Parseval identity*. The identity illustrates that the variations in Y_t can be divided into frequencies, where $R_p^2/2$ is the contribution of the variance from the p'th harmonic.

By comparing (7.8), (7.9), and (7.3), it is seen that the periodogram may be determined based on a *harmonic analysis* by ($p \neq N/2$)

$$I_N(\omega_p) = \frac{NR_p^2}{8\pi}. \tag{7.11}$$

7.1.2 Properties of the periodogram

By applying (6.2) on page 147 we find

$$\begin{aligned}
\mathrm{E}[I(\omega)] &= \frac{1}{2\pi}\sum_{k=-(N-1)}^{N-1}\left(1-\frac{|k|}{N}\right)\gamma(k)\mathrm{e}^{-i\omega k} \\
&= \frac{1}{2\pi}\sum_{k=-\infty}^{\infty}\gamma(k)\mathrm{e}^{-i\omega k} \\
&= f(\omega), \quad \text{for } N \to \infty,
\end{aligned}$$

i.e., the estimator of the periodogram is *asymptotically unbiased.* Furthermore, we have an important theorem.

THEOREM 7.1 (PROPERTIES IN CASE OF WHITE NOISE)
Let $\{Y_t\}$ be normally distributed white noise having variance σ_Y^2. Then the following holds

i) $\{I(\omega_p)\}, p = 0, 1, \dots, [N/2]$ *are independent.*

ii) $\frac{I(\omega_p)4\pi}{\sigma_Y^2} \in \chi^2(2), p \neq 0, N/2$, *for N even.*

iii) $\frac{I(\omega_p)2\pi}{\sigma_Y^2} \in \chi^2(1), p = 0, N/2.$

If the assumption of normality does not hold, then the theorem is only an approximation.

Proof If we introduce

$$A(\omega_p) = \sum_{t=1}^{N} Y_t\cos(\omega_p t), \qquad B(\omega_p) = \sum_{t=1}^{N} Y_t\sin(\omega_p t) \tag{7.12}$$

we get

$$\operatorname{Var}[A(\omega_p)] = \begin{cases} \sigma_Y^2 \dfrac{N}{2} & p \neq 0, N/2\,, \quad (N \text{ even}) \\ \sigma_Y^2 N & p = 0 \end{cases}$$

$$\mathrm{E}[A(\omega_p)] = 0$$

$$\operatorname{Var}[B(\omega_p)] = \begin{cases} \sigma_Y^2 \dfrac{N}{2} & p \neq 0, N/2\,, \quad (N \text{ even}) \\ 0 & p = 0 \end{cases}$$

$$\mathrm{E}[B(\omega_p)] = 0$$

It is seen that $A(\omega_p)$ and $B(\omega_p)$ are uncorrelated. Furthermore, both $A(\omega_p)$ and $B(\omega_p)$ will be normally distributed and, thus, also independent. When rewriting (7.3) on page 188 as

$$\frac{I_N(\omega_p)4\pi}{\sigma_Y^2} = \left[\left[\frac{A(\omega_p)}{\sigma_Y \sqrt{N/2}} \right]^2 + \left[\frac{B(\omega_p)}{\sigma_Y \sqrt{N/2}} \right]^2 \right]$$

it is seen that $I_N(\omega_p)4\pi\,/\,\sigma_Y^2$ is a sum of two mutually independent, squared, $\mathrm{N}(0,1)$-distributed variables and, thus, that $I_N(\omega_p)4\pi/\sigma_Y^2$ is distributed as χ^2 with 2 degrees of freedom; cf. Section 2.7 on page 23. ■

As a special case we have

$$\operatorname{Var}\left[\frac{I_N(\omega_p)4\pi}{\sigma_Y^2} \right] = 4,$$

and thus

$$\operatorname{Var}[I_N(\omega_p)] = \sigma_Y^4/4\pi^2. \tag{7.13}$$

It is seen that the *variance of the estimate does not go to zero for* $N \to \infty$, *and thus the estimator is not consistent.* By using more observations we obtain more values in the periodogram, whereas the precision of the estimates does not improve.

The previous results are based on the assumption that $\{Y_t\}$ is white noise. It is easy to illustrate a consistency problem without this assumption. The periodogram is a linear combination of N samples of the autocovariance. We have seen previously that $\operatorname{Var}[\mathrm{C}(k)] = O(1/N)$, and since the term is included N times in $I(\omega)$, we get

$$\operatorname{Var}[I_N(\omega)] = O(1). \tag{7.14}$$

This result shows that the variance does not go to zero for $N \to \infty$ and, thus, that the *periodogram* $I(\omega)$ *is not a consistent estimator* for the spectrum $f(\omega)$. The problem of large variation of the values in the periodogram is illustrated

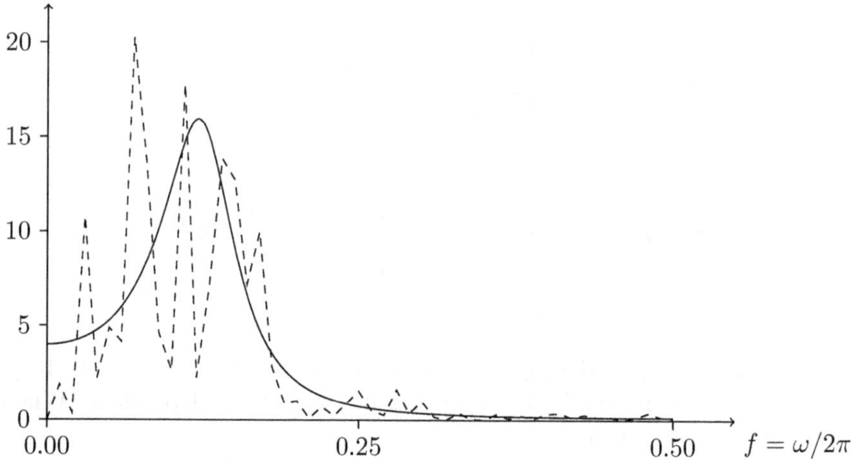

Figure 7.1: *The periodogram for the AR(2)-process: $Y_t - 1.131Y_{t-1} + 0.64Y_{t-2} = \varepsilon_t$ (100 observations). The smooth curve is the theoretical spectrum.*

in Figure 7.1. In the following section we will introduce some modifications which will imply consistent estimates of the spectrum.

If we assume a *weak dependency* (which implies that values having large time-distances only are weakly correlated), then Theorem 7.1 on page 190 can be extended to the following.

THEOREM 7.2 (PROPERTIES OF PERIODOGRAM ESTIMATOR)
Let $\{Y_t\}$ be a weakly dependent process with the spectral density $f(\omega)$. It then holds that

i) $\{I(\omega_p)\}, p = 0, 1, \ldots, \lfloor N/2 \rfloor$, *are approximately independent.*

ii) $\frac{2I(\omega_p)}{f(\omega_p)} \in_{\text{approx.}} \chi^2(2), p \neq 0, N/2,$ *for N even.*

iii) $\frac{I(\omega_p)}{f(\omega_p)} \in_{\text{approx.}} \chi^2(1), p = 0, N/2.$

Proof Omitted. See, e.g., Brillinger (1981) or Shumway (1988). ■

▸ **Remark 7.1**

The theorem can be applied in assessing the uncertainty of the periodogram. Especially, it should be noted that $\text{Var}[2I(\omega_p)/f(\omega_p)] \approx 2 \cdot 2 = 4$, when the endpoints are not considered, and thereby $\text{Var}[I(\omega_p)/f(\omega_p)] = 1$. On the contrary, we have for the endpoints $\text{Var}[I(\omega_p)/f(\omega_p)] = 2$, which implies that the uncertainty is twice as large for the endpoints as for the rest of the points. ◂

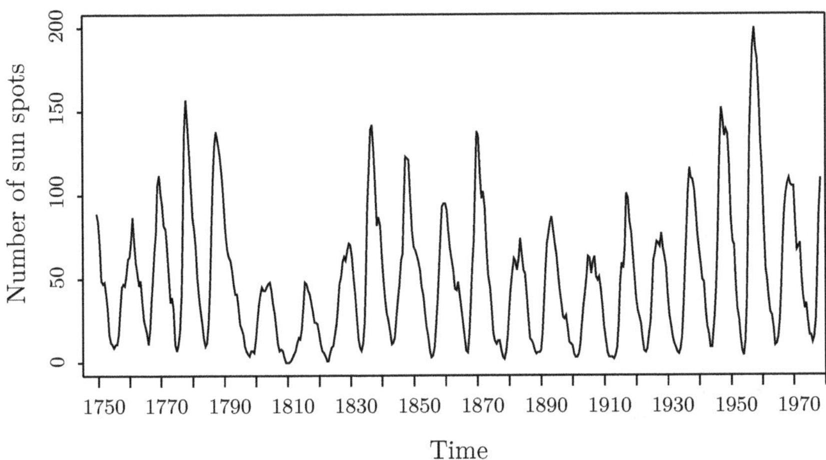

Figure 7.2: *Number of observed sun spots per half year in the period from June 1749 to December 1978.*

▸ **Remark 7.2**
In the literature, it is sometimes seen that instead of $I(\omega)$ ($f(\omega)$, respectively) $\omega \in [-\pi, \pi]$ the mapping of $I(\nu)$ ($f(\nu)$, respectively) $\nu \in [-0.5, 0.5]$, where $\omega = 2\pi\nu$, is applied. Introduce

$$f(\omega)d\omega = f(\nu)d\nu\,, \quad \omega = 2\pi\nu$$

and the following *relation between* $f(\omega)$ *and* $f(\nu)$ is obtained

$$f(\omega)2\pi = f(\nu)\,, \quad \omega = 2\pi\nu \tag{7.15}$$

Obviously, $I(\omega)2\pi = I(\nu)\,, \omega = 2\pi\nu$. ◂

Example 7.1 (Periodogram for the number of sun spots)
The time series shown in Figure 7.2 is often seen in the literature. It shows a series of sun spots. An obvious periodicity corresponding to about 11 years (or 22 half years) is seen in the plot. The periodogram for the series in Figure 7.2 is shown in Figure 7.3 on the following page. The figure shows that there presumably exist several periodicities other than the 11 year cycle.

Due to the large variability in the estimates of the periodogram, it is difficult to determine which of the values are actual cycles. In order to calculate a 90% confidence interval for some of the peak values in Figure 7.3, we apply Theorem 7.2. Since $\chi^2(2)_{0.05} = 0.103$ and $\chi^2(2)_{0.95} = 5.99$, we obtain the confidence interval shown in Table 7.1 on the next page.

The large confidence intervals are a result of the large variability of the χ^2-distributed stochastic variable having only 2 degrees of freedom. Some

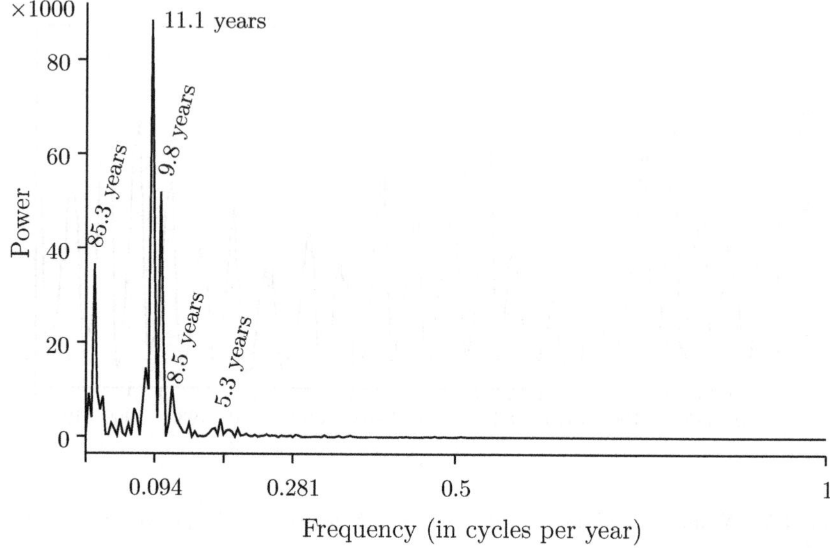

Figure 7.3: *The periodogram for the sun spots series.*

Table 7.1: *90% confidence intervals for $f(\nu)$ corresponding to the peak values of the sun spot.*

Frequency (cycle/year)	Period (year)	I(ν) $\times 10^4$	90% confidence-interval
0.012	85.3	3.66	$[1.2, 71.1]$
0.090	11.1	8.84	$[2.9, 171.6]$
0.102	9.8	5.19	$[1.7, 100.8]$
0.117	8.5	1.07	$[0.4, 20.8]$

methods to improve on the number of degrees of freedom will be given in the following section. The method consists of averaging the estimate of the periodogram with neighbor frequencies for which $f(\omega)$ (or $f(\nu)$) do not vary too strongly.

7.2 Consistent estimates of the spectrum

The previous section illustrated that the periodogram $I_N(\omega)$ is not a consistent estimator for the spectrum $f(\omega)$ because $I_N(\omega)$ contains "too many" values of the estimated autocorrelation function. In this section we will consider consistent estimates of the spectrum. Therefore, the variance will be decreased, but at the cost of an increase in bias (non-centrality).

7.2.1 The truncated periodogram

The problem with the periodogram, (7.1) on page 188, is that it contains too many values of the estimated autocovariance function. Thus, it is obvious to apply the *truncated periodogram*

$$\widehat{f}(\omega) = \frac{1}{2\pi} \sum_{k=-M}^{M} \mathrm{C}(k) \mathrm{e}^{-i\omega k}, \quad M < (N-1), \tag{7.16}$$

where M is the *truncation point.* The truncated periodogram is a linear combination of $M+1$ values of $\mathrm{C}(k)$, and thus (Priestley 1981, p. 433)

$$\mathrm{Var}[\widehat{f}(\omega)] = O(M/N). \tag{7.17}$$

At the same time we have

$$\begin{aligned} \mathrm{E}[\widehat{f}(\omega)] &= \frac{1}{2\pi} \sum_{k=-M}^{M} \mathrm{E}[\mathrm{C}(k)] \mathrm{e}^{i\omega k} \\ &= \frac{1}{2\pi} \sum_{k=-M}^{M} \left(1 - \frac{|k|}{N}\right) \gamma(k) \mathrm{e}^{-i\omega k}, \quad \text{for } N \to \infty \\ &= f(\omega), \quad \text{for } M \to \infty, \end{aligned} \tag{7.18}$$

i.e., $\widehat{f}(\omega)$ is an asymptotically unbiased estimator for $f(\omega)$.

From (7.17) it follows that if $N \to \infty$ and $M \to \infty$ in such a way that $M/N \to 0$ for $N \to \infty$, the variance of $\widehat{f}(\omega)$ will go to zero for $N \to \infty$. Thus for any value of ω, we have that *$\widehat{f}(\omega)$ is a consistent estimator for $f(\omega)$*. It is easy to find values of M which fulfill the above statement. An example is to choose $M = pN$, or more general $M = N^\alpha$, $(0 < \alpha < 1)$.

The estimate of $\widehat{f}(\omega)$, determined by (7.16), is a special case of the more general estimate

$$\widehat{f}(\omega) = \frac{1}{2\pi} \sum_{k=-(N-1)}^{N-1} \lambda_k \, \mathrm{C}(k) \mathrm{e}^{-i\omega k}, \tag{7.19}$$

since (7.16) yields the *rectangular lag-window*

$$\lambda_k = \begin{cases} 1 & |k| \le M \\ 0 & |k| > M \end{cases} \tag{7.20}$$

The window can also be formulated in the frequency domain and is then called the *spectral window.* The spectral window is obtained by taking the Fourier transform of the time window, i.e., (cf. Appendix C on page 361)

$$W(\theta) = \frac{1}{2\pi} \sum_{k=-M}^{M} \mathrm{e}^{-i\theta k} = \frac{1}{2\pi} \left\{ \frac{\sin((M+\frac{1}{2})\theta)}{\sin(\theta/2)} \right\} = D_M(\theta), \tag{7.21}$$

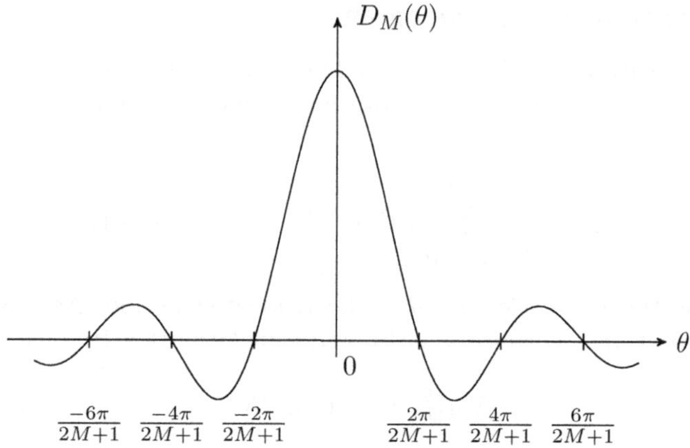

Figure 7.4: *Dirichlet kernel. The spectral window for the truncated periodogram.*

where $D_M(\theta)$ is the so-called *Dirichlet kernel* of order M, see Figure 7.4.

It should be noted that $D_M(\theta)$ has a peak in zero as well as some additional minor peaks. Furthermore, it is seen that $D_M(\theta)$ may take negative values. Since multiplication with the lag-window in (7.19) corresponds to folding in the frequency domain (see next section), the negative values imply that $\widehat{f}(\omega)$ can become negative. This is obviously not desirable because $f(\omega)$ is a non-negative function.

7.2.2 Lag- and spectral windows

In this section we will consider both *lag-windows* (time domain) and *spectral windows* (frequency domain).

A lag-window is identified with a sequence of $\{\lambda_k\}$, which fulfills

i) $\lambda_0 = 1$

ii) $\lambda_k = \lambda_{-k}$

iii) $\lambda_k = 0\,, |k| > M$,

where M is the *truncation point.*

Corresponding to a lag-window $\{\lambda_k\}$ we have the *smoothed spectrum*

$$\widehat{f}(\omega) = \frac{1}{2\pi} \sum_{k=-(N-1)}^{N-1} \lambda_k \, \mathrm{C}(k) \mathrm{e}^{-i\omega k}. \tag{7.22}$$

Furthermore we introduce the *spectral window* corresponding to $\{\lambda_k\}$ by

$$W(\theta) = \frac{1}{2\pi} \sum_{k=-(N-1)}^{N-1} \lambda_k \mathrm{e}^{-i\theta k}, \tag{7.23}$$

i.e., as the Fourier transform of $\{\lambda_k\}$.

Substituting (7.5) from page 188 for $C(k)$ in (7.22) yields

$$\begin{aligned}\widehat{f}(\omega) &= \frac{1}{2\pi}\sum_{k=-(N-1)}^{N-1}\lambda_k\int_{-\pi}^{\pi}I_N(\theta)e^{ik\theta}\,d\theta e^{-i\omega k}\\ &= \int_{-\pi}^{\pi}I_N(\theta)\left\{\frac{1}{2\pi}\sum_{k=-(N-1)}^{N-1}\lambda_k e^{-ik(\omega-\theta)}\right\}d\theta,\end{aligned}$$

or

$$\widehat{f}(\omega) = \int_{-\pi}^{\pi}I_N(\theta)W(\omega-\theta)\,d\theta. \tag{7.24}$$

Thus, the smoothed spectrum can be obtained either by *applying weights to the estimated autocovariance function* (7.22) or as a *convolution*, (7.24), which implies that $f(\omega)$ corresponds to a "locally" weighted mean of the periodogram in a neighbourhood of ω.

Some of the most commonly used lag-windows are given in Table 7.2 on page 199. For practical applications these lag-windows yield more or less the same result, perhaps with the exception of the truncated periodogram. The spectral windows are shown in Figure 7.5 on the following page.

The Daniell window has a rectangular spectral window.

$$W(\theta) = \begin{cases} M/2\pi & -\pi/M \le \theta \le \pi/M \\ 0 & \text{otherwise}\end{cases} \tag{7.25}$$

The corresponding lag-window is obtained by an inverse Fourier transform of (7.25)

$$\lambda_k = \frac{M}{2\pi}\int_{-\pi/M}^{\pi/M}e^{ik\theta}\,d\theta = \frac{\sin(\pi k/M)}{(\pi k/M)}. \tag{7.26}$$

The asymptotical variances for the different lag-windows are given in the last column of Table 7.2. It is seen that if a *small variance is desired, then M must be small.* The consequence is an increase in bias (non-centrality) for the estimators.

The *skewness* $b(\omega)$ for the windows, disregarding the truncated periodogram and the Bartlett-window, is given in Table 7.2.

$$b(\omega) = E[\widehat{f}(\omega)] - f(\omega) \simeq \frac{c}{M^2}f''(\omega) \tag{7.27}$$

where the value of c is given in Table 7.2. It is seen that for values of ω where $f(\omega)$ has a (possibly local) maximum, $\widehat{f}(\omega)$ tends to underestimate $f(\omega)$, since $f''(\omega)$ is negative (and opposite for values of ω where $f(\omega)$ has a [possible local] minimum).

It is also seen that if *a small bias is desired, then M should be large.*

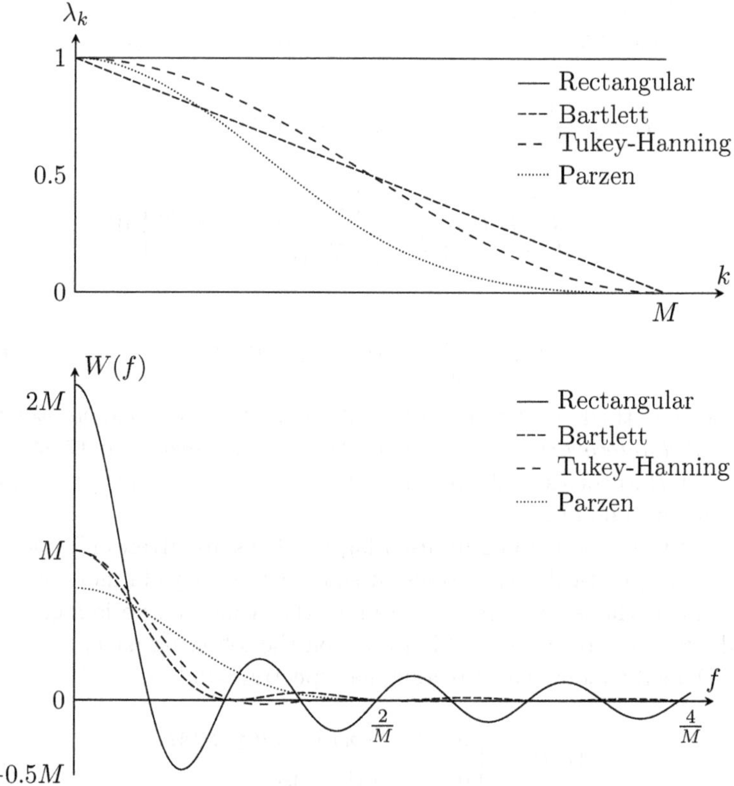

Figure 7.5: *Some lag- and spectral windows.*

Example 7.2 (The general Tukey window)

This example illustrates the relation between the truncated periodogram and an important class of windows, given by the *general Tukey window.* In the time domain the latter can be written as

$$\lambda_k = \begin{cases} 1 - 2a + 2a\cos\left(\dfrac{\pi k}{M}\right) & |k| \leq M \\ 0 & |k| > M, \end{cases} \tag{7.28}$$

where $a \leq 1/4$ ensures that $\lambda_k \geq 0$ for all values of k.

The general Tukey window and the truncation point M yield the spectral estimate

$$\widehat{f}(\omega) = \frac{1}{2\pi} \sum_{k=-M}^{M} \left(1 - 2a + 2a\cos\left(\frac{\pi k}{M}\right)\right) \mathrm{C}(k) e^{-i\omega k} \tag{7.29}$$

Table 7.2: *Commonly used lag-windows. The last column displays the asymptotic variances (for $\omega \neq 0, \pm\pi$). For $\omega = 0, \pm\pi$ the values must be doubled. These values are in parentheses—see (7.27) on page 197.*

Estimator	*Lag-window*	$N\,\mathrm{Var}[\widehat{f}(\omega)]/\left(Mf^2(\omega)\right)$
Truncated periodogram (Rectangular $\{\lambda_k\}$)	$\lambda_k = \begin{cases} 1 & \lvert k\rvert \le M \\ 0 & \lvert k\rvert > M \end{cases}$	2
Bartlett	$\lambda_k = \begin{cases} 1 - \lvert k\rvert/M & \lvert k\rvert \le M \\ 0 & \lvert k\rvert > M \end{cases}$	2/3
Daniell (Rectangular $W(\theta)$)	$\lambda_k = \dfrac{\sin(\pi k/M)}{(\pi k/M)}$	1 $(c = \pi^2/6)$
Tukey-Hanning	$\lambda_k = \begin{cases} \frac{1}{2}\left[1 + \cos\left(\frac{\pi k}{M}\right)\right] & \lvert k\rvert \le M \\ 0 & \lvert k\rvert > M \end{cases}$	3/4 $(c = \pi^2/4)$
Tukey-Hamming	$\lambda_k = \begin{cases} 0.54 + 0.46\cos(\pi k/M) & \lvert k\rvert \le M \\ 0 & \lvert k\rvert > M \end{cases}$	0.7948 $(c = 0.23\pi^2)$
Parzen	$\lambda_k = \begin{cases} 1 - 6\left(\frac{k}{M}\right)^2 + 6\left(\frac{\lvert k\rvert}{M}\right)^3 & \lvert k\rvert \le \frac{M}{2} \\ 2\left(1 - \frac{\lvert k\rvert}{M}\right)^3 & \frac{M}{2} \le \lvert k\rvert \le M \\ 0 & \lvert k\rvert > M \end{cases}$	0.5393 $(c = 6)$

or

$$\begin{aligned}\widehat{f}(\omega) &= \frac{1}{2\pi}\Bigg(\sum_{k=-M}^{M}(1-2a)\,\mathrm{C}(k)\mathrm{e}^{-i\omega k} \\ &\qquad + \sum_{k=-M}^{M} a(\mathrm{e}^{i\frac{\pi k}{M}} + \mathrm{e}^{-i\frac{\pi k}{M}})\,\mathrm{C}(k)\mathrm{e}^{-i\omega k}\Bigg) \\ &= \frac{1}{2\pi}\Bigg(a\sum_{k=-M}^{M}\mathrm{C}(k)\mathrm{e}^{-i(\omega k-\frac{\pi}{M}k)} \\ &\qquad + (1-2a)\sum_{k=-M}^{M}\mathrm{C}(k)\mathrm{e}^{-i\omega k} \\ &\qquad + a\sum_{k=-M}^{M}\mathrm{C}(k)\mathrm{e}^{-i(\omega k+\frac{\pi}{M}k)}\Bigg) \\ &= a\widehat{f}_1\left(\omega-\frac{\pi}{M}\right) + (1-2a)\widehat{f}_1(\omega) + a\widehat{f}_1\left(\omega+\frac{\pi}{M}\right),\end{aligned} \tag{7.30}$$

where $\widehat{f}_1$ is the truncated periodogram defined by (7.16) on page 195. This illustrates that the spectrum can be determined either by applying a lag-window in the time domain (7.29) or by a locally weighted mean of the truncated periodogram (7.30).

▸ **Remark 7.3**
If we select $a = 1/4$ for the general Tukey window we get the *Tukey-Hanning window*, and selecting $a = 0.23$ yields the *Tukey-Hamming window.* ◂

7.2.3 Approximative distributions for spectral estimates

From Theorem 7.1 on page 190 we have that the estimates of the periodogram are $\chi^2(2)$-distributed. In the following theorem, the corresponding results for the smoothed spectrum are given.

THEOREM 7.3 (PROPERTIES FOR A SMOOTHED SPECTRUM)
Let there be given a smoothed spectrum

$$\widehat{f}(\omega) = \frac{1}{2\pi}\sum_{k=-M}^{M}\lambda_k\,\mathrm{C}(k)\mathrm{e}^{-i\omega k}.$$

It holds that

$$\frac{\nu\widehat{f}(\omega)}{f(\omega)} \in_{\text{approx.}} \chi^2(\nu), \tag{7.31}$$

where

$$\nu = \frac{2N}{\sum_{k=-M}^{M}\lambda_k^2}. \tag{7.32}$$

Table 7.3: *Equivalent degrees of freedom for some windows.*

Estimation	Equivalent degrees of freedom (ν)
Truncated periodogram	N/M
Bartlett	$3N/M$
Daniell	$2N/M$
Tukey-Hanning	$8N/3M$
Tukey-Hamming	$2.516N/M$
Parzen	$3.709N/M$

Here, ν is called the equivalent degrees of freedom *(see Table 7.3).*

Proof Omitted—see, e.g., Priestley (1981). ∎

A $100(1-\alpha)\%$ confidence interval is obtained by

$$\mathrm{P}\left\{\chi^2(\nu)_{\alpha/2} \le \frac{\nu\widehat{f}(\omega)}{f(\omega)} \le \chi^2(\nu)_{1-\alpha/2}\right\} \simeq 1-\alpha,$$

or

$$\mathrm{P}\left\{\frac{\nu\widehat{f}(\omega)}{\chi^2(\nu)_{1-\alpha/2}} \le f(\omega) \le \frac{\nu\widehat{f}(\omega)}{\chi^2(\nu)_{\alpha/2}}\right\} \simeq 1-\alpha, \tag{7.33}$$

or

$$\begin{aligned}\mathrm{P}\Bigg\{\log\widehat{f}(\omega) + \log\left(\frac{\nu}{\chi^2(\nu)_{1-\alpha/2}}\right) &\le \log f(\omega)\\ &\le \log\widehat{f}(\omega) + \log\left(\frac{\nu}{\chi^2(\nu)_{\alpha/2}}\right)\Bigg\} \simeq 1-\alpha.\end{aligned} \tag{7.34}$$

By plotting the spectrum on a logarithmic scale, we achieve that adding and subtracting a constant to the smoothed spectrum yields the confidence intervals—see also Exercise 7.4 on page 212.

Example 7.3 (Variance/bias for the general Tukey window)
The general Tukey window is defined in Example 7.2 on page 198. From (7.31) it follows that (approximately)

$$\mathrm{Var}\left[\frac{\nu\widehat{f}(\omega)}{f(\omega)}\right] = 2\nu,$$

which implies that the variance proportion (between bias and variance) becomes

$$\mathrm{Var}\left[\frac{\widehat{f}(\omega)}{f(\omega)}\right] = 2/\nu \tag{7.35}$$

The degree of freedom ν is calculated from (7.32). We get

$$\begin{aligned}\sum_{k=-M}^{M} \lambda_k^2 &= \sum_{k=-M}^{M} \left(1 - 2a + 2a\cos\left(\frac{\pi k}{M}\right)\right)^2 \\ &= \sum_{k=-M}^{M} \left((1-2a)^2 + 4a^2\cos^2\left(\frac{\pi k}{M}\right)\right. \\ &\qquad\qquad \left. + (1-2a)4a\cos\left(\frac{\pi k}{M}\right)\right) \\ &= (1-2a)^2(2M+1) \\ &\quad + 4a^2\left[1 + 2\sum_{k=1}^{M} \frac{\cos(\frac{2\pi k}{M}) + 1}{2}\right] \\ &\quad + (1-2a)4a\left(-1 + 2\sum_{k=0}^{M}\cos\left(\frac{\pi k}{M}\right)\right) \\ &= (1-2a)^2(2M+1) + 4a^2(1+M) + (1-2a)4a(-1) \\ &= (2 + 12a^2 - 8a)M + (1 + 16a^2 - 8a),\end{aligned} \tag{7.36}$$

and the variance proportion becomes

$$\mathrm{Var}\left[\frac{\widehat{f}(\omega)}{f(\omega)}\right] = \sum_{k=-M}^{M} \frac{\lambda_k^2}{N} = \frac{(2+12a^2-8a)M + (1+16a^2-8a)}{N}.$$

The Tukey-Hanning window is obtained for $a = 1/4$, and the variance proportion for this window becomes

$$\mathrm{Var}\left[\frac{\widehat{f}(\omega)}{f(\omega)}\right] = \frac{3M}{4N}.$$

The equivalent degrees of freedom for the Tukey-Hanning window is $8N/3M$ (calculated from (7.32) and (7.36)), which is the same value as listed in Table 7.3 on the previous page.

Example 7.4 (Retaining the variance proportion)
In a given case we have applied a Tukey-Hanning window using the truncation point $M_T = 40$. We decide to apply a Parzen window instead and want to determine the truncation point so that the variance proportion remains the same. We denote the truncation point for the Parzen and the Tukey-Hanning window, M_P and M_T, respectively. The claim for equal variance proportion implies the following relation between M_P and M_T:

$$\frac{\frac{2}{3.709}M_P}{N} = \frac{\frac{2}{8/3}M_T}{N},$$

and the claim is hereby obtained for $M_P \simeq 56$.

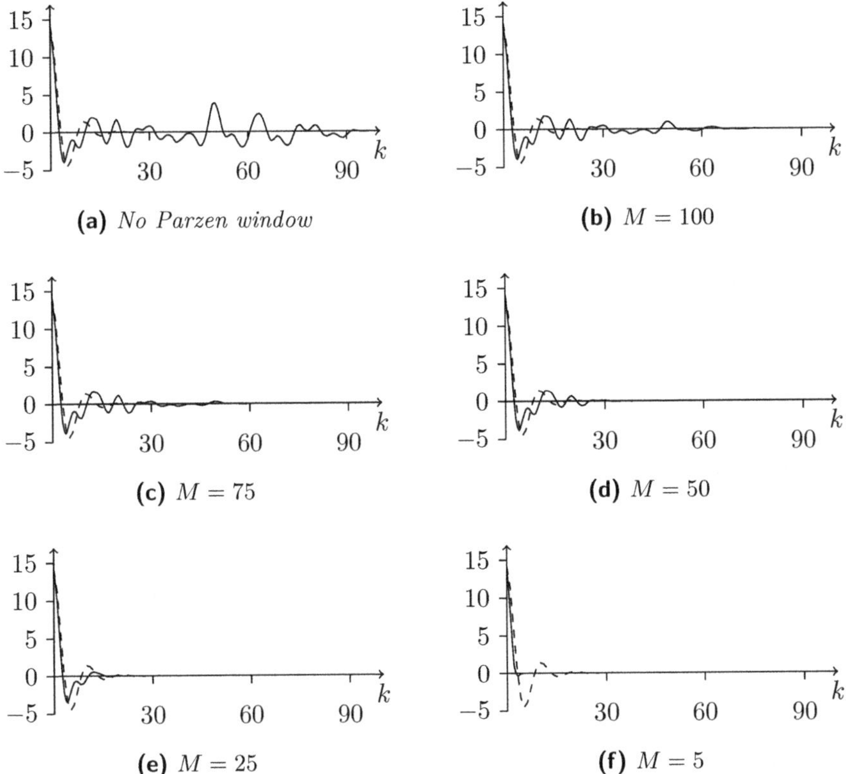

(a) *No Parzen window* (b) $M = 100$

(c) $M = 75$ (d) $M = 50$

(e) $M = 25$ (f) $M = 5$

Figure 7.6: *The estimated and truncated autocovariance function for various values of the truncation point M. A Parzen window is used.*

Example 7.5 (Estimation of spectra)
This example originates from Olbjer, Holst, and Holst (2005). We consider the process $Y_t - 1.31Y_{t-1} + 0.64Y_{t-2} = \varepsilon_t$. In Figure 7.6 the theoretical autocovariance function, the estimated autocovariance function, and the estimated autocovariance function multiplied by a Parzen window having different values for M (number of observations = 100), are shown.

The periodogram (which corresponds to Figure 7.1 on page 192 apart from a different realization) and the smoothed spectrum corresponding to the Parzen windows in Figure 7.6 (but for higher values of M) are shown in Figure 7.7 on the following page.

7.3 The cross-spectrum

The cross-spectrum is used to described the *relations between two processes in the frequency domain* and corresponds to the cross-covariance function in

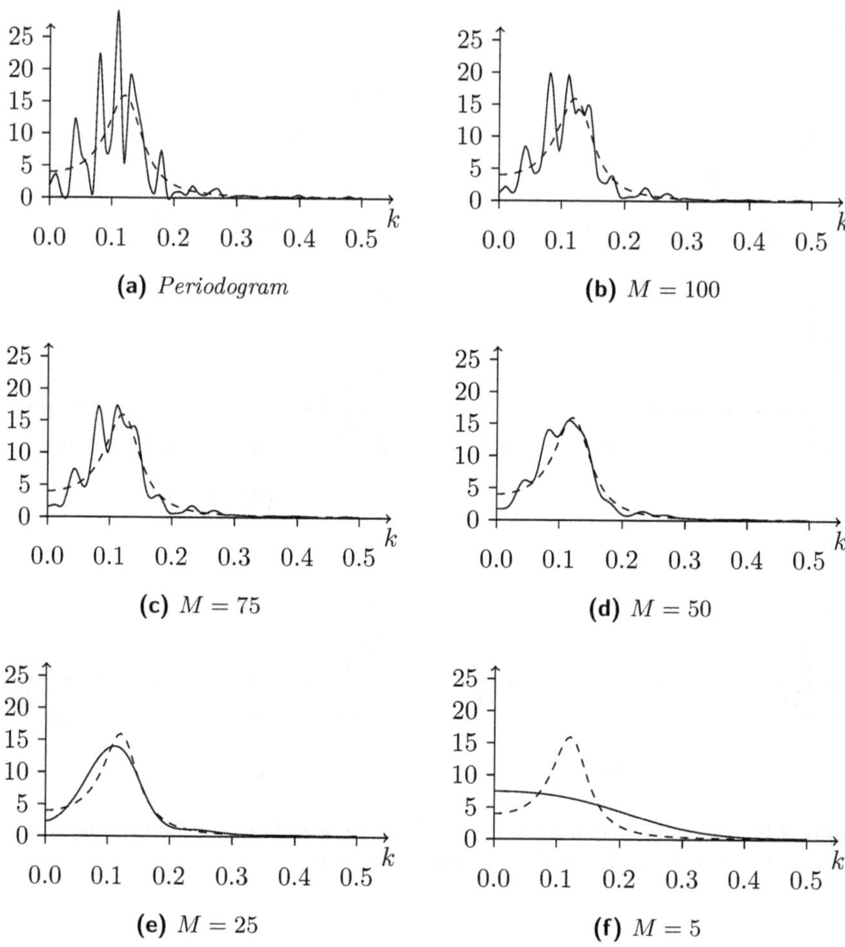

Figure 7.7: *Estimation of the spectrum: (a): The periodogram. (b–f): Smoothed spectrum (Parzen window).*

the time domain. In this section we consider the cross-spectrum for a given stochastic process, while the estimation of the cross-spectrum is discussed in Section 7.4.

We consider the two stochastic processes $\{X_t\}$ and $\{Y_t\}$ for $t = 0, \pm 1, \pm 2, \ldots$. Since the processes are observed at the same time instances, we have a *bivariate stochastic process*, $\{\boldsymbol{Z}_t\} = \{(X_t, Y_t)^T\}$. If the bivariate process is stationary, then the properties of the covariance for the bivariate process are described by a sequence of matrices, where the matrix at lag k is given by

$$\boldsymbol{\Gamma}_k = \begin{pmatrix} \gamma_{XX}(k) & \gamma_{XY}(k) \\ \gamma_{YX}(k) & \gamma_{YY}(k) \end{pmatrix} \tag{7.37}$$

$\boldsymbol{\Gamma}_k$ is referred to as the *covariance matrix* at lag k.

The cross-spectrum is defined as the Fourier transform of the cross-covariance function, i.e.,

$$f_{XY}(\omega) = \frac{1}{2\pi} \sum_{k=-\infty}^{\infty} \gamma_{XY}(k) \mathrm{e}^{-i\omega k}, \tag{7.38}$$

where $\omega \in [-\pi, \pi]$. Note that in general $\gamma_{XY}(k)$ is *not* an even function and, thus, $f_{XY}(\omega)$ is *a complex function.* The inverse relation is

$$\gamma_{XY}(k) = \int_{-\pi}^{\pi} \mathrm{e}^{i\omega k} f_{XY}(\omega)\,\mathrm{d}\omega. \tag{7.39}$$

The usual *Fourier transformation pair* is given by Equations (7.38) and (7.39). Alternative formulations are sometimes used in the literature (the factor 2π and the interval considered for ω). Please note, that $f_{XY}(-\omega)$ is the complex conjugate of $f_{XY}(\omega)$.

The spectral properties of the bivariate process are described by the *spectral matrix*:

$$\boldsymbol{f}(\omega) = \begin{pmatrix} f_{XX}(\omega) & f_{XY}(\omega) \\ f_{YX}(\omega) & f_{YY}(\omega) \end{pmatrix} \tag{7.40}$$

where $f_{XX}(\omega)$ and $f_{YY}(\omega)$ are the spectra for $\{X_t\}$ and $\{Y_t\}$, respectively. Since $\gamma_{XY}(k) = \gamma_{YX}(-k)$, we have

$$f_{YX}(\omega) = \frac{1}{2\pi} \sum_{k=-\infty}^{\infty} \gamma_{XY}(-k) \mathrm{e}^{-i\omega k} = f^*_{XY}(\omega), \tag{7.41}$$

where f^* is the complex conjugate of f. The previous results can be generalized to *multivariate processes.* Consider the m-dimensional process $\{X_{1,t}, X_{2,t}, \ldots, X_{m,t}\}$, having the covariance matrix at lag k

$$\boldsymbol{\Gamma}_k = \{\gamma_{ij}(k)\}, \quad i, j = 1, \ldots, m. \tag{7.42}$$

Here, the Fourier transform can be formulated in terms of matrices

$$\boldsymbol{f}(\omega) = \frac{1}{2\pi} \sum_{k=-\infty}^{\infty} \boldsymbol{\Gamma}_k \mathrm{e}^{-ik\omega} \tag{7.43}$$

$$\boldsymbol{\Gamma}_k = \int_{-\pi}^{\pi} \mathrm{e}^{ik\omega} \boldsymbol{f}(\omega)\,\mathrm{d}\omega, \tag{7.44}$$

where $\boldsymbol{f}(\omega) = \{f_{ij}(\omega)\}$, for $\omega \in [-\pi, \pi]$.

7.3.1 The co-spectrum and the quadrature spectrum

Since $\gamma_{XY}(k)$ is generally not an even function, the corresponding cross-spectrum becomes a complex function

$$\begin{aligned} f_{XY}(\omega) &= \frac{1}{2\pi}\sum_{k=-\infty}^{\infty}\gamma_{XY}(k)\mathrm{e}^{-i\omega k} \\ &= \frac{1}{2\pi}\sum_{k=-\infty}^{\infty}\gamma_{XY}(k)\cos(\omega k) - i\frac{1}{2\pi}\sum_{k=-\infty}^{\infty}\gamma_{XY}(k)\sin(\omega k) \\ &= c_{XY}(\omega) - iq_{XY}(\omega), \end{aligned} \tag{7.45}$$

where $c_{XY}(\omega)$ and $-q_{XY}(\omega)$ are the real and complex part of $f_{XY}(\omega)$.

Here, $c_{XY}(\omega)$ is called the *co-spectrum*, whereas the complex part (with opposite sign) $q_{XY}(\omega)$, is called the *quadrature spectrum*. Please note that $c_{XY}(\omega)$ is an even function while $q_{XY}(\omega)$ is an odd function. Thus, common practice is to plot the spectrum only in the interval $[0, \pi]$.

7.3.2 Cross-amplitude spectrum, phase spectrum, coherence spectrum, gain spectrum

The polar form of $f_{XY}(\omega)$ is determined as follows:

$$f_{XY}(\omega) = \alpha_{XY}(\omega)\mathrm{e}^{i\phi_{XY}(\omega)}, \tag{7.46}$$

where

$$\alpha_{XY}(\omega) = |f_{XY}(\omega)| = \sqrt{c_{XY}^2(\omega) + q_{XY}^2(\omega)}, \tag{7.47}$$

and

$$\phi_{XY}(\omega) = \arctan[-q_{XY}(\omega)/c_{XY}(\omega)]. \tag{7.48}$$

The function $\alpha_{XY}(\omega)$ is called the *cross-amplitude spectrum*, and $\phi_{XY}(\omega)$ is called the *phase spectrum*. It follows from (7.48) that $\phi_{XY}(\omega)$ is undetermined from an additive multiple of 2π, but usually we define $\phi_{XY}(\omega)$ in the interval $[-\pi, \pi]$ and the value is unique by defining $\phi_{XY}(\omega)$ as the angle (in the interval $[-\pi, \pi]$) between the positive region of the c_{XY} axis and the line that connects zero with the point $(c_{XY}, -q_{XY})$.

The cross-amplitude spectrum, $\alpha_{XY}(\omega)$, may be interpreted as the *mean product of the amplitudes* of the components at the frequency ω in $\{X_t\}$ and $\{Y_t\}$. The phase spectrum may be interpreted as the average *change in phase* between the same components.

Furthermore, the *complex coherency* is defined by

$$w_{XY}(\omega) = \frac{f_{XY}(\omega)}{\sqrt{f_{XX}(\omega)f_{YY}(\omega)}}. \tag{7.49}$$

The *coherency* (at the frequency ω) or the *coherence spectrum* is defined as $|w_{XY}(\omega)|$. (Some authors refer to $|w_{XY}(\omega)|^2$ as the coherency.) It is also possible to interpret $|w_{XY}(\omega)|$ as the correlation coefficient between the stochastic components at the frequency ω in $\{X_t\}$ and $\{Y_t\}$, and thereby we have that

$$0 \leq |w_{XY}(\omega)| \leq 1. \tag{7.50}$$

As for the usual correlation coefficient, the value of $|w_{XY}(\omega)|$ is a measure for how strongly the components are linearly related at the frequency ω.

Finally, we define the *gain spectrum* by

$$G_{XY}(\omega) = \frac{|f_{XY}(\omega)|}{f_X(\omega)} = \frac{\alpha_{XY}(\omega)}{f_X(\omega)}, \tag{7.51}$$

which is the proportion between the cross-amplitude spectrum and the spectrum for the input process. The gain spectrum has already been described in Chapter 4 and will be further discussed in Chapter 8, which considers linear stochastic systems.

Here G_{XY} is the *gain* in the system from the input $\{X_t\}$ to the output $\{Y_t\}$, and it is shown that

$$G_{XY}(\omega) = |\mathcal{H}_{XY}(\omega)|, \tag{7.52}$$

i.e., $G_{XY}(\omega)$ *equals the numerical value of the frequency response function*; compare with (4.24) on page 74 in Chapter 4.

Example 7.6 (A time delay with noise)
Consider the bivariate process defined by

$$\begin{aligned} X_t &= \varepsilon_{1,t} \\ Y_t &= X_{t-d} + \varepsilon_{2,t}\,, \qquad (d \in \mathbb{N}_0) \end{aligned}$$

where $\{\varepsilon_{1,t}\}$ and $\{\varepsilon_{2,t}\}$ are mutually independent white noise processes with mean value 0 and variance σ_ε^2.

The cross-covariance function is

$$\begin{aligned} \gamma_{XY}(k) &= \text{Cov}[X_t, Y_{t+k}] \\ &= \text{Cov}[\varepsilon_{1,t}, \varepsilon_{1,t-d+k} + \varepsilon_{2,t+k}] \\ &= \begin{cases} \sigma_\varepsilon^2 & \text{for } k = d \\ 0 & \text{otherwise} \end{cases} \end{aligned}$$

and, hereby, the *cross-spectrum* becomes

$$\begin{aligned} f_{XY}(\omega) &= \frac{1}{2\pi} \sum_{k=-\infty}^{\infty} \gamma_{XY}(k) \mathrm{e}^{-i\omega k} \\ &= \frac{1}{2\pi} \sigma_\varepsilon^2 \mathrm{e}^{-i\omega d} \\ &= \frac{1}{2\pi} \sigma_\varepsilon^2 \cos(\omega d) - i \frac{1}{2\pi} \sigma_\varepsilon^2 \sin(\omega d) \end{aligned}$$

i.e., we obtain

Co-spectrum: $$c_{XY}(\omega) = \frac{1}{2\pi} \sigma_\varepsilon^2 \cos(\omega d)$$

Quadrature spectrum: $$q_{XY}(\omega) = \frac{1}{2\pi} \sigma_\varepsilon^2 \sin(\omega d)$$

Furthermore, we obtain

Cross-amplitude spectrum: $$\alpha_{XY}(\omega) = \frac{1}{2\pi} \sigma_\varepsilon^2$$

Phase spectrum: $$\phi_{XY}(\omega) = \tan^{-1}\left(\frac{-q_{XY}(\omega)}{c_{XY}(\omega)} \right) = -\omega d$$

In order to determine the coherence spectrum and the gain, the spectrum for $\{X_t\}$ and $\{Y_t\}$ have to be calculated. It is seen that

$$\gamma_{XX}(k) = \begin{cases} \sigma_\varepsilon^2 & \text{for } k = 0 \\ 0 & \text{otherwise} \end{cases} \qquad \gamma_{YY}(k) = \begin{cases} 2\sigma_\varepsilon^2 & \text{for } k = 0 \\ 0 & \text{otherwise,} \end{cases}$$

and thus,

$$f_X(\omega) = \frac{1}{2\pi} \sum_{k=-\infty}^{\infty} \gamma_{XX}(k) \mathrm{e}^{-i\omega k} = \frac{\sigma_\varepsilon^2}{2\pi} \qquad f_Y(\omega) = \frac{\sigma_\varepsilon^2}{\pi}$$

Thereby we get the *coherence spectrum*

$$|w_{XY}(\omega)| = \frac{\frac{1}{2\pi}\sigma_\varepsilon^2}{\sqrt{(\sigma_\varepsilon^2/2\pi)(\sigma_\varepsilon^2/\pi)}} = \frac{\sqrt{2}}{2}$$

and the *gain*

$$G_{XY}(\omega) = \frac{\frac{1}{2\pi}\sigma_\varepsilon^2}{\frac{1}{2\pi}\sigma_\varepsilon^2} = 1$$

The coherency reveals that there is not a total linear dependency and this is due to the noise component $\{\varepsilon_{2,t}\}$. The gain shows that all input frequencies pass undisturbed through the system. Compare with Example 4.5 on page 77.

7.4 Estimation of the cross-spectrum

Consider N observations of the bivariate process

$$\left\{ \begin{pmatrix} X_t \\ Y_t \end{pmatrix} \right\}, \quad t = 1, 2, \ldots, N.$$

As discussed in Section 7.2 on page 194, a consistent estimate of the spectrum (the smoothed spectrum) is obtained either by a weighted estimate of the autocovariance function or as a weighted integral of the periodogram. Both procedures can be applied in estimating the cross-spectrum.

The first method is based on the *estimated cross-covariance function*, C_{XY}, where the estimate of the cross-spectrum is found by ($M < N$)

$$\widehat{f}_{XY}(\omega) = \frac{1}{2\pi} \sum_{k=-M}^{M} \lambda_k \, \mathrm{C}_{XY}(k) e^{-i\omega k}, \quad \omega \in [-\pi, \pi], \tag{7.53}$$

or

$$\begin{aligned} \widehat{f}_{XY}(\omega) &= \frac{1}{2\pi} \sum_{k=-M}^{M} \lambda_k \, \mathrm{C}_{XY}(k) \cos(\omega k) \\ &\quad - i \frac{1}{2\pi} \sum_{k=-M}^{M} \lambda_k \, \mathrm{C}_{XY}(k) \sin(\omega k) &(7.54) \\ &= \widehat{c}_{XY}(\omega) - i \widehat{q}_{XY}(\omega). &(7.55) \end{aligned}$$

The lag-window $\{\lambda_k\}$ and truncation point M are chosen as for the (auto-)spectrum; cf. Section 7.2. From (7.54) and (7.55) it is clear how the co-spectrum, $c_{XY}(\omega)$, and the quadrature spectrum, $q_{XY}(\omega)$, are estimated.

Based on $\widehat{c}_{XY}(\omega)$ and $\widehat{q}_{XY}(\omega)$ the estimates of the cross-amplitude and phase spectrum are found by using the theoretical relations

$$\widehat{\alpha}_{XY}(\omega) = \sqrt{\widehat{c}^2_{XY}(\omega) + \widehat{q}^2_{XY}(\omega)}, \tag{7.56}$$

and

$$\tan \widehat{\phi}_{XY}(\omega) = \frac{-\widehat{q}_{XY}(\omega)}{\widehat{c}_{XY}(\omega)}. \tag{7.57}$$

In order to estimate the coherency spectrum, we need an estimate of the (auto-)spectrum, since

$$\widehat{w}_{XY}(\omega) = \frac{\widehat{f}_{XY}(\omega)}{\sqrt{\widehat{f}_{XY}(\omega) \widehat{f}_{Y}(\omega)}}. \tag{7.58}$$

The gain can be estimated by

$$\widehat{G}_{XY}(\omega) = \frac{|\widehat{f}_{XY}(\omega)|}{\widehat{f}_{X}(\omega)}. \tag{7.59}$$

A lag-window, and especially the truncation point, should ideally be chosen so that it "matches" the decay in the covariance function. Thus, it is not mandatory that the same truncation point is used in the estimation of the cross-spectrum as in the estimation of the (auto-)spectrum.

The estimate of the cross-spectrum will be biased if the maximum of the cross-covariance function is not close to lag 0. This bias may be removed using a *time alignment* of one of the time series so that the maximum of the cross-covariance function is at lag 0.

The second method is based on the *cross-periodogram*, which is defined by

$$I_{XY,N}(\omega) = \frac{1}{2\pi} \sum_{k=-(N-1)}^{N-1} C_{XY}(k) e^{-i\omega k}. \tag{7.60}$$

As in Section 7.1 on page 187 the cross-periodogram may be written as

$$I_{XY,N}(\omega) = \frac{1}{2\pi N} \left(\sum_{t=1}^{N} X_t e^{-i\omega t} \right) \left(\sum_{t=1}^{N} Y_t e^{i\omega t} \right) \tag{7.61}$$

$$= \xi_X(\omega) \xi_Y^*(\omega), \tag{7.62}$$

where

$$\xi_X(\omega) = \frac{1}{\sqrt{2\pi N}} \sum_{t=1}^{N} X_t e^{-i\omega t}$$

and $\xi_Y^*(\omega)$ is the complex conjugated of $\xi_Y(\omega)$.

As for the expression in the time domain, (7.53) on the previous page, we can calculate the cross-spectrum by

$$\widehat{f}_{XY}(\omega) = \int_{-\pi}^{\pi} I_{XY,N}(\theta) W_N(\omega - \theta) \, d\theta, \tag{7.63}$$

where $W_N(\theta)$ is the spectral window corresponding to the lag-window $\{\lambda_k\}$; see Section 7.2.2. Now the estimates of the co-spectrum, amplitude spectrum etc., can easily be found as described for the first method.

For a detailed description of the properties of the estimators, see Priestley (1981).

7.5 Problems

Exercise 7.1

Question 1 Consider the moving average

$$X_t = \frac{\varepsilon_t + \varepsilon_{t-1} + \cdots + \varepsilon_{t-(k-1)}}{k}$$

where $\{\varepsilon_t\}$ is assumed to be white noise with mean value 0 and variance σ_ε^2.

Show that the spectral density for $\{X_t\}$ can be written as

$$f_x(\omega) = \frac{\sigma_\varepsilon^2}{2\pi} \frac{\sin^2(k\omega/2)}{k^2 \sin^2(\omega/2)}.$$

The function

$$D_k(\omega) = \frac{\sin(k\omega/2)}{k \sin(\omega/2)}$$

is known as a Dirichlet kernel. This function occurs for example when working with rectangular time windows.

Note that if $\omega \to 0$ and $k \to \infty$ in such a way that $k\omega =$ constant, then:

$$D_k(\omega) \to \frac{\sin(k\omega/2)}{k\omega/2}$$

Question 2 Consider the rectangular window in the time domain

$$g(t) = \begin{cases} a/T & \text{for } |t| \le T \\ 0 & \text{otherwise} \end{cases}$$

Find an expression for this window specified in the frequency domain.

Exercise 7.2

In the time domain the general Tukey window can be written

$$\lambda_k = \begin{cases} 1 - 2a + 2a\cos(\frac{\pi k}{M}) & |k| \le M \\ 0 & |k| > M \end{cases}$$

Question 1 Find the values of a which ensure that $\lambda_k \ge 0$ for all values of k.

Question 2 The truncated periodogram is given by

$$\widehat{f}_1(\omega) = \frac{1}{2\pi} \sum_{k=-M}^{M} c_k \cos(\omega k), \quad \omega = \pi j/M, j = 0, 1, \ldots, M$$

Determine how the spectral estimate can be written as a moving average of values in the truncated periodogram using the general Tukey window.

Notice that the Tukey-Hanning window is obtained for $a = 1/4$ while the Tukey-Hamming window is obtained for $a = 0.23$.

Question 3 The spectrum is estimated as

$$\widehat{f}(\omega) = \frac{1}{2\pi} \sum_{k=-N}^{N} \lambda_k c_k \cos(\omega k)$$

where λ_k is the general Tukey window. Determine the variance relation $\mathrm{Var}[\widehat{f}(\omega)/f(\omega)]$.

Find the variance relation for the Tukey-Hanning window, and compare with Table 7.2 on page 199.

Question 4 In a particular case a Tukey-Hanning window with $M = 40$ has been used. Now a Parzen window is wanted instead. Determine for which value of M the variance relation is unchanged.

Exercise 7.3

For processes in continuous time it is well known that the autocovariance function $\gamma(\tau)$ is a continuous function and the spectral density can be determined using

$$f(\omega) = \frac{1}{2\pi}\int_{-\infty}^{\infty} \gamma(\tau)e^{-i\omega\tau}\,d\tau$$

For continuous time processes the lag-window (the window in the time domain) is a continuous function, $\lambda(\tau)$, and the window in the spectral domain is given by

$$k(\omega) = \frac{1}{2\pi}\int_{-\infty}^{\infty} \lambda(\tau)e^{-i\omega\tau}\,d\tau$$

An example is the Bartlett window:

$$\lambda(\tau) = \begin{cases} 1 - |\tau|/T & |\tau| \leq T \\ 0 & |\tau| > T \end{cases}$$

Let $T = 1$ and find the corresponding spectral window.

Exercise 7.4

Figure 7.8 shows a smoothed estimate of the spectrum of an AR(2) process with $\phi_1 = -1.0$ and $\phi_2 = 0.5$ based on 400 observations (simulated). A Parzen window has been used to smoothen the estimate, and the correlation function is considered until lag 48. Furthermore, the theoretical spectrum is plotted.

Deduce whether the theoretically spectrum is inside an 80% confidence interval. (Hint: The distance between 1 and 2 on the y-axis corresponds to $\log_{10}(2) - \log_{10}(1) \approx 0.3010$ on the logarithmic scale.)

Exercise 7.5

Consider the bivariate process

$$\begin{aligned} X_{1t} &= \varepsilon_{1,t} \\ X_{2t} &= \beta_1 X_{1,t} + \beta_2 X_{1,t-1} + \varepsilon_{2,t} \end{aligned}$$

where $\{\varepsilon_{1,t}\}$ and $\{\varepsilon_{2,t}\}$ are mutually uncorrelated white noise processes with variance σ_1^2 and σ_2^2, respectively.

Question 1 Find the cross-covariance function of the process.

Question 2 Calculate the co-, quadrature-, cross-amplitude, and phase spectral density functions.

Question 3 Sketch the cross-amplitude and phase spectral density for $\beta_2/\beta_1 = 1$ and $\beta_2/\beta_1 = -1$. Comment on the results.

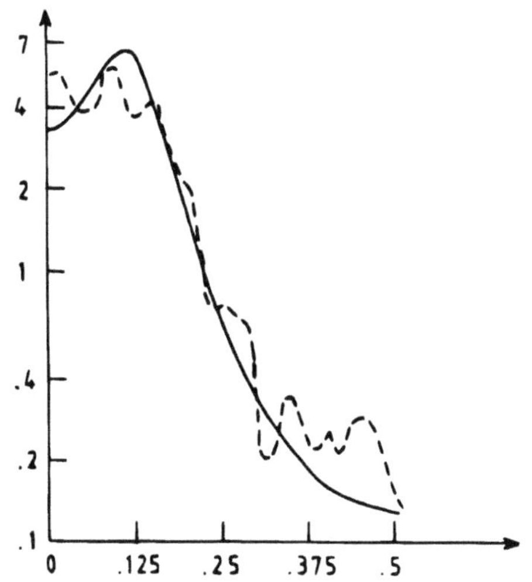

Figure 7.8: *The estimated smoothed spectrum for the AR(2) process is shown with the dashed line. The solid line corresponds to the theoretical spectrum.*

CHAPTER 8

Linear systems and stochastic processes

In Chapter 5 it was shown that a general linear process can be characterized as a linear system having white noise as input. In this chapter, we will consider systems where the input is not necessarily white noise but, in general, a stationary stochastic process. The chapter is based on the theory discussed in Chapter 5 but also on the discussion on linear dynamic systems presented in Chapter 4.

In the first sections we will restrict our attention to single-input, single-output (SISO) models. In Section 8.5 this will be generalized to multiple-input, single-output (MISO) models. Chapters 9 and 10 consider multivariate or vector processes, which contain multiple-input, multiple-output (MIMO) processes.

Considering the introductory examples in Chapter 1, the case on page 3 describing the indoor air temperature in a test building relates to the types of models considered in this chapter. The output is the indoor air temperature and the inputs are the heat supply, the ambient air temperature, and the solar radiation.

These types of models are extensively described in Ljung (1987), Söderström and Stoica (1989), and Johansson (1993). A classical reference to input-output (or transfer function) modeling is Åström and Eykhoff (1971). An introduction is also found in Chatfield (2003).

8.1 Relationship between input and output processes

Let us consider a *single-input, single-output system (SISO system) without noise.*

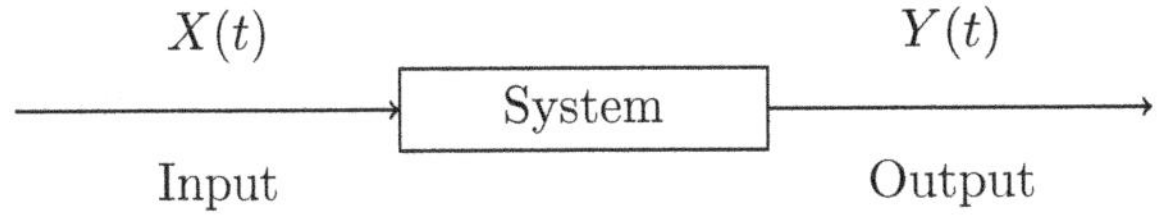

Figure 8.1: *Single-input, single-output system without noise.*

We will assume that the system is *linear and time-invariant*; cf. Chapter 4.

Such a system in *discrete time* may, according to Theorem 4.1 on page 70, be characterized by the impulse response function $\{h_k\}$, and the output is obtained as the convolution

$$Y_t = \sum_{i=-\infty}^{\infty} h_i X_{t-i}. \tag{8.1}$$

Especially for causal systems,

$$Y_t = \sum_{i=0}^{\infty} h_i X_{t-i},$$

which demonstrates, that the present value of the output does not depend on future values of the input.

Furthermore, we will assume that the system is stable and that $\{X_t\}$ does not depend on $\{Y_t\}$, i.e., $\{X_t\}$ is an *exogenous* variable. The case where $\{Y_t\}$ has feedback to $\{X_t\}$ will be discussed in Chapter 9, which deals with multivariate time series.

8.1.1 Moment relations

We now assume that $\{X_t\}$ is a *stationary process*, and thus the *second moment representation* for $\{X_t\}$ is given by the mean value μ_X and the autocovariance function $\gamma_{XX}(k)$. The following then holds.

THEOREM 8.1 (MOMENT RELATIONS—DISCRETE TIME SYSTEM)
The output from the stable system in (8.1) *is a stationary process* $\{Y_t\}$ *characterized by the second moment representation*

$$\mu_Y = \mu_X \sum_{i=-\infty}^{\infty} h_i, \tag{8.2}$$

$$\gamma_{YY}(k) = \sum_{i=-\infty}^{\infty} \sum_{j=-\infty}^{\infty} h_i h_j \gamma_{XX}(k-j+i). \tag{8.3}$$

Furthermore, the cross-covariance function between input and output is given by

$$\gamma_{XY}(k) = \sum_{i=-\infty}^{\infty} h_i \gamma_{XX}(k-i). \tag{8.4}$$

Proof Equation (8.2) follows directly. Equation (8.3) follows from

$$\begin{aligned}\gamma_{YY}(k) &= \text{Cov}[Y_t, Y_{t+k}]\\ &= \text{Cov}\left[\sum_{i=-\infty}^{\infty} h_i X_{t-i}, \sum_{j=-\infty}^{\infty} h_j X_{t+k-j}\right]\\ &= \sum_{i=-\infty}^{\infty}\sum_{j=-\infty}^{\infty} h_i h_j \,\text{Cov}[X_{t-i}, X_{t+k-j}]\\ &= \sum_{i=-\infty}^{\infty}\sum_{j=-\infty}^{\infty} h_i h_j \gamma_{XX}(k-j+i).\end{aligned}$$

Equation (8.4) follows from the same arguments. The stationarity follows from the fact that the system is stable and that the input process is stationary. ■

▸ **Remark 8.1**
It follows from (8.4) that if $\{X_t\}$ is *white noise*, then

$$\gamma_{XY}(k) = h_k \sigma_X^2, \tag{8.5}$$

i.e., the estimate of the cross-covariance function may be applied as an estimate for the impulse response function, conditioned that the input process is white noise. ◂

In *continuous time* we get the following.

THEOREM 8.2 (MOMENT RELATIONS—CONTINUOUS TIME SYSTEM)
Let the input process $\{X(t)\}$ be a stationary process with mean μ_X and autocovariance $\gamma_{XX}(\tau)$. The output from a continuous, linear, time-invariant, and stable system, with impulse response function $h(u)$ and output $\{Y(t)\}$, is characterized by the second moment representation

$$\mu_Y = \mu_X \int_{-\infty}^{\infty} h(u)\,du, \tag{8.6}$$

$$\gamma_{YY}(\tau) = \int_{-\infty}^{\infty}\int_{-\infty}^{\infty} h(u)h(v)\gamma_{XX}(\tau - v + u)\,du\,dv. \tag{8.7}$$

Furthermore, the cross-covariance function between input and output is given by

$$\gamma_{XY}(\tau) = \int_{-\infty}^{\infty} h(u)\,\gamma_{XX}(\tau - u)\,du. \tag{8.8}$$

Proof Omitted. Follows from the same arguments as applied for Theorem 8.1, or see, e.g., Chatfield (2003). ■

▸ **Remark 8.2**

From Theorems 8.1 and 8.2 it follows that if the input process is a normal process, then the output process will also be a normal process. ◂

8.1.2 Spectral relations

Let us first consider systems in *discrete time.* From Theorem 4.7 on page 83 the system can be written in the z domain as

$$Y(z) = H(z)X(z), \tag{8.9}$$

where $H(z)$ is the *transfer function* of the system. Equation (8.9) corresponds to the convolution (8.1) on page 216 in the time domain. By considering the definition of the z-transform, it is seen that (8.2) can be written as

$$\mu_Y = H(1)\mu_X. \tag{8.10}$$

If we instead choose to apply the *frequency response function*

$$\mathcal{H}(\omega) = \sum_{k=-\infty}^{\infty} h_k \mathrm{e}^{-i\omega k} = H(\mathrm{e}^{i\omega}), \tag{8.11}$$

then (8.10) can be written

$$\mu_Y = \mathcal{H}(0)\mu_X. \tag{8.12}$$

According to Definition 5.6 on page 115, we have the following relation between the spectrum $f_{YY}(\omega)$ and the autocovariance function $\gamma_{YY}(k)$:

$$f_{YY}(\omega) = \frac{1}{2\pi} \sum_{k=-\infty}^{\infty} \gamma_{YY}(k)\mathrm{e}^{-i\omega k}. \tag{8.13}$$

If we substitute $\gamma_{YY}(k)$, given by (8.3) on page 216, we obtain

$$\begin{aligned} f_{YY}(\omega) &= \frac{1}{2\pi} \sum_{\ell,j,k=-\infty}^{\infty} h_\ell h_j \gamma_{XX}(k-j+\ell)\mathrm{e}^{-i\omega k} \\ &= \frac{1}{2\pi} \sum_{\ell,j,k=-\infty}^{\infty} h_\ell \mathrm{e}^{i\omega\ell} h_j \mathrm{e}^{-i\omega j} \gamma_{XX}(k-j+\ell)\mathrm{e}^{-i\omega(k-j+\ell)}. \end{aligned}$$

Since $H(\mathrm{e}^{i\omega}) = \sum_{j=-\infty}^{\infty} h_j \mathrm{e}^{-i\omega j}$, we get

$$f_{YY}(\omega) = H(\mathrm{e}^{-i\omega})H(\mathrm{e}^{i\omega})f_{XX}(\omega), \tag{8.14}$$

which expresses the *relation between the spectrum for the input process and the spectrum for the output process.* Equation (8.14) thus corresponds to Equation (8.7) on the preceding page.

Correspondingly, we have the following relation between the cross-spectrum, $f_{XY}(\omega)$, and the cross-covariance function, $\gamma_{XY}(k)$:

$$f_{XY}(\omega) = \frac{1}{2\pi} \sum_{k=-\infty}^{\infty} \gamma_{XY}(k) e^{-i\omega k}. \tag{8.15}$$

Substituting (8.4) yields

$$\begin{aligned} f_{XY}(\omega) &= \frac{1}{2\pi} \sum_{k=-\infty}^{\infty} \sum_{\ell=-\infty}^{\infty} h_\ell \gamma_{XX}(k-\ell) e^{-i\omega k} \\ &= \frac{1}{2\pi} \sum_{\ell=-\infty}^{\infty} h_\ell e^{-i\omega \ell} \sum_{k=-\infty}^{\infty} \gamma_{XX}(k-\ell) e^{-i\omega(k-\ell)}. \end{aligned}$$

This implies that the cross-spectrum between input and output is given by

$$f_{XY}(\omega) = H(e^{i\omega}) f_{XX}(\omega). \tag{8.16}$$

If $\{X_t\}$ is *white noise* then $f_{XX}(\omega) = \sigma_X^2/2\pi$, and

$$f_{XY}(\omega) = H(e^{i\omega}) \frac{\sigma_X^2}{2\pi}. \tag{8.17}$$

In this case an estimate of $f_{XY}(\omega)$ (possibly normalized) can be applied as an estimate for the frequency response function $\mathcal{H}(\omega) = H(e^{i\omega})$.

We will now present the corresponding results in *continuous time*. According to Theorem 4.16 on page 92, the system can be formulated in the s domain as

$$Y(s) = H(s)X(s), \tag{8.18}$$

where $H(s)$ is the transfer function for the system. The frequency response becomes

$$\mathcal{H}(\omega) = H(i\omega) \tag{8.19}$$

and

$$\mu_Y = H(0)\mu_X = \mathcal{H}(0)\mu_X. \tag{8.20}$$

The spectrum for the output becomes

$$f_{YY}(\omega) = H(i\omega)H(-i\omega) f_{XX}(\omega), \tag{8.21}$$

and the cross-spectrum between input and output is given by

$$f_{XY}(\omega) = H(i\omega) f_{XX}(\omega). \tag{8.22}$$

It should be noted that the above formulas in conjunction with the formulas for transfer functions for systems in series, parallel, and feedback (Theorem 4.11

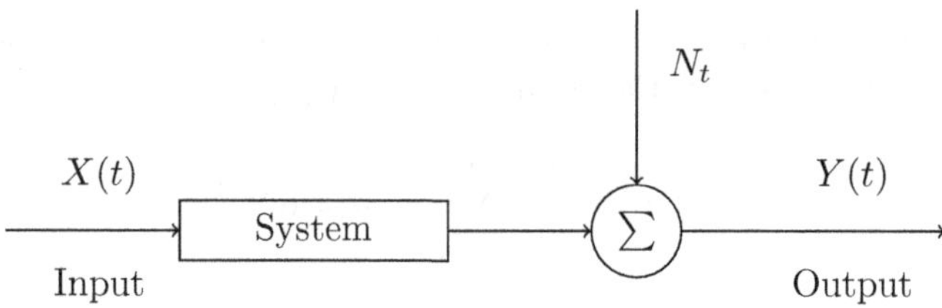

Figure 8.2: *Single-input, single-output system with noise.*

on page 85) are useful in calculating the theoretical spectra for more complex systems.

Finally it should be noticed that Theorem 5.7 on page 117, which is used to determine the spectrum for an ARMA process based on the transfer function, is only a special case of the above, namely, where the input process is white noise.

The frequency response function, which is a complex function, is usually split into a modulus and argument

$$\mathcal{H}(\omega) = |\mathcal{H}(\omega)|e^{i\arg\{\mathcal{H}(\omega)\}} = G(\omega)e^{i\phi(\omega)}, \tag{8.23}$$

where $G(\omega)$ and $\phi(\omega)$ are the *gain* and *phase*, respectively, of the system at the frequency ω from the input $\{X_t\}$ to the output $\{Y_t\}$.

8.2 Systems with measurement noise

Now we consider a system where the output is affected by noise, as indicated in Figure 8.2. Apart from the *observation noise* N_t, the system in Figure 8.2 corresponds to the system in Section 8.1. Since we have added N_t to the system, we get

$$Y_t = \sum_{i=-\infty}^{\infty} h_i X_{t-i} + N_t. \tag{8.24}$$

We assume that N_t *and* X_t *are mutually uncorrelated* stationary stochastic processes and we want to examine the noise influence on the input-output relations from Section 8.1.

In this section we will consider only systems (and processes) in *discrete time.* In this case the moment relations are as follows.

THEOREM 8.3 (MOMENT RELATIONS—SYSTEM WITH MEASUREMENT NOISE)
Let us consider a linear system with measurement noise as specified above. Since $\{X_t\}$ and $\{N_t\}$ are assumed to be mutually uncorrelated stationary

processes, we have

$$\gamma_{YY}(k) = \sum_{i=-\infty}^{\infty}\sum_{j=-\infty}^{\infty} h_i h_j \gamma_{XX}(k-j+i) + \gamma_{NN}(k), \tag{8.25}$$

$$\gamma_{XY}(k) = \sum_{i=-\infty}^{\infty} h_i \gamma_{XX}(k-i). \tag{8.26}$$

Proof Since X_t and N_t are mutually independent we get

$$\begin{aligned}\gamma_{YY}(k) &= \mathrm{Cov}[Y_t, Y_{t+k}] \\ &= \mathrm{Cov}\Big[\sum_i h_i X_{t-i} + N_t, \sum_j h_j X_{t+k-j} + N_{t+k}\Big] \\ &= \sum_i\sum_j h_i h_j \,\mathrm{Cov}[X_{t-i}, X_{t+k-j}] + \mathrm{Cov}[N_t, N_{t+k}],\end{aligned}$$

The results follow directly

$$\begin{aligned}\gamma_{XY}(k) &= \mathrm{Cov}[X_t, Y_{t+k}] \\ &= \mathrm{Cov}\Big[X_t, \sum_i h_i X_{t+k-i} + N_{t+k}\Big] \\ &= \mathrm{Cov}\Big[X_t, \sum_i h_i X_{t+k-i}\Big] = \sum_i h_i \gamma_{XX}(k-i).\end{aligned}$$

■

If we compare with Theorem 8.1 on page 216 we see that the formula for the cross-covariance is unchanged.

The corresponding spectral relations are given in the following.

THEOREM 8.4 (SPECTRAL RELATIONS—SYSTEM WITH MEASUREMENT NOISE)
Let the system $\{X_t\}$, $\{N_t\}$ be specified as in Theorem 8.3. Then we have

$$\begin{aligned}f_{YY}(\omega) &= H(e^{-i\omega})H(e^{i\omega})f_{XX}(\omega) + f_{NN}(\omega) & (8.27)\\ &= G^2(\omega)f_{XX}(\omega) + f_{NN}(\omega), & (8.28)\end{aligned}$$

$$f_{XY}(\omega) = H(e^{i\omega})f_{XX}(\omega) = \mathcal{H}(\omega)f_{XX}(\omega). \tag{8.29}$$

Proof Follows the same arguments as in Section 8.1.2. ■

In (8.28) we have applied $G^2(\omega) = \overline{\mathcal{H}}(\omega)\mathcal{H}(\omega) = \overline{H}(e^{i\omega})H(e^{i\omega})$. It is thus seen that when measurement noise is present, it holds

$$\mathcal{H}(\omega) = \frac{f_{XY}(\omega)}{f_{XX}(\omega)} \tag{8.30}$$

(for all ω where $f_{XX}(\omega) \neq 0$), i.e., *the frequency response function is the ratio between the cross-spectrum and the spectrum for the input process.* This is one of the most important results within the field of linear systems as it allows for the estimation of the gain and phase of the system by substituting estimates for $f_{XY}(\omega)$ and $f_{XX}(\omega)$ in (8.30).

8.3 Input-output models

The model (8.24) may contain an arbitrarily large number of parameters. Box and Jenkins (1970/1976) have, in analogy to the ARMA models, proposed writing the causal version of (8.24) on page 220 on the form

$$\begin{aligned} Y_t + \delta_1 Y_{t-1} + \cdots + \delta_r Y_{t-r} = \omega_0 X_{t-b} + \omega_1 X_{t-b-1} + \cdots \\ + \omega_s X_{t-b-s} + N_t^{'}, \end{aligned} \tag{8.31}$$

where b is an integer valued *time delay.*

In this section we will consider different types of *input-output models,* namely, *transfer function models, difference equation models,* and *output-error models.* First, we give an introduction to the most general model, the transfer function model.

8.3.1 Transfer function models

When we consider transfer function models, we assume that the noise, N_t, can be described by an ARMA model (possibly with seasonal components), i.e.,

$$N_t + \varphi_1 N_{t-1} + \cdots + \varphi_p N_{t-p} = \varepsilon_t + \theta_1 \varepsilon_{t-1} + \cdots + \theta_q \varepsilon_{t-q}. \tag{8.32}$$

By introducing

$$\delta(\mathrm{B}) = 1 + \delta_1 \mathrm{B} + \cdots + \delta_r \mathrm{B}^r, \tag{8.33}$$

$$\omega(\mathrm{B}) = \omega_0 + \omega_1 \mathrm{B} + \cdots + \omega_s \mathrm{B}^s, \tag{8.34}$$

$$\varphi(\mathrm{B}) = 1 + \varphi_1 \mathrm{B} + \cdots + \varphi_p \mathrm{B}^p, \tag{8.35}$$

$$\theta(\mathrm{B}) = 1 + \theta_1 \mathrm{B} + \cdots + \theta_q \mathrm{B}^q, \tag{8.36}$$

we can write (8.31) as

$$Y_t = \frac{\omega(\mathrm{B})}{\delta(\mathrm{B})} \mathrm{B}^b X_t + \frac{\theta(\mathrm{B})}{\varphi(\mathrm{B})} \varepsilon_t, \tag{8.37}$$

since $N_t^{'} = \delta(\mathrm{B}) N_t$. The polynomials (8.36) may be expanded to contain seasonal components, e.g., $\theta(\mathrm{B})$ may be replaced by $\theta(\mathrm{B})\Theta(\mathrm{B}^S)$. The term $\omega(\mathrm{B})\,\mathrm{B}^b \,/\, \delta(\mathrm{B})$ is called the *transfer function component.*

Box and Jenkins (1970/1976) refer to this model as a *transfer function model.* The term *transfer function* is often used for linear systems in the z or s domain. The model is also referred to as the *Box-Jenkins model.*

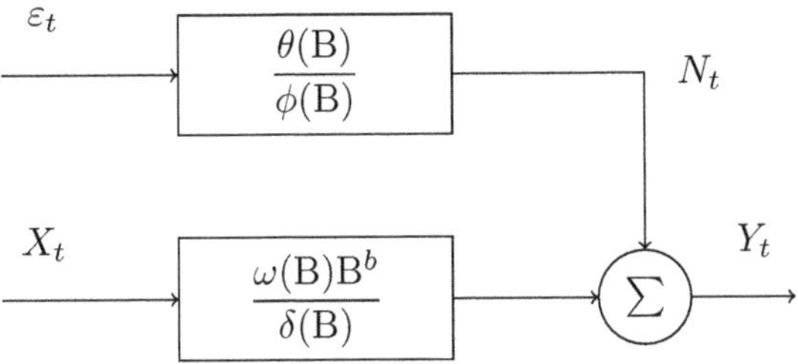

Figure 8.3: *Diagram for the transfer function model* (8.37).

8.3.2 Difference equation models

We often simplify model (8.37) by setting $\delta(B) = \varphi(B)$. Thereby we obtain the *ARMAX model* (AutoRegressive-Moving Average with eXogenous input)

$$\varphi(B)Y_t = \omega(B)X_{t-b} + \theta(B)\varepsilon_t, \tag{8.38}$$

where $\varphi(B)$, $\omega(B)$ and $\theta(B)$ are p'th, s'th and q'th order polynomials, respectively, in the shift operator B. Here b is an integer valued time delay.

Special cases of the ARMAX model are the ARX model ($\theta(B) = 1$) and FIR (Finite Impulse Response) ($\varphi(B) = \theta(B) = 1$). The ARMAX model is sometimes referred to as the CARMA model (Controlled ARMA).

8.3.3 Output error models

An *output error model (OE model)* is a model in the form

$$Y_t = \frac{\omega(B)}{\delta(B)} X_{t-b} + N_t \tag{8.39}$$

where there is no model for the noise term. The model is most often applied in estimation by the output error method (OEM). This will be elaborated in the section on estimation.

8.4 Identification of transfer function models

The identification of an input-output model (transfer function model) is done in a *two stage procedure.* First, a transfer function from $\{X_t\}$ to $\{Y_t\}$ (the transfer function component) is identified, and then a model for the noise is identified.

In this section we will consider single-input, single-output transfer function models. In Section 8.5 multiple-input systems will be considered.

The procedure for identification of a causal (physically realizable) linear system in the following is the *Box-Jenkins approach.* Notice, that it is assumed that both input and output series are mean-corrected.

Let us consider an input-output model of the type (8.31). The first problem is to *identify* r, s, and b. This can be done based on the estimated impulse response function, since, after such an estimation, we *select* r, s, and b so that we get a description in the form

$$\widehat{h}(\mathrm{B}) = \frac{\widehat{\omega}(\mathrm{B})\,\mathrm{B}^b}{\widehat{\delta}(\mathrm{B})}. \tag{8.40}$$

The estimation of the impulse response function is based on the fact that if the input is white noise, then the impulse response function can be found based on the estimated cross-covariance function; cf. (8.5) on page 217.

In other words, the impulse response function may be estimated by a method called *pre-whitening*, i.e., based on "pre-whitening" of the input.

Pre-whitening procedure for estimating h_k

a) A suitable ARMA model is applied to the input series:

$$\varphi(\mathrm{B})X_t = \theta(\mathrm{B})\alpha_t,$$

where α_t is white noise.

b) We perform a *pre-whitening* of the input series

$$\alpha_t = \theta(\mathrm{B})^{-1}\varphi(\mathrm{B})X_t$$

c) The output-series $\{Y_t\}$ is filtered with the same model, i.e.,

$$\beta_t = \theta(\mathrm{B})^{-1}\varphi(\mathrm{B})Y_t.$$

d) Now the *impulse response function is estimated* by

$$\widehat{h}_k = \mathrm{C}_{\alpha\beta}(k)/\,\mathrm{C}_{\alpha\alpha}(0) = \mathrm{C}_{\alpha\beta}(k)/S_\alpha^2.$$

Proof It holds that

$$Y_t = h(\mathrm{B})X_t + N_t$$

and it follows

$$\begin{aligned}\beta_t &= \theta(\mathrm{B})^{-1}\varphi(\mathrm{B})Y_t\\ &= \theta(\mathrm{B})^{-1}\varphi(\mathrm{B})[h(\mathrm{B})X_t + N_t]\\ &= h(\mathrm{B})\alpha_t + \theta(\mathrm{B})^{-1}\varphi(\mathrm{B})N_t\,.\end{aligned}$$

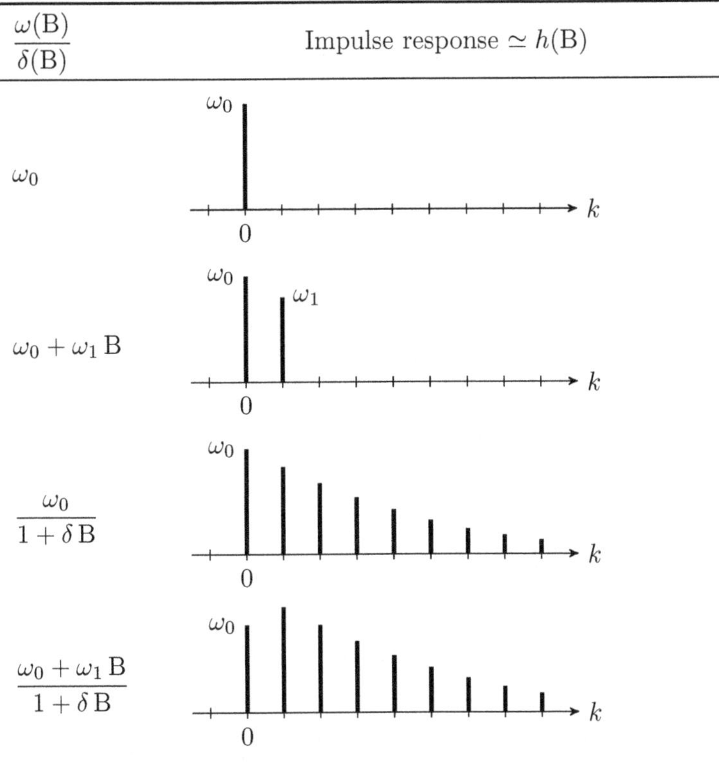

Figure 8.4: *Some examples of the transfer function:* $\frac{\omega(\text{B})\,\text{B}^b}{\delta(\text{B})}$ *and the corresponding impulse response function* $h(\text{B})$.

Multiplying by α_{t-k} and taking expected values yields

$$\gamma_{\alpha\beta}(k) = h_k \gamma_{\alpha\alpha}(0) = h_k \sigma_\alpha^2.$$

See (8.5) on page 217. ∎

The identification of r, s, and b is built on (8.40) and the corresponding theoretical relations. Some examples are shown in Figure 8.4.

If we have a time delay, b, the transfer estimate will start at $k = b$. Having determined r, s, and b we can estimate the remaining parameters in the model

$$Y_t = \frac{\omega(\text{B})\,\text{B}^b}{\delta(\text{B})} X_t + N_t, \tag{8.41}$$

e.g., by the *method of moments* or the *prediction error method.*

For the "residuals," $\{\widehat{N}_t\}$, we subsequently estimate a traditional ARMA model, as described in Chapter 6. After this identification, the *total model is estimated.*

▸ **Remark 8.3**

The Box-Jenkins approach described above is not adequate in the following situations:

- *Multiple-input systems.* In these cases the input processes are most often correlated, and the correlation between the input processes makes use of the pre-whitening method wery problematic.

- *Input process is not of the ARMA type.* In this case no reasonable ARMA model can be found for the input process, and hence the described method for prewhitening can not be used.

Procedures for estimating the impulse response functions in these cases are described in Section 8.5. ◂

8.5 Multiple-input models

Until now we have considered single-input, single-output models. Let us now consider the multiple-input, single-output (MISO) model

$$Y_t = \sum_{i=-\infty}^{\infty} h_{1,i} X_{1,t-i} + \cdots + \sum_{i=-\infty}^{\infty} h_{m,i} X_{m,t-i} + N_t \tag{8.42}$$

where $\{X_{1,t}\}, \ldots, \{X_{m,t}\}$ are m (exogenous) input variables, which are assumed not to depend on $\{Y_t\}$. The observational noise process $\{N_t\}$ is assumed to be mutually uncorrelated with all m input variables.

8.5.1 Moment relations

Multiplying (8.42) by $\{X_{1,t}\}, \ldots, \{X_{m,t}\}$ and taking the expectation leads to the moment relations

$$\begin{aligned}
\gamma_{X_1 Y}(k) &= \sum_{i=-\infty}^{\infty} h_{1,i}\gamma_{x_1 x_1}(k-i) + \cdots + \sum_{i=-\infty}^{\infty} h_{m,i}\gamma_{x_1 x_m}(k-i) \\
&\vdots \\
\gamma_{X_m Y}(k) &= \sum_{i=-\infty}^{\infty} h_{1,i}\gamma_{x_m x_1}(k-i) + \cdots + \sum_{i=-\infty}^{\infty} h_{m,i}\gamma_{x_m x_m}(k-i)
\end{aligned} \tag{8.43}$$

which is seen to be a generalization of (8.4) on page 216 and (8.26) on page 221.

8.5.2 Spectral relations

Since the Fourier transformation is a linear operator, it is seen from (8.43) that the corresponding spectral relations are

$$\begin{aligned} f_{X_1 Y}(\omega) &= H_1(\omega) f_{X_1 X_1}(\omega) + \cdots + H_m(\omega) f_{X_1 X_m}(\omega) \\ &\vdots \\ f_{X_m Y}(\omega) &= H_1(\omega) f_{X_m X_1}(\omega) + \cdots + H_m(\omega) f_{X_m X_m}(\omega) \end{aligned} \tag{8.44}$$

or on matrix form

$$\boldsymbol{f}_{XY}(\omega) = \boldsymbol{f}_{XX}(\omega)\boldsymbol{H}(\omega) \tag{8.45}$$

where $\boldsymbol{f}_{XX}(\omega)$ is an $m \times m$ spectral matrix (a straightforward generalization of (7.40) on page 205), and $\boldsymbol{H}(\omega)$ is a vector containing the m frequency response functions $H_1, \ldots, H_m$.

Notice that

$$\boldsymbol{H}(\omega) = \boldsymbol{f}_{XX}^{-1} \boldsymbol{f}_{XY}(\omega) \tag{8.46}$$

which enables a framework for estimating the transfer functions based on estimates of the (auto-)spectra and cross-spectra. Equation (8.46) facilitates a frequency domain based method for finding the impulse response functions for multiple-input systems. By using the inverse Fourier transformation, we are able to find the impulse response function from $\{X_{i,t}\}$ to $\{Y_t\}$:

$$h_{i,k} = \mathfrak{F}^{-1}(H_i(\omega)) = \frac{1}{2\pi} \int_{-\pi}^{\pi} H_i(\omega) e^{i\omega k}\, d\omega, \quad k = \ldots, -1, 0, 1, \ldots \tag{8.47}$$

8.5.3 Identification of multiple-input models

The pre-whitening method described in Section 8.4 on page 223 is very useful in many cases, but it is not suitable for multiple-input systems. In that case we can use the frequency domain method described in Section 8.5.2, i.e., based on the estimated spectral matrix $\widehat{\boldsymbol{f}}_{XX}(\omega)$ and the estimated vector of cross-spectra $\widehat{\boldsymbol{f}}_{XY}$. We find an estimate of the frequency response function as

$$\widehat{\boldsymbol{H}}(\omega) = \widehat{\boldsymbol{f}}_{XX}^{-1}(\omega) \widehat{\boldsymbol{f}}_{XY}(\omega) \tag{8.48}$$

and hence the impulse response function is found by (8.47).

Alternatively the impulse response function can be found by linear regression. In practice only the first q values of the impulse response function are of importance. We have

$$\begin{aligned} Y_t &= h_1(\mathrm{B}) X_{1,t} + \cdots + h_m(\mathrm{B}) X_{m,t} + N_t \\ &\simeq h_{10} X_{1,t} + h_{11} X_{1,t-1} + \cdots + h_{1q} X_{1,t-q} + \cdots \\ &\quad + h_{m0} X_{m,t} + h_{m1} X_{m,t-1} + \cdots + h_{mq} X_{m,t-q} + N_t \\ &= \boldsymbol{X}_t^T \boldsymbol{\theta} + e_t \end{aligned} \tag{8.49}$$

which is a general linear model with

$$\boldsymbol{X}_t^T = (X_{1,t}, \dots, X_{m,t-q}) \tag{8.50}$$
$$\boldsymbol{\theta}^T = (h_{10}, \dots, h_{mq}) \tag{8.51}$$
$$e_t = N_t\,. \tag{8.52}$$

Now the impulse response values can be found using the LS method (see Equation (6.37) on page 160).

Given N observations, the linear relation in (8.49) can be written for all observations as

$$\boldsymbol{Y} = \boldsymbol{X}\boldsymbol{\theta} + \boldsymbol{e} \tag{8.53}$$

where

$$\boldsymbol{Y}^T = (Y_{q+1}, \dots, Y_N) \tag{8.54}$$
$$\boldsymbol{X}^T = (\boldsymbol{X}_{q+1}, \dots, \boldsymbol{X}_N) \tag{8.55}$$
$$\boldsymbol{e}^T = (e_{q+1}, \dots, e_N) \tag{8.56}$$

and hence, the values of the impulse response functions can be found by solving the normal equation (6.37).

This leads to *LS estimates of the impulse response function* values, i.e.,

$$\widehat{\theta} = (\boldsymbol{X}^T\boldsymbol{X})^{-1}\boldsymbol{X}^T\boldsymbol{Y}. \tag{8.57}$$

However, in many applications the values of the input processes are highly autocorrelated, and this leads to near singularity (due to collinearity) of $\boldsymbol{X}^T\boldsymbol{X}$ and, hence, a large variance of $\widehat{\boldsymbol{\theta}}$. This problem can be treated by instead considering the Ridge estimator

$$\widehat{\boldsymbol{\theta}} = (\boldsymbol{X}^T\boldsymbol{X} + \lambda\boldsymbol{I})^{-1}\boldsymbol{X}^T\boldsymbol{Y} \tag{8.58}$$

where $\lambda \in [0, 1]$ is the *Ridge parameter*. The optimal value for λ is the smallest value λ_{min} for which $\widehat{\boldsymbol{\theta}}(\lambda)$ is reasonably constant.

This method is in general useful as an alternative to least squares estimation where the explanatory variables are highly correlated. The term *collinearity* is often used to describe the situation where the explanatory variables are so highly correlated that it is impossible to come up with reliable estimates using the least squares approach; see also Christensen (2002) or Hastie, Tibshirani, and Friedman (2001).

8.6 Estimation

The estimation problem has been discussed in Chapter 6—also for models having input variables. In this section aspects related to input-output models will be discussed. In addition, we will discuss a method used for input-output models, namely the output error method (OEM).

8.6.1 Moment estimates

The values dominating the impulse response function, $\widehat{h}_u$, can be used to formulate the equations that are used to determine the estimates for the parameters in the transfer function component. Hereby, we obtain the so-called moment estimates.

For example, we have

$$\begin{aligned}\frac{\omega_0}{(1+\delta\,\mathrm{B})} &= \omega_0(1-\delta\,\mathrm{B}+\delta^2\,\mathrm{B}^2-\delta^3\,\mathrm{B}^3+\cdots)\\ &= h_0+h_1\,\mathrm{B}+\cdots,\end{aligned}$$

and from this we get $\widehat{\omega_0}=\widehat{h_0}$ and $\widehat{\delta}=-\widehat{h_1}/\widehat{h_0}$.

8.6.2 LS estimates

The LS estimator for dynamic models that was derived in Chapter 6 may be applied to all models, which can be written in the form

$$Y_t = \boldsymbol{X}_t^T\boldsymbol{\theta}+\varepsilon_t \tag{8.59}$$

This is the case for, for example, the ARX and FIR models. The method and the properties of the estimator are discussed in Section 6.4.2 on page 159.

8.6.3 Prediction error method

The prediction error estimate may be found as discussed in Section 6.4.3 on page 163:

$$\widehat{\boldsymbol{\theta}} = \arg\min_{\theta}\left\{S(\boldsymbol{\theta})=\sum_{t=t_0}^{N}\varepsilon_t^2(\boldsymbol{\theta})\right\}. \tag{8.60}$$

where

$$\varepsilon_t(\boldsymbol{\theta}) = Y_t-\widehat{Y}_{t|t-1}(\boldsymbol{\theta}). \tag{8.61}$$

and t_0 is selected appropriately. The input process $\{X_t\}$ is usually assumed to be deterministic. If this is not the case, we have to condition on the observed values of $\{X_t\}$. The *conditional mean* is calculated as given by Equation (6.86) on page 169 for the *transfer function model* and by (6.89) for the *ARMAX model.*

8.6.4 ML estimates

Assuming that $\{\varepsilon_t\}$ is *normally distributed* white noise, we obtain the ML estimates of the parameters in the same way as for the prediction error estimates, see Chapter 6.

8.6.5 Output error method

When applying the *output error method* we consider the output error model, (8.39) on page 223, and the parameters are estimated by

$$\widehat{\boldsymbol{\theta}} = \arg\min_{\theta} \left\{ S(\boldsymbol{\theta}) = \sum_{t=}^{N} N_t^2(\boldsymbol{\theta}) \right\}. \tag{8.62}$$

where

$$N_t(\boldsymbol{\theta}) = Y_t - \frac{\omega(\mathrm{B})\,\mathrm{B}^b}{\delta(\mathrm{B})} X_t. \tag{8.63}$$

Please note that $N_t(\boldsymbol{\theta})$ is the *simulation error* since it represents the deviation between Y_t and the output from the model where X_t is input without considering the noise.

The variance matrix for the parameter estimates are often very difficult to calculate, see Söderström and Stoica (1989). Furthermore, the most common validation techniques cannot be applied, e.g., test for white noise. Therefore, the method should be used only under special conditions.

8.7 Model checking

As a final step we perform a *model check* (or model diagnostics) as described in Chapter 6, but we extend the diagnostics with a *test for whether the cross-correlations between* $\{\varepsilon_t\}$ *and* $\{X_t\}$ *are zero.* If $\{\varepsilon_t\}$ is white noise, we have according to Theorem 6.2 on page 151 that

$$\widehat{\rho}_{\varepsilon X}(k) \underset{\text{approx.}}{\in} \mathrm{N}\left(0, \frac{1}{N}\right). \tag{8.64}$$

An approximate 5% level test is done by accepting the hypothesis $\rho_{\varepsilon X}(k) = 0$ (against a two-sided alternative) if $\widehat{\rho}_{\varepsilon X}(k)$ is within the 2σ-levels $\pm 2/\sqrt{N}$.

The *Portmanteau test*, which includes more values of the cross-correlation function, is based on the test statistics

$$Q_X^2 = N \sum_{k=0}^{m} \widehat{\rho}_{\varepsilon X}(k)^2, \tag{8.65}$$

which under a hypothesis for $\rho_{\varepsilon X}(k) = 0$ and $\rho_{\varepsilon}(k) = 0$ is approximately χ^2-distributed with $m+1-(r+s+1) = m-r-s$ degrees of freedom. Often, the model control indicates that a re-identification is needed.

The tests for whether a parameter is zero and for lower model order are unchanged from the discussion in Section 6.5 on page 170.

In the discussion we have considered models with only one input series; but it is formally straightforward to extend the models to include *several input*

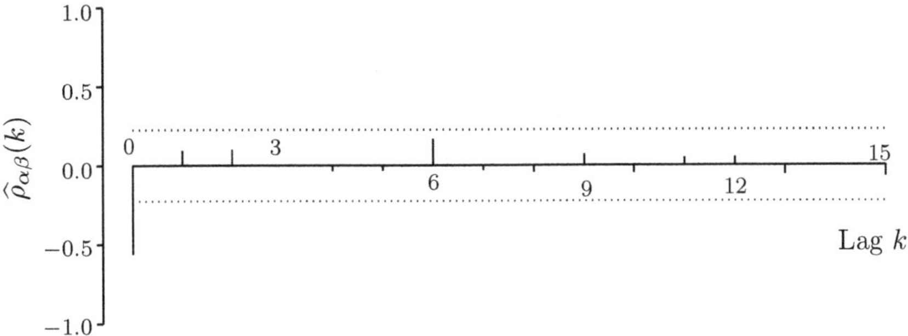

Figure 8.5: *The empirical cross-correlation function between the pre-whitened input series $\{\alpha_t\}$ and the filtered output series, $\{\beta_t\}$.*

series. However, often there will be a cross-correlation between input series, which implies that the identification is more uncertain, and the separately estimated impulse response functions (by pre-whitening) may give only a weak indication of the order of each of the transfer function components. In such a case it is necessary to apply several different models and base the model selection on tests, e.g., tests for single parameters and tests for lower model order.

Example 8.1 (Identification of a transfer function model)
This example is a continuation of the example in Section 6.7 on page 179. In that example, we found that the electricity consumption could be described by the following one-dimensional model:

$$\nabla\nabla_{12} \ln Y_t = (1 + \theta\,\mathrm{B})(1 + \Theta\,\mathrm{B}^{12})\varepsilon_t. \tag{8.66}$$

Since we want to exploit the information contained in the observations of the temperature, we consider a transfer function model. In order to identify the transfer function we pre-whiten the input series, $\{X_t\}$. The model obtained is

$$(1 - 0.27\,\mathrm{B})\nabla_{12} X_t = (1 - 0.88\,\mathrm{B}^{12})\alpha_t. \tag{8.67}$$

The output series is filtered using the same model, i.e.,

$$(1 - 0.27\,\mathrm{B})\nabla_{12} Y_t = (1 - 0.88\,\mathrm{B}^{12})\beta_t. \tag{8.68}$$

Finally, the cross-correlation function between $\{\alpha_t\}$ and $\{\beta_t\}$ is estimated. The result is shown in Figure 8.5.

The cross-correlation function is proportional to the impulse response function, since

$$\widehat{h}_k = \widehat{\rho}_{\alpha\beta}(k)\frac{\widehat{\sigma}_\beta}{\widehat{\sigma}_\alpha}. \tag{8.69}$$

The estimated cross-correlation in Figure 8.5 on the previous page has a large negative value at lag 0, whereas the remaining values are within the 2σ levels. This implies that an increasing temperature will result in reduced electricity consumption and vice versa. This can be described (cf. Figure 8.4 on page 225) by the following model

$$\ln Y_t = \omega_0 X_t + N_t. \tag{8.70}$$

As a first guess of the structure for the noise term, N_t, the one-dimensional model (8.66) for $\ln Y_t$ may be applied

$$\nabla\nabla_{12} N_t = (1+\theta\,\mathrm{B})(1+\Theta\,\mathrm{B}^{12})\varepsilon_t. \tag{8.71}$$

Combining (8.70) and (8.71) yields the following transfer function model

$$\nabla\nabla_{12}\ln Y_t = \omega_0\nabla\nabla_{12}X_t + (1-\theta\,\mathrm{B})(1-\Theta\,\mathrm{B}^{12})\varepsilon_t. \tag{8.72}$$

Based on $\widehat{\rho}_{\alpha\beta}(k)$, $\widehat{\sigma}_\alpha$, and $\widehat{\sigma}_\beta$, we can calculate a moment estimate for ω_0. With this estimate as a starting value in a prediction error method, the parameters in the total model (8.72) can be estimated. We find (for $N = 96$) that

$$\begin{aligned}\nabla\nabla_{12}\ln Y_t &= -0.00222^{\pm 0.00020}\nabla\nabla_{12}X_t \\ &\quad + (1-0.64\,\mathrm{B})^{\pm 0.09}\cdot(1-0.75\,\mathrm{B}^{12})^{\pm 0.09}\varepsilon_t, \\ \widehat{\sigma}_\varepsilon^2 &= 0.0174^2.\end{aligned} \tag{8.73}$$

For the one-dimensional model in Section 6.7, we found that $\widehat{\sigma}_\varepsilon^2 = 0.0255$, and it is thus seen that by using the information in the temperature series, we may further reduce the residual variance.

If we write the model between $\{X_t\}$ and $\{Y_t\}$ in the form

$$\ln Y_t = -0.00222^{\pm 0.00020}X_t + N_t, \tag{8.74}$$

the model, due to the logarithmic transform, may be interpreted as follows. An increase of 1 °C in the temperature in a given month implies a decrease in the energy consumption of 0.22% ± 0.02% in that same month. (If we put $Y = f(X)$ we have the linear approximation $Y - Y_0 = f'(X_0)(X - X_0)$. With $Y = \exp(\alpha X)$ we get $(Y - Y_0)/Y_0 = \alpha(X - X_0)$.

In order to check the model, we have plotted the residuals, $\{\varepsilon_t\}$, the autocorrelation for the residuals, $\widehat{\rho}_\varepsilon$, and the cross-correlation between the residuals and the pre-whitened input series, $\widehat{\rho}_{\alpha\varepsilon}(k)$, in Figure 8.6.

When comparing the empirical correlation functions with the 2σ levels, it is seen that we cannot reject that $\{\varepsilon_t\}$ may be white noise or that $\rho_{\varepsilon\alpha}(k) = 0$. Based on this we conclude the model is satisfactory.

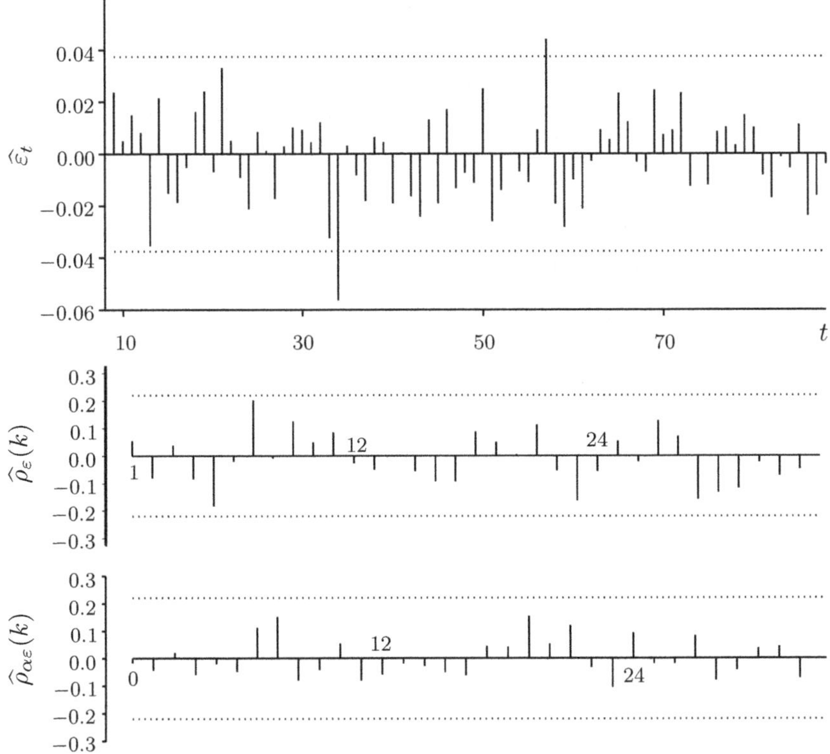

Figure 8.6: *Model control. Plot of* $\{\widehat{\varepsilon}_t\}$, $\widehat{\rho}_\varepsilon(k)$ *and* $\widehat{\rho}_{\alpha\varepsilon}(k)$ *for the model in* (8.73).

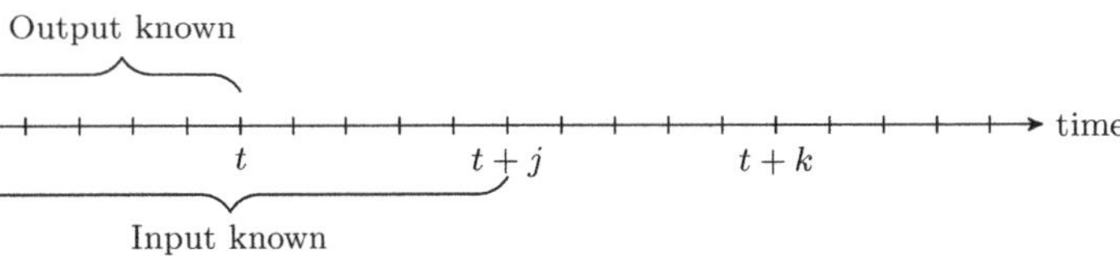

Figure 8.7: *General assumptions in prediction using input-output models.*

8.8 Prediction in transfer function models

The optimal prediction, in terms of minimizing the square of the prediction errors, is found, as in Chapter 3 and Chapter 5, as the conditional mean.

Let us assume that the time of prediction is t and that we want to predict the output at time Y_{t+k}. Further we assume that the values of the input process are available up until time $t+j$, see Figure 8.7. The *prediction horizon* is thus k.

The general assumptions in Figure 8.7 have some important special cases:

i) $\{X_t\}$ may be a controllable input (which we determine ourselves). Therefore it is possible to predict Y_{t+k} under certain controlling strategies, i.e., given the choice of future values for the controllable input.

ii) $\{X_t\}$ is an input variable which is known up until time t (i.e., the knowledge of X_t and Y_t follow each other in time, e.g., as for the temperature in Example 8.1 on page 231).

In order to make the prediction as indicated in Figure 8.7, we write the model (8.37) in the form

$$Y_{t+k} = \sum_{i=0}^{\infty} h_i X_{t+k-i} + N_{t+k}, \tag{8.75}$$

where $\{N_t\}$ is an ARMA process, which can be written in MA form

$$N_t = \sum_{i=0}^{\infty} \psi_i \varepsilon_{t-i}, \tag{8.76}$$

where $\{\varepsilon_t\}$ is white noise with variance, σ_ε^2. Furthermore, we assume that $\{X_t\}$ can be described by an ARMA process. An ARMA process can always be written as

$$X_t = \sum_{i=0}^{\infty} \overline{\psi}_i \eta_{t-i}, \tag{8.77}$$

where $\{\eta_t\}$ is white noise with variance σ_η^2.

By taking the conditional means in (8.75), we get the *prediction formula*:

$$\begin{aligned}
\widehat{Y}_{t+k|t} &= \mathrm{E}[Y_{t+k}|Y_t, Y_{t-1}, \dots, X_{t+j}, X_{t+j-1}, \dots] \\
&= \sum_{i=0}^{\infty} h_i \, \mathrm{E}[X_{t+k-i}| \quad \cdots \quad] + \mathrm{E}[N_{t+k}| \quad \cdots \quad] \\
&= \sum_{i=0}^{k-j-1} h_i \widehat{X}_{t+k-i|t+j} + \sum_{i=k-j}^{\infty} h_i X_{t+k-i} + \widehat{N}_{t+k|t}\,.
\end{aligned} \tag{8.78}$$

The prediction errors become

$$Y_{t+k} - \widehat{Y}_{t+k|t} = \sum_{i=0}^{k-j-1} h_i (X_{t+k-i} - \widehat{X}_{t+k-i|t+j}) + N_{t+k} - \widehat{N}_{t+k|t} \tag{8.79}$$

When $Y_t, Y_{t-1}, \ldots$ and $X_{t+j}, X_{t+j-1}, \ldots$ are known, then also $\varepsilon_t, \varepsilon_{t-1}, \ldots$ and $\eta_{t+j}, \eta_{t+j-1}, \ldots$ are known. Thereby we get the following expression for the prediction errors:

$$N_{t+k} - \widehat{N}_{t+k|t} = \sum_{i=0}^{k-1} \psi_i \varepsilon_{t+k-i}, \tag{8.80}$$

and

$$X_{t+k} - \widehat{X}_{t+k|t+j} = \sum_{i=0}^{k-j-1} \overline{\psi}_i \eta_{t+k-i}, \quad (k > j). \tag{8.81}$$

Substituting this into (8.79) we get the prediction error for Y_{t+k}:

$$\begin{aligned} Y_{t+k} - \widehat{Y}_{t+k|t} &= \sum_{i_1=0}^{k-j-1} h_{i_1} \sum_{i_2=0}^{k-i_1-j-1} \overline{\psi}_{i_2} \eta_{t+k-i_1-i_2} + \sum_{i=0}^{k-1} \psi_i \varepsilon_{t+k-i} \\ &= \sum_{\ell=0}^{k-j-1} \left(\sum_{i_1+i_2=\ell} h_{i_1} \overline{\psi}_{i_2} \right) \eta_{t+k-\ell} + \sum_{i=0}^{k-1} \psi_i \varepsilon_{t+k-i} . \end{aligned}$$

Since $\{\eta_t\}$ and $\{\varepsilon_t\}$ are assumed to be mutually uncorrelated, we get the following *variance of the prediction error*:

$$\text{Var}[Y_{t+k} - \widehat{Y}_{t+k|t}] = \sigma_\eta^2 \sum_{\ell=0}^{k-j-1} \left(\sum_{i_1+i_2=\ell} h_{i_1} \overline{\psi}_{i_2} \right)^2 + \sigma_\varepsilon^2 \sum_{i=0}^{k-1} \psi_i^2. \tag{8.82}$$

An important condition for the above statement is that $\{X_t\}$ can be described by an ARMA model.

In some applications we observe a time delay between input and output. If the time delay is b time units then $h_0, h_1, \ldots, h_{b-1}$ will be zero, and from (8.79) we get the prediction error

$$Y_{t+k} - \widehat{Y}_{t+k|t} = N_{t+k} - \widehat{N}_{t+k|t}, \quad \text{for } k \le b + j. \tag{8.83}$$

Thus, the prediction errors depend on the noise process only when the prediction horizon is less than or equal to the time delay plus j. In the important special case, where $j = 0$ and the information on $\{Y_t\}$ and $\{X_t\}$ follows each other (see Figure 8.7 on page 233), we get from (8.83) that the prediction error depends solely on the noise process, $\{N_t\}$, when the prediction horizon is less than or equal to the time delay.

Example 8.2 (Prediction in a transfer function model)
This is a continuation of Example 8.1 on page 231 for the model which describes the relation between the monthly electricity consumption, Y_t, and the monthly mean air temperature, X_t.

Table 8.1: *Predictions of the electricity consumption with the prediction horizon (lead time) of* $1, 2, \ldots, 12$ *hours from the time* $t = 96$ *based on the one-dimensional model for the electricity consumption (1) and the transfer function model (3).*

Observation number	(1) Univariate prediction of the el. consumption	(2) Univariate prediction of the temperature	(3) Transfer prediction of el. consumption by using (2)
97	129.1	11.0	132.0
98	118.8	12.4	121.8
99	111.9	13.7	115.6
100	102.2	15.9	105.5
101	97.7	17.0	101.6
102	106.7	17.6	109.6
103	110.4	16.5	113.6
104	119.9	13.7	123.4
105	135.6	11.4	138.8
106	147.8	9.1	152.5
107	154.6	8.6	157.3
108	147.0	9.4	151.1

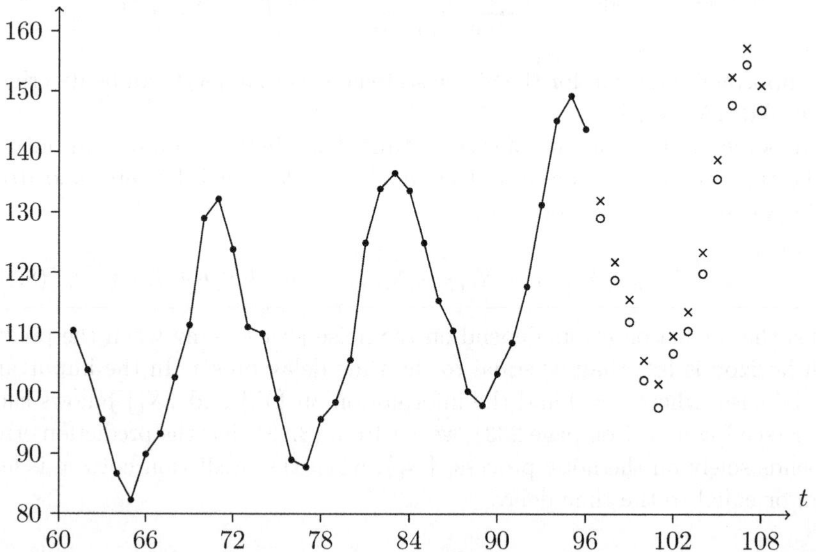

Figure 8.8: *Predictions in transfer function model. The one-dimensional model (1) in Table 8.1 is shown with circles and the transfer function model is shown with crosses.*

Table 8.2: *The standard deviation, given in percent of the expected electricity consumption, at different prediction horizons and models.*

Prediction horizon	Univariate model	Transfer function model (prediction of temperature)	Transfer function model (temperature known)
1	2.55	2.49	1.72
2	2.63	2.62	1.84
3	2.71	2.70	1.95
4	2.79	2.78	2.06
5	2.87	2.86	2.16
6	2.94	2.93	2.25
7	3.02	3.00	2.35
8	3.09	3.07	2.44
9	3.16	3.14	2.52
10	3.23	3.21	2.60
11	3.30	3.27	2.68
12	3.36	3.34	2.76
13	3.54	3.51	2.96
14	3.62	3.60	3.07
15	3.70	3.69	3.17
16	3.78	3.78	3.27
17	3.86	3.87	3.37
18	3.94	3.95	3.47

In Table 8.1, we have given the predictions of the electricity consumption starting from time $t = 96$ using different prediction horizons. The one-dimensional prediction of the energy consumption (1) is based on the model (8.66), the prediction of the temperature is based on the model (8.67), and the prediction by the transfer function model is based on (8.73). The predictions of the temperature are used to predict the electricity consumption, cf. (8.78). The predictions are shown in Figure 8.8.

In Table 8.2 we compare the standard deviations of the prediction error at different time horizons and models. Due to the logarithmic transform, the standard deviation is given as a percentage of the expected electricity consumption.

In spite of the transfer function model producing residuals with significantly smaller variance (see Example 8.1), we see that the standard deviation of the prediction error for the one-dimensional model and transfer function model are almost identical. This is because the uncertainty of the temperature prediction enters the model—cf. (8.82). In the last column of Table 8.2 we have shown the standard deviation of the prediction error assuming that we know the temperature in the period without uncertainty. Here, only the uncertainty from the noise process, $\{N_t\}$, is present.

8.8.1 Minimum variance controller

In this section we consider the important special case of predictions in transfer function models where we are able to select the input signal and thereby control the output signal. By solving the prediction equations with respect to the control variable (the input in this case), we obtain the so called minimum variance controller, which leads to a controller where the variance around the wanted reference value is minimized.

We consider the following model with a time delay k from input to output

$$\phi(\mathrm{B})Y_t = \omega(\mathrm{B})\,\mathrm{B}^k X_t + \theta(\mathrm{B})\varepsilon_t, \tag{8.84}$$

where we furthermore assume that $\{X_t\}$ is a *controllable signal* which we can determine at time t ($j = 0$, shown in Figure 8.7 on page 233).

If we write (8.84) at time $t+k$ we have

$$\phi(\mathrm{B})Y_{t+k} = \omega(\mathrm{B})X_t + \theta(\mathrm{B})\varepsilon_{t+k}. \tag{8.85}$$

By rearranging we get

$$\begin{aligned} Y_{t+k} &= \frac{\omega(\mathrm{B})}{\phi(\mathrm{B})}X_t + \frac{\theta(\mathrm{B})}{\phi(\mathrm{B})}\varepsilon_{t+k} \\ &= \frac{\omega(\mathrm{B})}{\phi(\mathrm{B})}X_t + (1 + \cdots + \psi_{k-1}\,\mathrm{B}^{k-1})\varepsilon_{t+k} + \mathrm{B}^k\frac{\gamma(\mathrm{B})}{\phi(\mathrm{B})}\varepsilon_{t+k} \end{aligned} \tag{8.86}$$

where the latter is found by polynomial division.

Alternatively Y_{t+k} can be expressed as

$$Y_{t+k} = \frac{\omega(\mathrm{B})}{\phi(\mathrm{B})}X_t + \psi^*(\mathrm{B})\varepsilon_{t+k} + \mathrm{B}^k\frac{\gamma(\mathrm{B})}{\phi(\mathrm{B})}\varepsilon_{t+k}, \tag{8.87}$$

where $\psi^*(\mathrm{B})$ and $\gamma(\mathrm{B})$ can be determined from the *identity*

$$\theta(\mathrm{B}) = \psi^*(\mathrm{B})\phi(\mathrm{B}) + \mathrm{B}^k\gamma(\mathrm{B}), \tag{8.88}$$

with

$$\psi^*(\mathrm{B}) = 1 + \cdots + \psi_{k-1}\,\mathrm{B}^{k-1}, \tag{8.89}$$

$$\gamma(\mathrm{B}) = \gamma_0 + \cdots + \gamma_{p-1}\,\mathrm{B}^{p-1}, \tag{8.90}$$

where we assume that $q \le k+p-1$ (otherwise the order of $\gamma(\mathrm{B})$ equals $q-k$). The identity (8.88) is called the *diophantine equation.*

The optimal prediction of Y_{t+k}, given $Y_t, Y_{t-1}, \ldots, X_t, X_{t-1}, \ldots$, is given by (8.87) as

$$\widehat{Y}_{t+k|t} = \frac{\omega(\mathrm{B})}{\phi(\mathrm{B})}X_t + \frac{\gamma(\mathrm{B})}{\phi(\mathrm{B})}\varepsilon_t. \tag{8.91}$$

Since

$$\varepsilon_t = \frac{\phi(\mathrm{B})}{\theta(\mathrm{B})} Y_t - \frac{\omega(\mathrm{B})}{\theta(\mathrm{B})} \mathrm{B}^k X_t,$$

we get

$$\begin{aligned} \widehat{Y}_{t+k|t} &= \left(\frac{\omega(\mathrm{B})}{\phi(\mathrm{B})} - \mathrm{B}^k \frac{\gamma(\mathrm{B})}{\phi(\mathrm{B})} \frac{\omega(\mathrm{B})}{\theta(\mathrm{B})} \right) X_t + \frac{\gamma(\mathrm{B})}{\theta(\mathrm{B})} Y_t \\ &= \frac{\omega(\mathrm{B})\psi^*(\mathrm{B})}{\theta(\mathrm{B})} X_t + \frac{\gamma(\mathrm{B})}{\theta(\mathrm{B})} Y_t, \end{aligned} \tag{8.92}$$

where the identity (8.88) is applied.

It is well-known that the optimal prediction is characterized by minimizing the variance of the prediction error. Since we are able to determine the input signal we can select X_t so that the variance of the output with respect to a given reference point is minimized. Without loss of generality, we put the reference point to zero, and we get the *minimum variance controller*, characterized by X_t being chosen such that $\widehat{Y}_{t+k|t} = 0$, i.e., *the minimum variance controller* (the control strategy) is determined by

$$\omega(\mathrm{B})\psi^*(\mathrm{B}) X_t = -\gamma(\mathrm{B}) Y_t, \tag{8.93}$$

and from (8.86) it is seen that the error is an MA($k-1$) process, and thus that the *variance of Y_t using the control error* (8.93) becomes

$$\mathrm{Var}[Y_t | X_t \text{ determined from (8.93)}] = (1 + \psi_1^2 + \cdots + \psi_{k-1}^2)\sigma_\varepsilon^2, \tag{8.94}$$

i.e., equal to the variance of the k-step prediction in the ARMA process $\phi(\mathrm{B})Y_t = \theta(\mathrm{B})\varepsilon_t$.

Instead of applying (8.93) it is often simpler to directly write the k-step prediction and subsequently isolate X_t. This is illustrated in the following.

Example 8.3 (Minimum variance controller)

We consider the model

$$(1 + \phi\,\mathrm{B})Y_t = (\omega_0 + \omega_1\,\mathrm{B})\,\mathrm{B}^2 X_t + \varepsilon_t, \tag{8.95}$$

where $\{\varepsilon_t\}$ is white noise with variance σ_ε^2. We find

$$\begin{aligned} \widehat{Y}_{t+1|t} &= \mathrm{E}[Y_{t+1} | Y_t, Y_{t-1}, \ldots, X_t, X_{t-1}, \ldots] \\ &= -\phi Y_t + \omega_0 X_{t-1} + \omega_1 X_{t-2}, \end{aligned} \tag{8.96}$$

$$\widehat{Y}_{t+2|t} = -\phi \widehat{Y}_{t+1|t} + \omega_0 X_t + \omega_1 X_{t-1}. \tag{8.97}$$

By substituting (8.96) into (8.97) we get

$$\widehat{Y}_{t+2|t} = \phi^2 Y_t + \omega_0 X_t + (\omega_1 - \phi\omega_0) X_{t-1} - \phi\omega_1 X_{t-2}.$$

Since we want to minimize the variance of the output (around the reference point 0), we require that $\widehat{Y}_{t+2|t} = 0$. Thus, the minimal variance regulator becomes

$$X_t = -\frac{1}{\omega_0}\left(\phi^2 Y_t + (\omega_1 - \phi\omega_0)X_{t-1} - \phi\omega_1 X_{t-2}\right). \tag{8.98}$$

Using this control strategy the variance of the output becomes

$$\sigma_Y^2 = (1 + \psi_1^2)\sigma_\varepsilon^2 = (1 + \phi^2)\sigma_\varepsilon^2. \tag{8.99}$$

These types of regulators are further described in Åström (1970) and Holst and Poulsen (1988).

The next example shows that the famous PI-controller (proportional-integral) is obtained as a minimum variance controller for a particular model.

Example 8.4 (PI-controller)
The discrete time PI-controller can be written

$$X_t = X_{t-1} + K\left[\left(1 + \frac{h}{T}\right)Y_t - Y_{t-1}\right] \tag{8.100}$$

or

$$X_t = \frac{K\left[\left(1 + \frac{h}{T}\right) - \mathrm{B}\right]}{(1 - \mathrm{B})} Y_t \tag{8.101}$$

This controller is obtained as the minimum variance controller

$$X_t = \frac{\gamma(\mathrm{B})}{\omega(\mathrm{B})\psi^*(\mathrm{B})} Y_t \tag{8.102}$$

for a model where the time delay is $k = 1$. Hence, $\psi^*(\mathrm{B}) = 1$ and $\omega(\mathrm{B}) = (1 - \mathrm{B})$.

Since the order of $\gamma(\mathrm{B}) = p - 1 = 1$, we see that the order is $p = 1$, and from the identity $K[(1 + h/T) - \mathrm{B}] = \mathrm{B}^{-1}(\theta(\mathrm{B}) - \phi(\mathrm{B}))$, we find that

$$\theta_1 - \phi_1 = K\left(1 + \frac{h}{T}\right) \tag{8.103}$$

$$\theta_2 - \phi_2 = K \tag{8.104}$$

which can be used to determine (ϕ_1, ϕ_2) and (θ_1, θ_2).

The technique of writing down the prediction of future values as a function of the control variable may lead to very powerful approaches for advanced control. The predictions are based on a stochastic model for the considered system, and hence, these methods are often called *model based control.*

In Palsson, Madsen, and Søgaard (1994) advanced predictive controllers are developed. These controllers contain methods for control of time-varying systems and systems with unknown time delay.

8.9 Intervention models

When applying ARMA models the assumption that the time series is stationary is crucial. In some situations, however, non-stationarities may be explained by a certain happening—also called an intervention. For example, one observed an increase in the price of meat when Denmark entered the Common Market in January 1973. Another such example is the change in the weekly sales of a certain product due to an advertisement campaign. Here the advertisement campaign is the intervention.

An *intervention model* is a model which can describe the changes in an output process due to external phenomena, which are then considered as input to the system.

For intervention models the input is a *qualitative variable* whereas the input to a transfer function model is a *quantitative variable.*

Let us introduce the *intervention function*:

$$I_t = \begin{cases} 1 & t = t_0 \\ 0 & t \neq t_0 \end{cases} \tag{8.105}$$

In the literature it is also seen that the intervention function is defined as a step, i.e., as $\mathrm{S}\, I_t = I_t/(1 - \mathrm{B})$, where S is the summation operator.

The *intervention model* can be written in the form

$$Y_t = \frac{\omega(\mathrm{B})}{\delta(\mathrm{B})} I_t + \frac{\theta(\mathrm{B})}{\phi(\mathrm{B})} \varepsilon_t, \tag{8.106}$$

where $\{I_t\}$ is the intervention function. The polynomials $\phi(\mathrm{B})$, $\theta(\mathrm{B})$, $\delta(\mathrm{B})$, and $\omega(\mathrm{B})$ are defined as for the transfer function model.

In the same way as for the transfer function models, the polynomials can be extended to include seasonal components, e.g., we can substitute $\phi(\mathrm{B})$ in (8.106) with $\phi(\mathrm{B})\Phi(\mathrm{B}^s)$. The term $\omega(\mathrm{B})I_t/\delta(\mathrm{B})$ is called the *intervention component*, and the model (8.106) may be extended to include *several intervention components* and hereby account for several types of interventions that influence the process.

The similarity between the intervention model (8.106) and the transfer function model is considerable; however, for the issues of *identification* they are different because the intervention function takes only the values 0 and 1. For the transfer function, we estimated the impulse response function, and this was the foundation for the subsequent identification. For an intervention model, the input may be directly interpreted as an impulse response function and unless the interventions are very close in time, the (first) identification can be based directly on the observed time series. For this, Figure 8.4 on page 225 may be useful. For a further discussion, see Milhøj (1986) or Shao (1997).

Example 8.5 (An intervention model)
Let $\{Y_t\}$ be the observed price of meat in Denmark, and let I_t be the intervention that describes when Denmark joined the Common Market in 1973, i.e.,

$$I_t = \begin{cases} 1 & t = \text{January 1973} \\ 0 & \text{otherwise} \end{cases}$$

In Milhøj (1986) the following intervention model has been applied

$$(1 - \mathrm{B}) \ln Y_t = \frac{\omega_0}{1 - \delta_1 \mathrm{B}} I_t + \varepsilon_t - \theta_1 \varepsilon_{t-1},$$

where $\{\varepsilon_t\}$ is white noise. Furthermore, he found that $\widehat{\omega}_0 = 0.14, \widehat{\delta}_1 = 0.63, \widehat{\theta}_1 = -0.24$, and $\widehat{\sigma}_\varepsilon^2 = 0.033^2$.

Intervention models are useful also in more technical applications, e.g., in cases of some *anomalies* in the data which may look as outliers for several samples, an intervention model might be useful for estimating the model despite these anomalies. Since missing data most often is a huge problem in time series analysis, the use of intervention models in such cases is often a more reasonable solution. Examples of the use of intervention models can be found in Box and Tiao (1975).

The following example also illustrates the potentials of using intervention models in a technical applications.

Example 8.6 (Chewing gum flavor release)
In this example, a model for the release of flavor compounds will be formulated, in which an intervention model is applied to describe the characteristics. Chewing gum can be used to get rid of bad breath since the release of flavor gives good taste and the expired air becomes fresh. Information about this release has been gathered where an interface was used to sample the breath directly from the nose.

During the chewing period the release depends on several factors, which are different from the factors acting in the period after the gum is spit out, also called phasing-out period. Therefore, two different time windows are considered: the first one covering the chewing period, the second window covering the entire phasing-out period. For each window, the characteristics are the qualitative input, described by the intervention function in (8.105), where it is assumed that the intervention takes place at the start of each time window.

The intervention model applied is the so-called OE model, as described in (8.39), with X_{t-b} equal to I_{t-b}. By applying the output error estimation method, the parameter estimates are obtained as equation (8.62) indicates, where $N_t(\boldsymbol{\theta})$ is defined by (8.63). The components in the transfer function

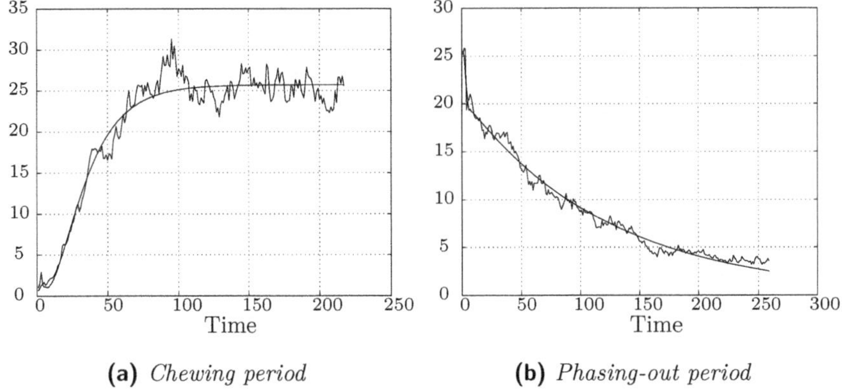

(a) *Chewing period* **(b)** *Phasing-out period*

Figure 8.9: *Measured and simulated intensities for chewing and phasing-out time windows.*

are described in (8.36) and for the OE estimation method, the best fit proved to be of 4th order for both polynomials and $b = 1$.

Assuming stability, which can be controlled by finding all the poles, the concentration Y_t in the model will approach a stationary (we are neglecting the noise term N_t) concentration given by

$$Y_\infty = h(1) = \frac{\omega(1)}{\delta(1)}. \tag{8.107}$$

which is directly seen from (8.2). The convergence rate is given by the time constants of the transfer function and the largest time constant determines the rate. The time constants are related to the roots of the transfer function. These roots are found as the roots of the denominator, $\delta(z^{-1}) = 0$, where the roots are either real or complex. For a real and positive root, $(p_i < 1)$, the time constant is found as

$$\tau_i = -\ln \frac{1}{p_i}. \tag{8.108}$$

Figures 8.9a and 8.9b show the measured release profile, along with a simulation for the estimated model, for both the chewing period and the phasing-out period, respectively. The estimated parameters in the model are the predicted maximum intensity and two time constants, one describing the rise process for the chewing period and another describing the phasing-out process. The parameters for the breathing, that is, the parameters in the transfer function, and the three estimated parameters can together provide an adequate description of the whole release process.

8.10 Problems

Exercise 8.1

Assume $\{X_t\}$ is a stationary process with spectral density $f_x(\omega)$. Now a new process $\{Y_t\}$ is defined by the linear filter:

$$Y_t = \sum_{p=0}^{k} a_p X_{t-p}$$

where the a_p's are real numbers.

Question 1 Show that the spectral density for $\{Y_t\}$ is

$$f_y(\omega) = \left[\sum_{q=0}^{k} \sum_{p=0}^{k} a_q a_p \cos(\omega(p-q)) \right] f_x(\omega)$$

Question 2 Set $a_p = 1/(k+1)$ for $p = 0, 1, \ldots, k$ and show that the spectral density for $\{Y_t\}$ becomes

$$f_y(\omega) = \frac{\sin^2(\omega(k+1)/2)}{\sin^2(\omega/2)} \frac{f_x(\omega)}{(k+1)^2}$$

Question 3 Given $\{Y_t\}$ (from Question 2) a new process $\{Z_t\}$ is defined by

$$Z_t = \frac{1}{k+1} \sum_{p=0}^{k} Y_{t-p}$$

Find the spectral density for $\{Z_t\}$ expressed by the spectral density for $\{X_t\}$.

Question 4 Take $k = 1$ and find the impulse response function for the composed filter from X_t to Z_t.

Question 5 Find the frequency response function corresponding to the impulse response function in Question 4. Sketch the amplitude and phase. Is the composed filter a low- or high-pass filter?

Question 6 State a simple change in the composed filter in order to avoid a phase shift, while the amplitude is maintained.

Exercise 8.2

The importer of a given commodity has investigated the sale of the commodity and especially the effect of a particular advertising campaign through weekly magazines.

Intervention analysis has shown that the weekly sales can be described by the following model:

$$(Y_t - \mu) = \frac{\omega_0 \mathrm{B}}{1 + \omega_1 \mathrm{B}} I_t + \frac{1 + \theta_1 \mathrm{B}}{1 + \omega_1 \mathrm{B}} \varepsilon_t$$

where ε_t is white noise with the mean value 0 and variance σ_ε^2. I_t is an intervention function defined, such that it is equal to 1, if in week t an advertising campaign is undertaken and 0 otherwise.

Based on the previous experiences of advertisement sales, it is found that

$$\begin{aligned} \mu &= 200 \text{ units/week} & \theta_1 &= 0.3 \text{ units/week} \\ \omega_0 &= 100 \text{ units/week} & \omega_1 &= -0.9 \text{ units/week} \\ \sigma_\varepsilon^2 &= 400 \text{ (units/week)}^2 \end{aligned}$$

During the past 7 weeks the following sales has been registered.

Table 8.3: *Sales figures prior to the advertising.*

Week	1	2	3	4	5	6	7
Sales	238	247	230	208	215	196	207

There has been no advertising campaign in the above mentioned period and the effect of the previous campaigns is assumed to be negligible.

Question 1 Calculate the expected sales in week 8, 9, 10, and 11 assuming that no sales campaign is undertaken during these weeks.

Question 2 The importer is now planning an advertising campaign in week 8 and therefore, he wants an assessment over the effect. Assuming that a sales campaign is undertaken in week 8, you should calculate the expected sales in weeks 8, 9, 10, and 11, and specify a 95% confidence band for the future sales.

CHAPTER 9

Multivariate time series

In previous chapters we have mostly considered univariate time series. However, in Chapter 7 we introduced the bivariate process and the related covariance in (7.37) on page 204 and spectral matrix in (7.40). In this section this is generalized to multivariate or vector time series, and we shall see that the methods used for multivariate time series closely correspond to what we have seen previously for univariate time series.

In Chapter 8 we introduced the transfer function model which describes the variation of the output time series $\{Y_t\}$ as a function of the input time series $\{X_t\}$. The formulation in Chapter 8 was based on the important assumption that $\{X_t\}$ does not depend on $\{Y_t\}$. In technical terms this implies that previously we did not allow for a feedback from $\{Y_t\}$ to $\{X_t\}$. In other words we assumed an *open-loop* system. For physical systems we say that there is a *causal relationship* between the input and the output.

However, in many cases $\{X_t\}$ does depend on $\{Y_t\}$, and we are not able to classify one of the variables as an input and the other as an output; the variables are just inter-related. In technical terms such systems are often called *closed-loop* systems. In this case the system should be described as a *multivariate time series.*

From a modeling perspective it is important to notice that in this chapter we allow for a simultaneous and non-causal variation of several variables. In the bivariate case, which is easy to illustrate, we now have the situation shown in Figure 9.1 on the next page.

As illustrated in Figure 9.1

$$Y_t = h_1(B)X_t + N_{1,t}. \tag{9.1}$$

$$X_t = h_2(B)Y_t + N_{2,t}. \tag{9.2}$$

Consider as an example the temperature in two attached rooms, Room A and Room B. Initially we assume that the temperature is the same in both rooms. If, for some reason, the temperature increases in Room A (e.g., due to heating), then the temperature will also start to increase slowly in Room B due to heat transfer between the rooms, and vice versa.

In econometrics most time series are non-causal. If you for instance consider the relation between the price of chicken meat and the amount of chickens on

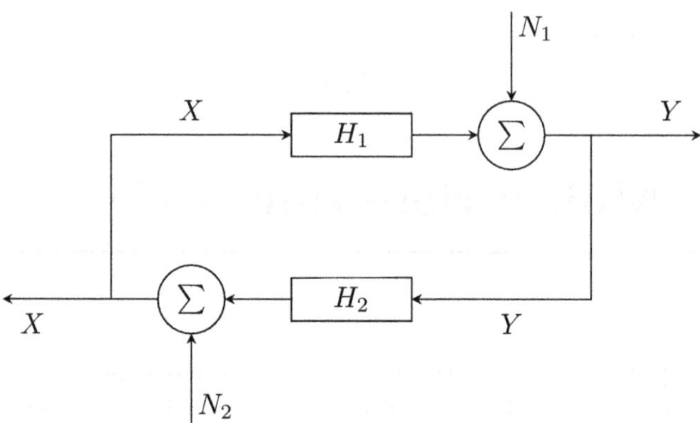

Figure 9.1: *Schematic diagram of a linear model for the bivariate process* $\{(Y_t, X_t)^T\}$ *defined in Equation* (9.1) *on the preceding page. The input to the linear system is the serial correlated process* $\{(N_{1,t}, N_{2,t})^T\}$.

the market, a multivariate approach is needed. Because an increase in price will motivate the farmers to produce more chickens, and, as a consequence, large amounts of chicken meat on the market will cause a decrease in price, and vice versa.

Example 9.1 (Closed-loop transfer functions)
Consider the model described by Equations (9.1) and (9.2). The input to this model is $(N_{1,t}, N_{2,t})^T$ and the output is $(Y_t, X_t)^T$. For each combination of input and output there exists a transfer function; hence, four (univariate) transfer functions exist.

Using the same technique as in Theorem 4.12 on page 86, while noticing the change in the sign, we are now able to find, e.g., the transfer function from $N_{1,t}$ to X_t:

$$H_1(z)H_2(z)X(z) + H_2(z)N_1(z) = X(z) \tag{9.3}$$

or

$$X(z) = \frac{H_2(z)}{1 - H_1(z)H_2(z)} N_1(z) \tag{9.4}$$

and similarly we could find the remaining three transfer functions.

However, it is often easier to find the transfer functions directly by using vector notation. Doing this, the model (9.1)–(9.2) can be written as

$$\begin{pmatrix} 1 & -h_1(B) \\ -h_2(B) & 1 \end{pmatrix} \begin{pmatrix} Y_t \\ X_t \end{pmatrix} = \begin{pmatrix} N_{1,t} \\ N_{2,t} \end{pmatrix} \tag{9.5}$$

or in the z domain as

$$\begin{pmatrix} 1 & -H_1(z) \\ -H_2(z) & 1 \end{pmatrix} \begin{pmatrix} Y(z) \\ X(z) \end{pmatrix} = \begin{pmatrix} N_1(z) \\ N_2(z) \end{pmatrix} \tag{9.6}$$

which implies that

$$\begin{aligned} \begin{pmatrix} Y(z) \\ X(z) \end{pmatrix} &= \frac{1}{1 - H_1(z)H_2(z)} \begin{pmatrix} 1 & H_1(z) \\ H_2(z) & 1 \end{pmatrix} \begin{pmatrix} N_1(z) \\ N_2(z) \end{pmatrix} \\ &= \boldsymbol{H}(z) \begin{pmatrix} N_1(z) \\ N_2(z) \end{pmatrix} \end{aligned} \tag{9.7}$$

where $\boldsymbol{H}$ is the *multivariate transfer function* from the input $(N_{1,t}, N_{2,t})^T$ to the output $(Y_t, X_t)^T$. Notice that the transfer function from $N_{1,t}$ to X_t is recognized as $\boldsymbol{H}_{21}$.

In this chapter we will introduce the concepts related to multivariate time series. We shall see that the theory in the multivariate case turns out to be a straightforward generalization of what we have seen previously for univariate time series. We shall assume that the system which generates the stochastic process is linear and time-invariant, and hence, it is known that there exists a multivariate impulse response function describing the dynamics. After introducing the linear multivariate (or vector) process, the multivariate ARMA (or VARMA) will be introduced in Section 9.3, and the concepts of stationarity and invertibility will be introduced. Then non-stationary models vector ARIMA (MARIMA or VARIMA) model will be treated. The very useful concept of covariance functions is introduced, and it is shown how the theoretical covariance matrix functions can be calculated in the multivariate case in Section 9.3.1. Section 9.5 is considering predictions in multivariate linear processes. Finally, identification, estimation and model validation are considered in Section 9.6. In general the subjects in this chapter are ordered such that the similarities with the univariate case in Chapter 5 and 6 can be seen rather easily.

9.1 Stationary stochastic processes and their moments

We will restrict our attention to stationary vector processes. Consider the m-dimensional jointly stationary vector process[1] $\{\boldsymbol{Z}_t = [Z_{1,t}, \dots, Z_{m,t}]^T, t = 0, \pm 1, \dots\}$. As in the univariate case, the stationarity implies that the mean

[1]A jointly stationary process implies that every univariate component process is stationary, but a vector of univariate stationary processes is not necessarily a jointly stationary process.

is constant, i.e.,

$$\mathrm{E}[\boldsymbol{Z}_t] = \boldsymbol{\mu} = \begin{pmatrix} \mu_1 \\ \vdots \\ \mu_m \end{pmatrix} \tag{9.8}$$

and that the cross-covariance between $Z_{i,t}$ and $Z_{j,s}$ for all values of (i,j) is a function of only the time difference $(s-t)$. Hence, the *lag k covariance matrix* is

$$\begin{aligned} \boldsymbol{\Gamma}(k) = \mathrm{C}[\boldsymbol{Z}_t, \boldsymbol{Z}_{t+k}] &= \mathrm{E}[(\boldsymbol{Z}_t - \boldsymbol{\mu})(\boldsymbol{Z}_{t+k} - \boldsymbol{\mu})^T] \\ &= \mathrm{E}[(\boldsymbol{Z}_{t-k} - \boldsymbol{\mu})(\boldsymbol{Z}_t - \boldsymbol{\mu})^T] \\ &= \boldsymbol{\Gamma}^T(-k) \end{aligned} \tag{9.9}$$

$\boldsymbol{\Gamma}(k)$ is called the *covariance matrix function* for the vector process.

Notice that

$$\boldsymbol{\Gamma}(k) = \begin{pmatrix} \gamma_{11}(k) & \gamma_{12}(k) & \cdots & \gamma_{1m}(k) \\ \gamma_{21}(k) & \gamma_{22}(k) & \cdots & \gamma_{2m}(k) \\ \vdots & \vdots & \ddots & \vdots \\ \gamma_{m1}(k) & \gamma_{m2}(k) & \cdots & \gamma_{mm}(k) \end{pmatrix} \tag{9.10}$$

where $\gamma_{ij}(k) = \mathrm{Cov}[Z_{i,t}, Z_{j,t+k}]$. Hence, for $i=j$, $\gamma_{ii}(k)$ is the autocovariance function for $\{Z_{i,t}\}$, and for $i \neq j$, $\gamma_{ij}(k)$ is the cross-covariance function between $\{Z_{i,t}\}$ and $\{Z_{j,t}\}$.

As introduced in Section 5.2.2 the cross-correlation function between $\{Z_{i,t}\}$ and $\{Z_{j,t}\}$ $(i \neq j)$ is given by

$$\rho_{ij}(k) = \frac{\gamma_{ij}(k)}{\sqrt{\gamma_{ii}(0)\gamma_{jj}(0)}} \tag{9.11}$$

whereas for $i=j$, we obtain the autocorrelation function.

The *lag k correlation matrix* is defined by

$$\boldsymbol{\rho}(k) = [\rho_{ij}(k)] \tag{9.12}$$

which as a function of k is called the *correlation matrix function*, and by defining $\boldsymbol{D} = \mathrm{diag}[\gamma_{11}(0), \ldots, \gamma_{mm}(0)]$, we have the relation

$$\boldsymbol{\rho}(k) = \boldsymbol{D}^{-1/2}\boldsymbol{\Gamma}(k)\boldsymbol{D}^{-1/2} \tag{9.13}$$

In the literature the covariance and correlation matrix functions are also called the autocovariance and autocorrelation matrix functions, respectively.

As in the univariate case we have the following.

THEOREM 9.1
The covariance and correlation matrix functions are positive semidefinite in the sense that

$$\sum_{i=1}^{k}\sum_{j=1}^{k} \boldsymbol{\alpha}_i^T \boldsymbol{\Gamma}(t_i - t_j)\boldsymbol{\alpha}_j \geq 0 \tag{9.14}$$

for any m-dimensional vectors $(\boldsymbol{\alpha}_1, \boldsymbol{\alpha}_2, \ldots, \boldsymbol{\alpha}_k)^T$, *and similarly for* $\boldsymbol{\rho}(k)$.

Proof Follows from the fact that for any set of real values $\boldsymbol{\alpha}_1, \boldsymbol{\alpha}_2, \ldots, \boldsymbol{\alpha}_k$, and any set of time points, we have

$$\text{Var}\left[\sum_{i=1}^{k} \boldsymbol{\alpha}_i^T \boldsymbol{Z}_{t_i}\right] \geq 0 \tag{9.15}$$

■

For two jointly stationary vector processes $\boldsymbol{Z}_t$ and $\boldsymbol{W}_t$ we define the *lag k cross-covariance matrix* between $\boldsymbol{Z}$ and $\boldsymbol{W}$ as

$$\boldsymbol{\Gamma}_{Z,W}(k) = \text{C}[\boldsymbol{Z}_t, \boldsymbol{W}_{t+k}] \tag{9.16}$$

which as a function of k is called the *cross-covariance matrix function.*

9.2 Linear processes

We shall focus on linear vector processes in the time domain. In general, a multivariate linear process can be interpreted as the output from a multivariate linear system with multivariate white noise input. Let us then first define the concept.

> DEFINITION 9.1 (MULTIVARIATE WHITE NOISE)
> A process $\{\boldsymbol{\varepsilon}_t\}$ is said to be m-dimensional *white noise*, if $\{\boldsymbol{\varepsilon}_t\}$ is a sequence of mutual uncorrelated identically distributed random variables with zero mean and constant covariance matrix $\boldsymbol{\Sigma}$.

It follows that for white noise, the covariance matrix function is

$$\boldsymbol{\Gamma}_{\varepsilon}(k) = \begin{cases} \boldsymbol{\Sigma} & \text{if } k = 0 \\ \boldsymbol{0} & \text{if } k \neq 0 \end{cases} \tag{9.17}$$

where

$$\boldsymbol{\Sigma} = \begin{pmatrix} \sigma_1^2 & \sigma_{12} & \cdots & \sigma_{1m} \\ \sigma_{21} & \sigma_2^2 & \cdots & \sigma_{2m} \\ \vdots & \vdots & \ddots & \vdots \\ \sigma_{m1} & \sigma_{m2} & \cdots & \sigma_m^2 \end{pmatrix} = \begin{pmatrix} \sigma_{11} & \sigma_{12} & \cdots & \sigma_{1m} \\ \sigma_{21} & \sigma_{22} & \cdots & \sigma_{2m} \\ \vdots & \vdots & \ddots & \vdots \\ \sigma_{m1} & \sigma_{m2} & \cdots & \sigma_{mm} \end{pmatrix} \tag{9.18}$$

Hence, the elements of the vector white noise are uncorrelated for different time points, but they may be correlated at the same time point (contemporaneously correlated).

The fundamental process, as for the univariate case, is the *linear process.*

DEFINITION 9.2 (THE LINEAR MULTIVARIATE PROCESS)
A multivariate linear process $\{\boldsymbol{Y}_t\}$ is a process that can be written in the form

$$\boldsymbol{Y}_t - \boldsymbol{\mu} = \sum_{i=0}^{\infty} \boldsymbol{\psi}_i \boldsymbol{\varepsilon}_{t-i}, \tag{9.19}$$

where $\{\boldsymbol{\varepsilon}_t\}$ is white noise and $\boldsymbol{\mu}$ is the mean of the process. Furthermore, $\boldsymbol{\psi}_0 = \boldsymbol{I}$ is the $m \times m$ identity matrix.

Hence, as in Chapter 5, the linear process can be defined as the output from a linear convolution between a matrix weight function and the vector white noise.

The sequence of matrix weight functions $\{\boldsymbol{\psi}_k\}$ is often referred to as the $\boldsymbol{\psi}$ *weights*, and (9.19) is referred to as the *random shock form.*

A linear process defined as in Equation (9.19) is also called a purely stochastic process (or purely non-deterministic process)—compare with Definition 5.8 on page 103.

In order to ease the notation but without loss of generality, we will assume that $\boldsymbol{\mu} = \boldsymbol{0}$ in the following.

By introducing the linear operator

$$\boldsymbol{\psi}(\mathrm{B}) = \boldsymbol{I} + \sum_{i=1}^{\infty} \boldsymbol{\psi}_i \, \mathrm{B}^i, \tag{9.20}$$

equation (9.19) can be formulated (for $\boldsymbol{\mu} = \boldsymbol{0}$):

$$\boldsymbol{Y}_t = \boldsymbol{\psi}(\mathrm{B})\boldsymbol{\varepsilon}_t. \tag{9.21}$$

Here $\boldsymbol{\psi}(\mathrm{B})$ is referred to as the *transfer function* of the multivariate process.

THEOREM 9.2 (STATIONARITY FOR LINEAR MULTIVARIATE PROCESSES)
The linear process $\boldsymbol{Y}_t = \boldsymbol{\psi}(\mathrm{B})\boldsymbol{\varepsilon}_t$ is stationary if

$$\boldsymbol{\psi}(z) = \sum_{i=0}^{\infty} \boldsymbol{\psi}_i z^{-i} \tag{9.22}$$

converges for $|z| \geq 1$.

Proof Omitted. Since stationary processes are characterized by the fact that the variance is finite, one can interpret the result by comparison with (5.39) on page 109 for each sequence of ψ_{ij}. In the multivariate case the criterion in (5.39) must be fulfilled for each of the sequences of ψ_{ij}. ∎

Stationarity ensures that the influence of historical values of the noise process goes sufficiently fast to zero.

Example 9.2 (Bivariate linear process)
In the bivariate case ($m = 2$), the random shock form is

$$\begin{pmatrix} Y_{1,t} \\ Y_{2,t} \end{pmatrix} = \begin{pmatrix} \psi_{11}(B) & \psi_{12}(B) \\ \psi_{21}(B) & \psi_{22}(B) \end{pmatrix} \begin{pmatrix} \varepsilon_{1,t} \\ \varepsilon_{2,t} \end{pmatrix} \tag{9.23}$$

where $\boldsymbol{\varepsilon}_t = [\varepsilon_{1,t}, \varepsilon_{2,t}]^T$ is white noise with the covariance

$$\boldsymbol{\Sigma} = \begin{pmatrix} \sigma_1^2 & \sigma_{12} \\ \sigma_{21} & \sigma_2^2 \end{pmatrix}$$

Compare with the transfer function in Example 9.1 on page 248.

Given the existence of an inverse operator $\boldsymbol{\pi}(\mathrm{B})$ so that

$$\boldsymbol{\pi}(\mathrm{B})\boldsymbol{\psi}(\mathrm{B}) = \boldsymbol{I} \Leftrightarrow \boldsymbol{\pi}(\mathrm{B}) = \boldsymbol{\psi}^{-1}(\mathrm{B}), \tag{9.24}$$

the linear process (9.21) can be written in the form

$$\boldsymbol{\pi}(\mathrm{B})\boldsymbol{Y}_t = \boldsymbol{\varepsilon}_t, \tag{9.25}$$

where

$$\boldsymbol{\pi}(\mathrm{B}) = \boldsymbol{I} + \sum_{i=1}^{\infty} \boldsymbol{\pi}_i \, \mathrm{B}^i, \tag{9.26}$$

$\boldsymbol{\pi}(\mathrm{B})$ is referred to as the $\boldsymbol{\pi}$ *weights* of the multivariate process. Equation (9.25) is called the *inverse form.*

DEFINITION 9.3 (INVERTIBILITY FOR LINEAR PROCESSES)
The linear process $\boldsymbol{\pi}(\mathrm{B})\boldsymbol{Y}_t = \boldsymbol{\varepsilon}_t$ is said to be invertible if

$$\boldsymbol{\pi}(z) = \sum_{i=0}^{\infty} \boldsymbol{\pi}_i z^{-i} \tag{9.27}$$

converges for $|z| \geq 1$.

Invertibility ensures that the influence of past values of the process goes sufficiently fast to zero.

9.3 The multivariate ARMA process

The linear processes considered in the previous section are useful for deriving some results about multivariate stochastic models, but typically they are not useful for fitting to observed data since they contain too many parameters.

As in the univariate case in Chapter 5, parsimony can be achieved by considering the mixed representation using AR and MA polynomials simultaneously.

DEFINITION 9.4 (MULTIVARIATE ARMA PROCESS)
The multivariate (or vector) ARMA process is obtained by generalizing (5.92) on page 125 in the process model

$$\boldsymbol{Y}_t + \boldsymbol{\phi}_1 \boldsymbol{Y}_{t-1} + \cdots + \boldsymbol{\phi}_p \boldsymbol{Y}_{t-p} = \boldsymbol{\varepsilon}_t + \boldsymbol{\theta}_1 \boldsymbol{\varepsilon}_{t-1} + \cdots + \boldsymbol{\theta}_q \boldsymbol{\varepsilon}_{t-q}, \tag{9.28}$$

where $\{\boldsymbol{\varepsilon}_t\}$ is white noise. We shall refer to (9.28) as an *ARMA*$(\boldsymbol{P}, \boldsymbol{Q})$ *process* where $\boldsymbol{P}$ is a matrix with elements p_{ij} and $\boldsymbol{Q}$ is a matrix with elements q_{ij}. Sometimes the process is called a Vector ARMA or VARMA process.

By using the shift operator B, the ARMA$(\boldsymbol{P}, \boldsymbol{Q})$ process can be written

$$\boldsymbol{\phi}(\mathrm{B}) \boldsymbol{Y}_t = \boldsymbol{\theta}(\mathrm{B}) \boldsymbol{\varepsilon}_t, \tag{9.29}$$

where $\boldsymbol{\phi}(\mathrm{B})$ is a matrix of autoregressive operators whose elements $\phi_{ij}(\mathrm{B})$ are polynomials in the backward shift operator B of degree p_{ij}, and $\boldsymbol{\theta}(\mathrm{B})$ is a matrix of moving average operators whose elements $\theta_{ij}(\mathrm{B})$ are polynomials in the backward shift operator B of degree q_{ij}. It is noted that only the diagonal operators $\phi_{ii}(\mathrm{B})$ and $\theta_{ii}(\mathrm{B})$ have leading terms which are unity, whereas the off-diagonal elements have leading terms which are some power of B.

Alternatively the maximum order of the polynomials is used such that the multivariate ARMA(p, q) process can be written as

$$\boldsymbol{\phi}(\mathrm{B}) \boldsymbol{Y}_t = \boldsymbol{\theta}(\mathrm{B}) \boldsymbol{\varepsilon}_t, \tag{9.30}$$

where

$$\boldsymbol{\phi}(\mathrm{B}) = \boldsymbol{I} + \boldsymbol{\phi}_1 \mathrm{B} + \cdots + \boldsymbol{\phi}_p \mathrm{B}^p \tag{9.31}$$

$$\boldsymbol{\theta}(\mathrm{B}) = \boldsymbol{I} + \boldsymbol{\theta}_1 \mathrm{B} + \cdots + \boldsymbol{\theta}_q \mathrm{B}^q \tag{9.32}$$

For $q = 0$ we obtain the vector AR(p) process, whereas for $p = 0$ we have the vector MA(q) process.

THEOREM 9.3 (STATIONARITY)
An ARMA(p, q) process is stationary if all roots of $\det(\boldsymbol{\phi}(z^{-1})) = 0$, *with respect to z, lie within the unit circle.*

Proof Omitted. However, the result is easily seen by considering the fact that for the ARMA(p, q) process (9.30)

$$\boldsymbol{Y}_t = \boldsymbol{\phi}(\mathrm{B})^{-1}\boldsymbol{\theta}(\mathrm{B}) = \frac{\operatorname{adj}\boldsymbol{\phi}(\mathrm{B})}{\det(\boldsymbol{\phi}(\mathrm{B}))}\boldsymbol{\theta}(\mathrm{B}) \tag{9.33}$$

■

Similarly we have the following.

THEOREM 9.4 (INVERTIBILITY)
An ARMA(p, q) process is invertible if all roots of $\det(\boldsymbol{\theta}(z^{-1})) = 0$, *with respect to z, lie within the unit circle.*

Proof Omitted. ■

Remember that for univariate time series a given autocovariance function corresponds to more than one ARMA(p, q) process. To ensure a unique representation the concept of identifiability is imposed in the model selection.

▸ **Remark 9.1 (Identifiability)**
For multivariate ARMA models, left-multiplying both sides of (9.30) by an arbitrary non-singular matrix or a matrix polynomial in B yields a class of process with identical covariance matrix structures. Therefore, the stationarity and invertibility criterion have to be somehow extended. The solution is to choose models with minimal moving average order and minimum autoregressive order. For a complete discussion we refer to Hannan (1970). ◂

9.3.1 Theoretical covariance matrix functions

Still we assume that the process $\boldsymbol{Y}_t$ is stationary and that the mean is zero. We will now show how to calculate $\boldsymbol{\Gamma}(k)$ for pure autoregressive, pure moving average, and mixed autoregressive-moving average models.

9.3.1.1 Pure autoregressive models

Let us first write the pure autoregressive model in the form

$$\boldsymbol{Y}_t = -\boldsymbol{\phi}_1\boldsymbol{Y}_{t-1} - \cdots - \boldsymbol{\phi}_p\boldsymbol{Y}_{t-p} + \boldsymbol{\varepsilon}_t \tag{9.34}$$

By right multiplying by $\boldsymbol{Y}_t^T$ and using the expectation operator, we easily see that

$$\boldsymbol{\Gamma}(0) = -\boldsymbol{\Gamma}(-1)\boldsymbol{\phi}_1^T - \cdots - \boldsymbol{\Gamma}(-p)\boldsymbol{\phi}_p^T + \boldsymbol{\Sigma} \tag{9.35}$$

By instead multiplying by $\boldsymbol{Y}_{t-k}^T$, $(k > 0)$ we get

$$\boldsymbol{\Gamma}(k) = -\boldsymbol{\Gamma}(k-1)\boldsymbol{\phi}_1^T - \cdots - \boldsymbol{\Gamma}(k-p)\boldsymbol{\phi}_p^T \tag{9.36}$$

where $\boldsymbol{\Gamma}(-j) = \boldsymbol{\Gamma}(j)^T$.

9.3.1.2 Pure moving average models

The pure moving average model is

$$\boldsymbol{Y}_t = \boldsymbol{\varepsilon}_t + \boldsymbol{\theta}_1 \boldsymbol{\varepsilon}_{t-1} + \cdots + \boldsymbol{\theta}_q \boldsymbol{\varepsilon}_{t-q} \tag{9.37}$$

For this model the covariance matrix function is given by

$$\boldsymbol{\Gamma}(0) = \boldsymbol{\Sigma} + \boldsymbol{\theta}_1 \boldsymbol{\Sigma} \boldsymbol{\theta}_1^T + \cdots + \boldsymbol{\theta}_q \boldsymbol{\Sigma} \boldsymbol{\theta}_q^T \tag{9.38}$$

$$\boldsymbol{\Gamma}(k) = \begin{cases} \boldsymbol{\Sigma}\boldsymbol{\theta}_k^T + \boldsymbol{\theta}_1 \boldsymbol{\Sigma} \boldsymbol{\theta}_{k-1}^T + \cdots + \boldsymbol{\theta}_{q-k} \boldsymbol{\Sigma} \boldsymbol{\theta}_q^T & |k| = 1, 2, \ldots, q \\ \boldsymbol{0} & |k| > q \end{cases} \tag{9.39}$$

Note the similarity with (5.65) on page 118.

9.3.1.3 Mixed autoregressive moving average models

Similarly by multiplying by $\boldsymbol{Y}_{t-k}^T$ and taking expectations we get

$$\begin{aligned} \boldsymbol{\Gamma}(k) = & -\boldsymbol{\Gamma}(k-1)\boldsymbol{\phi}_1^T - \cdots - \boldsymbol{\Gamma}(k-p)\boldsymbol{\phi}_p^T \\ & + \boldsymbol{\Gamma}_{Y\varepsilon}(k) + \boldsymbol{\Gamma}_{Y\varepsilon}(k-1)\boldsymbol{\theta}_1^T + \cdots \\ & + \boldsymbol{\Gamma}_{Y\varepsilon}(k-q)\boldsymbol{\theta}_q^T, \quad \text{for } k \geq 0 \end{aligned} \tag{9.40}$$

where $\boldsymbol{\Gamma}_{Y\varepsilon}(k) = \mathrm{C}[\boldsymbol{Y}_t, \boldsymbol{\varepsilon}_{t+k}]$.

It is seen that, as in the univariate case (compare with (5.97) on page 126), the identification of mixed ARMA models is complicated by the fact that the first q covariance matrices follow no fixed pattern, but for $k > q$, they satisfy

$$\boldsymbol{\Gamma}(k) = -\boldsymbol{\Gamma}(k-1)\boldsymbol{\phi}_1^T - \cdots - \boldsymbol{\Gamma}(k-p)\boldsymbol{\phi}_p^T \tag{9.41}$$

which is the same recursive equation as for a pure autoregressive model. This will be used later in the definition of the q-conditioned partial correlation matrix, which in theory can be used for identifying mixed processes.

Example 9.3 (Theoretical correlation structures)
In this example we will consider three models and their theoretical correlation structures. The models are first order moving average, first order autoregressive, and a mixed first order autoregressive—first order moving average model.

Model A: First order moving average

$$\begin{pmatrix} Z_{1,t} \\ Z_{2,t} \end{pmatrix} = \begin{pmatrix} 1+1.7B & 2.1B \\ -0.7B & 1-1.1B \end{pmatrix} \begin{pmatrix} \varepsilon_{1,t} \\ \varepsilon_{2,t} \end{pmatrix} \qquad \boldsymbol{\Sigma} = \begin{pmatrix} 1.0 & 0.5 \\ 0.5 & 1.0 \end{pmatrix}$$

Model B: First order autoregressive

$$\begin{pmatrix} 1+0.9B & 2.4B \\ -0.9B & 1-1.9B \end{pmatrix} \begin{pmatrix} Z_{1,t} \\ Z_{2,t} \end{pmatrix} = \begin{pmatrix} \varepsilon_{1,t} \\ \varepsilon_{2,t} \end{pmatrix} \qquad \boldsymbol{\Sigma} = \begin{pmatrix} 1.0 & 0.75 \\ 0.75 & 1.0 \end{pmatrix}$$

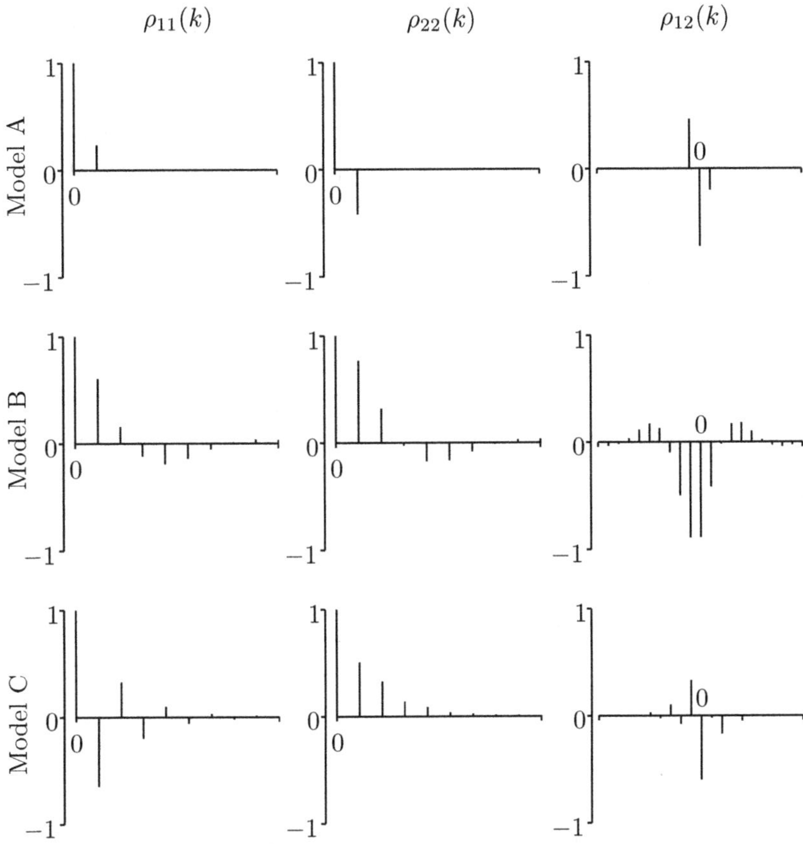

Figure 9.2: *Theoretical correlation functions corresponding to the three bivariate models.*

Model C: First order autoregressive—first order moving average

$$\begin{pmatrix} 1+0.4B & -0.3B \\ -0.3B & 1-0.4B \end{pmatrix} \begin{pmatrix} Z_{1,t} \\ Z_{2,t} \end{pmatrix} = \begin{pmatrix} 1-2.9B & -3.4B \\ 2.4B & 1+2.9B \end{pmatrix} \begin{pmatrix} \varepsilon_{1,t} \\ \varepsilon_{2,t} \end{pmatrix}$$

$$\boldsymbol{\Sigma} = \begin{pmatrix} 2 & -2 \\ -2 & 4 \end{pmatrix}$$

The theoretical correlation matrix functions for the three models are plotted in Figure 9.2. It can be seen in the figure that the three correlation functions have the following characteristics:

Model A Cut-off in the correlation functions at lags ± 1. In general a moving average process with maximum order q will show a cutoff for $k > q$.

Model B Damped sine wave behavior. In general for pure autoregressive processes the autocorrelation functions will behave as a mixture of real exponentials and damped sine waves. See also Example 9.4.

Model C For ρ_{22} an exponential decay is observed, but the decay starts in lag 1, rather than in lag 0. In general for mixed processes, the first few lagged correlations will follow no fixed pattern, but the tails will reflect the structure of the autoregressive part of the model.

It should be noticed that for multivariate processes the correlation between the individual signals will be reflected in the autocorrelation function for the individual series. This is illustrated in the following example.

Example 9.4 (Bivariate AR(1) behaves as univariate ARMA(2, 1))
This example illustrates that the autocorrelation for one of the signals in a two-dimensional AR(1) model behaves as the autocorrelation for a univariate ARMA(2, 1) model.

Consider the bivariate AR(1) model:

$$\begin{pmatrix} 1+\phi_{11}\,\mathrm{B} & \phi_{12}\,\mathrm{B} \\ \phi_{21}\,\mathrm{B} & 1+\phi_{22}\,\mathrm{B} \end{pmatrix} \begin{pmatrix} Y_{1,t} \\ Y_{2,t} \end{pmatrix} = \begin{pmatrix} \varepsilon_{1,t} \\ \varepsilon_{2,t} \end{pmatrix} \tag{9.42}$$

The covariance for the white noise process is

$$\mathbf{\Sigma} = \begin{pmatrix} \sigma_1^2 & \sigma_{12} \\ \sigma_{21} & \sigma_2^2 \end{pmatrix}$$

Consider now one of the two signals, $Y_t = Y_{1,t}$. The model for $\{Y_t\}$ can be written

$$\begin{pmatrix} Y_{1,t} \\ Y_{2,t} \end{pmatrix} = \begin{pmatrix} -\phi_{11} & -\phi_{12} \\ -\phi_{21} & -\phi_{22} \end{pmatrix} \begin{pmatrix} Y_{1,t-1} \\ Y_{2,t-1} \end{pmatrix} + \begin{pmatrix} \varepsilon_{1,t} \\ \varepsilon_{2,t} \end{pmatrix} \tag{9.43}$$

$$Y_t = \begin{pmatrix} 1 & 0 \end{pmatrix} \begin{pmatrix} Y_{1,t} \\ Y_{2,t} \end{pmatrix} \tag{9.44}$$

From (9.42) we see that

$$\begin{aligned} \begin{pmatrix} Y_{1,t} \\ Y_{2,t} \end{pmatrix} &= \frac{1}{(1+\phi_{11}\,\mathrm{B})(1+\phi_{22}\,\mathrm{B}) - \phi_{21}\phi_{12}\,\mathrm{B}^2} \\ &\quad \times \begin{pmatrix} 1+\phi_{22}\,\mathrm{B} & -\phi_{12}\,\mathrm{B} \\ -\phi_{21}\,\mathrm{B} & 1+\phi_{11}\,\mathrm{B} \end{pmatrix} \begin{pmatrix} \varepsilon_{1,t} \\ \varepsilon_{2,t} \end{pmatrix} \end{aligned} \tag{9.45}$$

Using (9.44) we see that

$$Y_t = \frac{(1+\phi_{22}\,\mathrm{B})\varepsilon_{1,t} - \phi_{12}\,\mathrm{B}\varepsilon_{2,t}}{1 + (\phi_{11}+\phi_{22})\,\mathrm{B} + (\phi_{11}\phi_{22} - \phi_{21}\phi_{12})\,\mathrm{B}^2} \tag{9.46}$$

From the denominator we see that the model for $\{Y_t\}$ must include a second order autoregressive part, i.e., $p = 2$.

A linear combination of two white noise processes is again a white noise process. Since then the numerator contains an MA(1) component plus a white noise component, the total process will be an MA(1).

In conclusion, (9.46) can be written as

$$(1 + \phi_1 \mathrm{B} + \phi_2 \mathrm{B}^2) Y_t = (1 + \theta_1 \mathrm{B}) \xi_t \,, \tag{9.47}$$

where $\{\xi_t\}$ is a white noise sequence with variance σ_ξ^2.

9.3.2 Partial correlation matrix

As with the univariate case the partial correlation matrix is a useful tool for identifying pure AR models, and we can use the following definition (see also Tiao and Box (1981)).

DEFINITION 9.5 (PARTIAL CORRELATION MATRIX)
Let us consider the multivariate AR(k) process

$$\boldsymbol{Y}_t = \boldsymbol{\phi}_{k1} \boldsymbol{Y}_{t-1} + \cdots + \boldsymbol{\phi}_{kk} \boldsymbol{Y}_{t-k} + \boldsymbol{\varepsilon}_t, \tag{9.48}$$

then the partial correlation matrix $\boldsymbol{\phi}_{kk}$ is defined as the last matrix coefficient in the AR(k) process (9.48).

Consider the following multivariate generalization of the Yule-Walker equations in unnormalized form.

$$\begin{pmatrix} \boldsymbol{\Gamma}(0) & \boldsymbol{\Gamma}(1)^T & \cdots & \boldsymbol{\Gamma}(k-1)^T \\ \boldsymbol{\Gamma}(1) & \boldsymbol{\Gamma}(0) & \cdots & \boldsymbol{\Gamma}(k-2)^T \\ \vdots & \vdots & \ddots & \vdots \\ \boldsymbol{\Gamma}(k-1) & \boldsymbol{\Gamma}(k-2) & \cdots & \boldsymbol{\Gamma}(0) \end{pmatrix} \begin{pmatrix} \boldsymbol{\phi}_{k1} \\ \boldsymbol{\phi}_{k2} \\ \vdots \\ \boldsymbol{\phi}_{kk} \end{pmatrix} = \begin{pmatrix} \boldsymbol{\Gamma}(1) \\ \boldsymbol{\Gamma}(2) \\ \vdots \\ \boldsymbol{\Gamma}(k) \end{pmatrix}. \tag{9.49}$$

The sequence of partial correlation matrices is thus found by solving (9.49) for successively higher orders of k, and a method for identifying the pure AR(k) process is built on the following theorem.

THEOREM 9.5 (PARTIAL AUTOCORRELATION MATRIX FOR AR PROCESSES)
For a multivariate AR(p) process it holds that

$$\begin{aligned} \boldsymbol{\phi}_{kk} &\neq \boldsymbol{0} \,, \quad k \leq p \\ \boldsymbol{\phi}_{kk} &= \boldsymbol{0} \,, \quad k > p. \end{aligned}$$

Proof Follows from the discussion above. ∎

In contrast the partial correlation matrix function for an MA(q) process will take values different from $\mathbf{0}$ for arbitrarily large k. A recursive formula for calculating the partial correlation matrix can be found in Wei (2006).

9.3.3 q-conditioned partial correlation matrix

As in the univariate case, for a mixed ARMA(p, q) process, neither the correlation matrices nor the partial correlation matrices have a cut-off. However, as shown in Section 9.3.1, for $k > q$, they satisfy

$$\mathbf{\Gamma}(k) = -\mathbf{\Gamma}(k-1)\boldsymbol{\phi}_1^T - \cdots - \mathbf{\Gamma}(k-p)\boldsymbol{\phi}_p^T \tag{9.50}$$

This is used for defining the q-conditioned partial correlation matrix as the last matrix coefficient in

$$\mathbf{\Gamma}(h) = \mathbf{\Gamma}(h-1)\boldsymbol{\phi}_{k1}^T + \cdots + \mathbf{\Gamma}(h-k)\boldsymbol{\phi}_{kk}^T \tag{9.51}$$

for $h = q+1, q+2, \ldots$.

By writing the generalized version of the Yule-Walker equations for $h = q+1, q+2, \ldots, q+k$, these equations can be solved for $\boldsymbol{\phi}_{kk}^T(q)$. It is seen that a mixed ARMA(p, q) process can be suggested in the identification step if the q-conditioned partial correlation matrix $\boldsymbol{\phi}_{kk}^T(q)$ has a "cut-off" at lag $k > p$, where q is the smallest value for which a "cut-off" is seen.

In practice, however, the use of the q-conditioned partial correlation matrices is not very useful due to approximations and estimation uncertainties.

9.3.4 VAR representation

All VARMA(p, q) processes can be represented by a VAR(1) process. Assume that $p \geq q+1$ then the VARMA(p, q) process

$$\boldsymbol{Y}_t + \boldsymbol{\phi}_1 \boldsymbol{Y}_{t-1} + \cdots + \boldsymbol{\phi}_p \boldsymbol{Y}_{t-p} = \boldsymbol{\varepsilon}_t + \boldsymbol{\theta}_1 \boldsymbol{\varepsilon}_{t-1} + \cdots + \boldsymbol{\theta}_q \boldsymbol{\varepsilon}_{t-q}$$

can be written as the VAR(1) process

$$\begin{pmatrix} \boldsymbol{Z}_{1,t} \\ \boldsymbol{Z}_{2,t} \\ \vdots \\ \boldsymbol{Z}_{p,t} \end{pmatrix} = \begin{pmatrix} -\boldsymbol{\phi}_1 & \boldsymbol{I} & \mathbf{0} & \cdots & \mathbf{0} \\ -\boldsymbol{\phi}_2 & \mathbf{0} & \boldsymbol{I} & \cdots & \mathbf{0} \\ \vdots & \vdots & \vdots & \ddots & \mathbf{0} \\ -\boldsymbol{\phi}_{p-1} & \mathbf{0} & \mathbf{0} & \cdots & \boldsymbol{I} \\ -\boldsymbol{\phi}_p & \mathbf{0} & \mathbf{0} & \cdots & \mathbf{0} \end{pmatrix} \begin{pmatrix} \boldsymbol{Z}_{1,t-1} \\ \boldsymbol{Z}_{2,t-1} \\ \vdots \\ \boldsymbol{Z}_{p,t-1} \end{pmatrix} + \begin{pmatrix} \boldsymbol{I} \\ \boldsymbol{\theta}_1 \\ \vdots \\ \boldsymbol{\theta}_{p-1} \end{pmatrix} \boldsymbol{\varepsilon}_t \tag{9.52}$$

where $\boldsymbol{Z}_{1,t} = \boldsymbol{Y}_t$.

9.4 Non-stationary models

In the analysis of time series, non-stationarities such as trends and seasonal behavior are often seen. As for univariate time series, trends are often dealt with using the difference operator (∇), whereas seasonal non-stationarity is dealt with by introducing the seasonal difference operator (∇_s).

9.4.1 The multivariate ARIMA process

For an m-dimensional multivariate time series some of the signals may show some sort of non-stationarity, whereas some other signals are stationary. This calls for an individual differencing of the various signals.

DEFINITION 9.6 (THE MULTIVARIATE ARIMA($\boldsymbol{P}, \boldsymbol{d}, \boldsymbol{Q}$) PROCESS)
The process $\{\boldsymbol{Y}_t\}$ is called a multivariate *integrated* (or summarised) *autoregressive moving average* or, in short, an *ARIMA*($\boldsymbol{P}, \boldsymbol{d}, \boldsymbol{Q}$) *process* if it can be written in the form

$$\boldsymbol{\phi}(B)\boldsymbol{W}_t = \boldsymbol{\theta}(B)\boldsymbol{\varepsilon}_t, \tag{9.53}$$

where $\{\boldsymbol{\varepsilon}_t\}$ is multivariate white noise. The ARMA part of the model is as described previously in (9.29), and $\boldsymbol{d}$ is a vector with elements d_i, where d_i is a non-negative integer describing the order of differencing on the i'th signal in $\boldsymbol{Y}_t$, i.e., the vector $\boldsymbol{W}_t$ has the elements $w_{i,t} = \nabla^{d_i} Y_{i,t}$.

Alternatively, if we define a multivariate differencing operator $\boldsymbol{D}(B)$:

$$\boldsymbol{D}(\mathrm{B}) = \begin{pmatrix} (1-\mathrm{B})^{d_1} & 0 & \cdots & 0 \\ 0 & (1-\mathrm{B})^{d_2} & \cdots & 0 \\ \vdots & \vdots & \ddots & 0 \\ 0 & 0 & \cdots & (1-\mathrm{B})^{d_2} \end{pmatrix} \tag{9.54}$$

we are able to write the ARIMA model as

$$\boldsymbol{\phi}(\mathrm{B})\boldsymbol{D}(\mathrm{B})\boldsymbol{Y}_t = \boldsymbol{\theta}(\mathrm{B})\boldsymbol{\varepsilon}_t, \tag{9.55}$$

where the zeros of $\det(\boldsymbol{\phi}(z^{-1}))$ and $\det(\boldsymbol{\theta}(z^{-1}))$ are inside the unit circle.

9.4.2 The multivariate seasonal model

Seasonal models are useful for describing, e.g., diurnal variations or annual variations in time series. Very often we focus on a single *seasonal period* s; however, in many cases more than one seasonal period is needed. For instance for hourly variations of heat consumption in a district heating system we most likely need a diurnal period, a weekly period, and an annual seasonal period.

Remember that in the univariate case, the multiplicative seasonal model is of the form

$$\phi(B)\Phi(B^s)\nabla^d\nabla_s^D Y_t = \theta(B)\Theta(B^s)\varepsilon_t \tag{9.56}$$

where $\{\varepsilon_t\}$ is white noise and ϕ and θ are polynomials of order p and q, respectively, and Φ and Θ are polynomials of order P and Q in the seasonal lag operator B^s.

As in the univariate case multivariate models are most often introduced as *multiplicative* models. The most important attraction of multiplicative models is that they very often achieve parsimony in representation (i.e., number of needed parameters) compared with *non-multiplicative* models in which individual parameters are introduced at all relevant lags.

In the multivariate case the seasonal components are often limited to some of the diagonal elements of the autoregressive or moving average matrices. Let us consider the following example of a multiplicative multivariate seasonal model with only one seasonal period s:

$$\begin{aligned}&\begin{pmatrix}(1-\phi_{11}\,\mathrm{B})(1-\Phi_{11}\,\mathrm{B}^s) & \phi_{12}\,\mathrm{B}\\ 0 & (1-\phi_{22}\,\mathrm{B})(1-\Phi_{22}\,\mathrm{B}^s)\end{pmatrix}\begin{pmatrix}\nabla Y_{1,t}\\ \nabla\nabla_s Y_{2,t}\end{pmatrix}\\ &\quad=\begin{pmatrix}(1-\theta_{11}\,\mathrm{B})(1-\Theta_{11}\,\mathrm{B}^s) & \theta_{12}\,\mathrm{B}\\ -\theta_{21}\,\mathrm{B} & (1-\theta_{22}\,\mathrm{B})(1-\Theta_{22}\,\mathrm{B}^s)\end{pmatrix}\begin{pmatrix}\varepsilon_{1,t}\\ \varepsilon_{2,t}\end{pmatrix}\end{aligned} \tag{9.57}$$

9.4.3 Time-varying models

For multivariate models a time-varying mean can be introduced as in Section 5.6.4 on page 134; this also includes models with covariates. Likewise, models with time-varying coefficients can be introduced as in Section 5.6.5 on page 135. This subject will not be covered further here; however, some time-varying models are introduced in Chapter 10, and adaptive methods for dealing with slowly varying models are described in Chapter 11.

9.5 Prediction

Now we shall consider how to predict future values of the multivariate stochastic process $\{\boldsymbol{Y}_t\}$ based on past observations. It is assumed that observations from the process up to time t, i.e., $\{\boldsymbol{Y}_s\,, s\le t\}$, are observed, and now we want to predict the value of $\boldsymbol{Y}_{t+k}$. Hence, the prediction horizon or lead time is k. Furthermore it is assumed that there are no missing observations. In the case of missing observations for multivariate time series, we refer to the procedures described in either Section 9.5.1 or Section 10.5.

Based on the observations $\boldsymbol{Y}_t, \boldsymbol{Y}_{t-1},\ldots$, we seek to predict $\boldsymbol{Y}_{t+k}$ $(k>0)$, and we thus introduce the *predictor*

$$\widehat{\boldsymbol{Y}}_{t+k|t} = \mathrm{g}(\boldsymbol{Y}_t, \boldsymbol{Y}_{t-1},\ldots). \tag{9.58}$$

The *prediction error* is

$$\boldsymbol{e}_{t+k|t} = \boldsymbol{Y}_{t+k} - \widehat{\boldsymbol{Y}}_{t+k|t} \tag{9.59}$$

Assuming that the prediction error is unbiased, the *covariance matrix of the prediction error* is

$$\boldsymbol{V}(k) = \mathrm{C}[\boldsymbol{e}_{t+k|t}, \boldsymbol{e}_{t+k|t}] = \mathrm{E}[\boldsymbol{e}_{t+k|t}\boldsymbol{e}_{t+k|t}^T] \tag{9.60}$$

In order to find an objective function we seek a reasonable scalar function of $\boldsymbol{V}(k)$. Assuming normality of the predictions, minimizing $\det \boldsymbol{V}(k)$ is equivalent to minimizing the volume of the hyper-ellipsoid defining the probability region for the prediction, which is minimized for

$$\mathrm{g}(\boldsymbol{Y}_t, \boldsymbol{Y}_{t-1}, \dots) = \mathrm{g}_0(\boldsymbol{Y}_t, \boldsymbol{Y}_{t-1}, \dots) = \mathrm{E}[\boldsymbol{Y}_{t+k}|\boldsymbol{Y}_t, \boldsymbol{Y}_{t-1}, \dots]. \tag{9.61}$$

Hence, the *optimal predictor* (or forecast) for $\boldsymbol{Y}_{t+k}$ becomes

$$\widehat{\boldsymbol{Y}}_{t+k|t} = \mathrm{E}[\boldsymbol{Y}_{t+k}|\boldsymbol{Y}_t, \boldsymbol{Y}_{t-1}, \dots] \tag{9.62}$$

The optimal predictor is thus the conditional mean. Furthermore, it is known that in the case of a normal process the conditional mean is linear—see also Theorem 5.15. Here we will limit our attention to the linear case, where the optimal prediction is assumed linear.

Let us consider any linear multivariate model AR, ARMA, ARIMA, etc. In order to find a formula for the covariance of the prediction error, it is useful to consider the MA form (or the $\boldsymbol{\psi}$ weight form) of the model

$$\boldsymbol{Y}_t = \boldsymbol{\varepsilon}_t + \boldsymbol{\psi}_1\boldsymbol{\varepsilon}_{t-1} + \boldsymbol{\psi}_2\boldsymbol{\varepsilon}_{t-2} + \cdots . \tag{9.63}$$

where $\{\boldsymbol{\varepsilon}_t\}$ is white noise with covariance $\boldsymbol{\Sigma}$.

As shown in Section 5.7.1 on page 137, knowledge about all previous values of $\boldsymbol{Y}_t$ is equivalent to knowledge of all previous values of $\boldsymbol{\varepsilon}_t$, and since $\{\boldsymbol{\varepsilon}_t\}$ is white noise, we have

$$\mathrm{E}[\boldsymbol{\varepsilon}_{t+k}|\boldsymbol{Y}_t, \boldsymbol{Y}_{t-1}, \dots] = \begin{cases} \boldsymbol{\varepsilon}_{t+k} & \text{for } k \le 0 \\ \mathbf{0} & \text{for } k > 0 \end{cases} \tag{9.64}$$

In order to find the prediction at *origin* t with *lead time* k, we write

$$\boldsymbol{Y}_{t+k} = \boldsymbol{\varepsilon}_{t+k} + \boldsymbol{\psi}_1\boldsymbol{\varepsilon}_{t+k-1} + \cdots + \boldsymbol{\psi}_k\boldsymbol{\varepsilon}_t + \boldsymbol{\psi}_{k+1}\boldsymbol{\varepsilon}_{t-1} + \cdots \tag{9.65}$$

Using the conditional mean on (9.65) while using (9.64) leads to the following *predictor*

$$\widehat{\boldsymbol{Y}}_{t+k|t} = \boldsymbol{\psi}_k\boldsymbol{\varepsilon}_t + \boldsymbol{\psi}_{k+1}\boldsymbol{\varepsilon}_{t-1} + \cdots . \tag{9.66}$$

By subtracting (9.66) from (9.65), we obtain the *prediction error*:

$$\begin{aligned} \boldsymbol{e}_{t+k|t} &= \boldsymbol{Y}_{t+k} - \widehat{\boldsymbol{Y}}_{t+k|t} \\ &= \boldsymbol{\varepsilon}_{t+k} + \boldsymbol{\psi}_1 \boldsymbol{\varepsilon}_{t+k-1} + \cdots + \boldsymbol{\psi}_{k-1} \boldsymbol{\varepsilon}_{t+1}. \end{aligned} \tag{9.67}$$

Now it is readily seen that the *covariance of the prediction error* becomes

$$\boldsymbol{V}(k) = \boldsymbol{\Sigma} + \boldsymbol{\psi}_1 \boldsymbol{\Sigma} \boldsymbol{\psi}_1^T + \cdots + \boldsymbol{\psi}_{k-1} \boldsymbol{\Sigma} \boldsymbol{\psi}_{k-1}^T. \tag{9.68}$$

Hence, the covariance of a one-step prediction is always $\boldsymbol{V}(1) = \boldsymbol{\Sigma}$.

▸ **Remark 9.2**

The values of the $\boldsymbol{\psi}$ matrices are often most easily found by a convenient generalization of Remark 4.1 on page 71. It is noticed that using the identity matrix as input in the convolution sum (9.21) on page 252 provides the needed values of the $\boldsymbol{\psi}$ matrices as output. Hence, we simply consider an appropriate formulation of the process and take

$$\boldsymbol{\varepsilon}_t = \begin{cases} \boldsymbol{I} & \text{for } t = 0 \\ \boldsymbol{0} & \text{for } t \neq 0 \end{cases} \tag{9.69}$$

as the input for the process. Then it is seen that the calculated output will be the needed sequence of $\boldsymbol{\psi}$ matrices. ◂

The practical calculation of the predictions (forecasts) is most easily carried out by applying the conditional mean operator directly to the process written as a (multivariate) difference equation for the point in time where the forecast is considered. Notice that in order to use the conditional mean operator all multiplicative polynomials in the backward shift operator must be turned into just one polynomial by multiplying out the polynomials.

The rules for evaluating the conditional means are

$$\begin{aligned} &\mathrm{E}[\boldsymbol{Y}_{t-j} | \boldsymbol{Y}_t, \boldsymbol{Y}_{t-1}, \dots] = \boldsymbol{Y}_{t-j} && j = 0, 1, 2, \dots \\ &\mathrm{E}[\boldsymbol{Y}_{t+j} | \boldsymbol{Y}_t, \boldsymbol{Y}_{t-1}, \dots] = \widehat{\boldsymbol{Y}}_{t+j|t} && j = 1, 2, \dots \\ &\mathrm{E}[\boldsymbol{\varepsilon}_{t-j} | \boldsymbol{Y}_t, \boldsymbol{Y}_{t-1}, \dots] = \boldsymbol{\varepsilon}_{t-j} && j = 0, 1, 2, \dots \\ &\mathrm{E}[\boldsymbol{\varepsilon}_{t+j} | \boldsymbol{Y}_t, \boldsymbol{Y}_{t-1}, \dots] = \boldsymbol{0} && j = 1, 2, \dots \end{aligned} \tag{9.70}$$

The rules in (9.70) are almost exactly as described for univariate processes in Section 5.7. Let us consider an example.

Example 9.5 (Predictions in the multivariate ARMA(1, 1) process)
Let us first consider predictions in the multivariate ARMA(1, 1) model. We assume that the process $\{\boldsymbol{Y}_t\}$ is observed up to and including time t. As almost always we assume that the actual time is t. In order to calculate the

prediction of the process at $t+1$, the model is written

$$\boldsymbol{Y}_{t+1} = -\boldsymbol{\phi}_1 \boldsymbol{Y}_t + \boldsymbol{\varepsilon}_{t+1} + \boldsymbol{\theta}_1 \boldsymbol{\varepsilon}_t$$

By using the conditional expectation operator and using the rules in (9.70), the predicted value for time $t+1$ is readily seen to be

$$\widehat{\boldsymbol{Y}}_{t+1|t} = -\boldsymbol{\phi}_1 \boldsymbol{Y}_t + \boldsymbol{\theta}_1 \boldsymbol{\varepsilon}_t \tag{9.71}$$

where $\boldsymbol{\varepsilon}_t$ can be calculated by starting up the one-step predictions at, say, $t-20$ by putting the first values of $\boldsymbol{\varepsilon}_t$ equal to zero and then successively estimating (approximating) $\boldsymbol{\varepsilon}_t$ by the prediction error

$$\boldsymbol{e}_{t+1|t} = \boldsymbol{Y}_{t+1} - \widehat{\boldsymbol{Y}}_{t+1|t} \tag{9.72}$$

From (9.68) it is seen that the covariance of the one-step prediction is $\boldsymbol{V}(1) = \boldsymbol{\Sigma}$.

For $k = 2, 3, \ldots$ it is readily seen that we obtain

$$\widehat{\boldsymbol{Y}}_{t+k|t} = -\boldsymbol{\phi}_1 \widehat{\boldsymbol{Y}}_{t+k-1|t} \tag{9.73}$$

Using the remark above it is seen that we obtain the following set of recurrence relationships for the $\boldsymbol{\psi}$ matrices:

$$\begin{aligned}
\boldsymbol{\psi}_1 &= -\boldsymbol{\phi}_1 + \boldsymbol{\theta}_1 \\
\boldsymbol{\psi}_2 &= -\boldsymbol{\phi}_1 \boldsymbol{\psi}_1 \\
&\ \vdots \\
\boldsymbol{\psi}_j &= -\boldsymbol{\phi}_1 \boldsymbol{\psi}_{j-1}
\end{aligned}$$

It should be noticed, however, that in most cases the prediction horizon k is only a few time steps.

9.5.1 Missing values for some signals

In prediction of multivariate time series very often the measurements of one or more of the individual signals are missing from time to time. In that case the linear projection from Theorem 2.6 can be used.

The following example illustrates the practical use of predictions and the *linear projection theorem* in a simple setting.

Example 9.6 (Missing signal values)
In most modern cities a monitoring of the air pollution is taken, and if a pollution above some limits is expected in the near future a warning is given to the public. In order to provide this warning, time series analysis is very useful.

In this example it is assumed that a model for the variation of the pollutants is established. Using the established model and the most recent data, predictions of the pollutants for the next hours are calculated. If the probability for the pollutants to become larger than some defined (critical) values is larger than, say, 20% a warning is given.

Carstensen (1990) has used time series analysis to set up models for NO and NO_2 for a particular location (Jagtvej) in Copenhagen.

In the following we will assume that measurements in general are available every third hour. Furthermore, we will assume that the measurements of NO and NO_2 are described by a simple first order bivariate autoregressive model.

Let us introduce $Y_{1,t} \sim NO_{2,t} - \mu_{NO_2}$ and $Y_{2,t} \sim NO_t - \mu_{NO}$. The model is then:

$$\begin{pmatrix} Y_{1,t} \\ Y_{2,t} \end{pmatrix} = \begin{pmatrix} 0.9 & -0.1 \\ 0.4 & 0.8 \end{pmatrix} \begin{pmatrix} Y_{1,t-1} \\ Y_{2,t-1} \end{pmatrix} + \begin{pmatrix} \varepsilon_{1,t} \\ \varepsilon_{2,t} \end{pmatrix}$$

or—using a matrix notation,

$$\boldsymbol{Y}_t = \boldsymbol{\Phi} \boldsymbol{Y}_{t-1} + \boldsymbol{\varepsilon}_t$$

where $\boldsymbol{\varepsilon}_t$ is white noise with the covariance

$$\mathrm{Var}[\boldsymbol{\varepsilon}_t] = \boldsymbol{\Sigma} = \begin{pmatrix} \sigma_1^2 & \sigma_{12} \\ \sigma_{21} & \sigma_2^2 \end{pmatrix} = \begin{pmatrix} 30 & 21 \\ 21 & 23 \end{pmatrix}$$

Finally we have $\mu_{NO_2} = 48\,\mu g/m^3$ and $\mu_{NO} = 79\,\mu g/m^3$.

Now we shall consider the predictions calculated shortly after 9 a.m. The most recent measurements are obtained at 9 a.m. (corresponding to time point t in the following), and the observations are $64\,\mu g/m^3 NO_2$ and $93\,\mu g/m^3 NO$.

The prediction of the concentration/pollution at 12 noon (i.e., time point $t+1$) becomes

$$\widehat{\boldsymbol{Y}}_{t+1|t} = \boldsymbol{\Phi} \boldsymbol{Y}_t = \begin{pmatrix} 0.9 & -0.1 \\ 0.4 & 0.8 \end{pmatrix} \begin{pmatrix} 16 \\ 14 \end{pmatrix} = \begin{pmatrix} 13.0 \\ 17.6 \end{pmatrix}$$

which corresponds to $61\,\mu g/m^3 NO_2$ and $96.6\,\mu g/m^3 NO$.

The covariance of the prediction is $\boldsymbol{\Sigma}$. This matrix contains the variance of the one-step predictions of the individual pollutants and the covariance between the predictions.

Now the time point is $(t+1) \sim$ 12:00 noon. The observation of $NO_{2,t+1} = 67\,\mu g/m^3 NO_2$, while the observation of NO_{t+1} is *missing* due to some troubles with the measurement equipment.

Now we will illustrate how the formulas for the linear projections can be used to estimate the missing concentration, i.e., to estimate NO_{t+1}. (In order to ease the notation we will avoid to condition on all previous observations).

$$\begin{aligned}
\mathrm{E}[Y_{2,t+1}|Y_{1,t+1}] &= \mathrm{E}[Y_{2,t+1}] + \frac{\sigma_{21}}{\sigma_1^2}(Y_{1,t+1} - \mathrm{E}[Y_{1,t+1}]) \\
&= 17.6 + \frac{21}{30}(19 - 13) = 21.8 \sim 100.8\,\mu\mathrm{g/m^3NO}.
\end{aligned}$$

The variance on this estimate is

$$\begin{aligned}
\mathrm{Var}[Y_{2,t+1} \mid Y_{1,t+1}] &= \mathrm{Var}[Y_{2,t+1}] - \frac{\sigma_{21}^2}{\sigma_1^2} \\
&= 23 - \frac{21^2}{30} = 8.3 \sim (2.9\,\mu\mathrm{g/m^3NO})^2
\end{aligned}$$

It is seen that the predicted value is above $100\,\mu\mathrm{g/m}^3$.

This example illustrates that the correlation between NO and NO_2 is used in the projection formulas to provide a rather precise estimation of the missing observation of NO_2. Since the estimate is at the actual point in time, this estimate is often called the *filter estimate* of the missing observation.

9.6 Identification of multivariate models

Until now we have considered the theory for multivariate stochastic processes; now we will consider the case of modeling, and it is assumed that an observed time series is available.

In principle modeling of multivariate time series is similar to the methods used for univariate models described in Chapter 6. It follows the same Box and Jenkins principles as sketched in Figure 6.1 on page 145, i.e., the principles for *identification, estimation, and model checking* are in general as for univariate time series. In this section we shall focus on the identification step. Later, methods for estimation and model checking of the identified models are considered, and then, if inadequacies are discovered, a further cycle of identification, estimation, and model checking is carried out, and so on.

Given an observed multivariate time series $\boldsymbol{Y}_1, \boldsymbol{Y}_2, \ldots, \boldsymbol{Y}_N$ the underlying model is identified from the pattern of its sample correlation and partial correlation matrices. However, initially a proper transformation and differencing of each signal in the multivariate time series is introduced according to exactly the rules mentioned previously in Sections 6.3.1 and 6.7. Hence, we will assume that an appropriate transformation and differencing is selected for each of the signals in the multivariate time series.

The key to the identification is the sample correlation and partial correlation matrices.

The *sample covariance matrix* at lag k, $\boldsymbol{C}(k)$, is

$$\boldsymbol{C}(k) = \frac{1}{N}\sum_{t=1}^{N-k}(\boldsymbol{Y}_t - \overline{\boldsymbol{Y}})(\boldsymbol{Y}_{t+k} - \overline{\boldsymbol{Y}})^T \tag{9.74}$$

for $k = 0, 1, \ldots, N-1$. Furthermore, $\overline{\boldsymbol{Y}} = (\sum_{t=1}^{N} \boldsymbol{Y}_t)/N$. Notice the similarity with the estimator for the univariate case in (6.1).

The *sample correlation matrix* at lag k, $\boldsymbol{R}(k)$, is then obtained by using the theoretical relation in (9.13) or alternatively

$$r_{ij}(k) = \frac{c_{ij}(k)}{\sqrt{c_{ii}(0)c_{jj}(0)}} \tag{9.75}$$

where c_{ij} and $r_{ij} = \widehat{\rho}_{ij}$ are the elements in $\boldsymbol{C}$ and $\boldsymbol{R}$, respectively.

As seen in Section 9.3.1, the theoretical correlation matrix of a moving average model with maximum order q has a cut-off after lag $|k| > q$. For an evaluation of a point at which a cut-off takes place, use may be made of an approximate formula which is embedded in the following theorem.

THEOREM 9.6
The estimated cross-correlation between two mutually uncorrelated time series (i.e., $\rho_{ij}(k) = 0$ for all values of k) is asymptotically normally distributed having

$$\mathrm{E}[\widehat{\rho}_{ij}(k)] \simeq 0 \tag{9.76}$$

$$\mathrm{Var}[\widehat{\rho}_{ij}(k)] \simeq \frac{1}{N}\left[1 + 2\sum_{h=1}^{\infty} \rho_{ii}(h)\rho_{jj}(h)\right], \tag{9.77}$$

where the theoretical autocorrelations $\rho_{ii}(h)$ may be estimated by their sample estimates $r_{ii}(h)$.

Proof Omitted. See Bartlett (1946). ∎

The *sample partial correlation matrix*, $\boldsymbol{S}_k$, is estimated as the estimate of the last coefficient matrix in the autoregressive model of order k as described previously in Section 9.3.2.

For identification of pure vector autoregressive models the fact that the partial correlation matrix for a vector autoregressive model with maximum order p has a cut-off after lag $|k| > p$ is used. To determine the point of cut-off, a useful result is that the estimates of the elements of the partial correlation matrix have an approximate standard error $1/\sqrt{N}$ given that $k > p$; this is the same as in Theorem 6.3 on page 154.

In principle for mixed models the q-conditioned partial correlation matrix can be used for identification. However, in practice it is rather difficult to use this quantity for identification, and hence, more iterative procedures are called for—exactly as for univariate time series.

9.6.1 Identification using pre-whitening

We suggest, as described in the previous section, that identification should be based on the correlation and partial correlation matrices of the stationary time series eventually obtained by differencing the original time series.

In the literature alternative methods for identification are suggested. One of the most frequently used approaches is first to fit univariate time series to each of the signals of the multivariate time series. By this approach we obtain white noise residuals ε_{it} of all the univariate time series. Following this *pre-whitening* of all the signals, the cross-correlations between the sequences of residuals are then used to find a multivariate MA part, which then links together the individual series. This method is suggested in, e.g., Jenkins and Alavi (1981) and Jenkins (1975). However, the method is not recommended. Let us illustrate the problems by considering the true multivariate ARMA(p, q) process:

$$\boldsymbol{\phi}(B)\boldsymbol{Y}_t = \boldsymbol{\theta}(B)\boldsymbol{\varepsilon}_t.$$

The model can be written on MA form

$$\boldsymbol{Y}_t = \boldsymbol{\phi}^{-1}(B)\boldsymbol{\theta}(B)\boldsymbol{\varepsilon}_t,$$

or

$$\det(\boldsymbol{\phi}(B))\boldsymbol{Y}_t = \operatorname{adj}\boldsymbol{\phi}(B)\boldsymbol{\theta}(B)\boldsymbol{\varepsilon}_t. \tag{9.78}$$

where $\operatorname{adj}\boldsymbol{\phi}(B)$ denotes the adjoint matrix. Hence, it is concluded that any multivariate ARMA model may be written as

$$\boldsymbol{\phi}_1(B)\boldsymbol{Y}_t = \boldsymbol{\theta}_1(B)\boldsymbol{\varepsilon}_t,$$

where

1. The autoregressive matrix $\boldsymbol{\phi}_1(B)$ is diagonal with identical elements, all equal to the determinant of $\boldsymbol{\phi}(B)$;

2. The moving average matrix $\boldsymbol{\theta}_1(B)$ is of maximum order $q_1 = q^* \times q$ where q^* is the maximum order of the adjoint matrix in (9.78).

Thus, it is seen that this method leads to a mis-specification of the AR part and an over-parameterisation of the MA part of the model.

The conclusion is that fitting models to the individual signals and then finding a model for the cross-correlation leads to a mis-specified model. This is also illustrated in Example 9.7 on page 275.

Instead, the method based on correlation matrices described in Section 9.6 should be used.

9.7 Estimation of parameters

Having identified the order of the model, we are now ready to estimate the model parameters. In this section we will describe both least squares and maximum likelihood methods for parameter estimation.

9.7.1 Least squares estimation

Let us consider the multivariate ARX(p) model

$$\boldsymbol{Y}_t + \boldsymbol{\phi}_1 \boldsymbol{Y}_{t-1} + \cdots + \boldsymbol{\phi}_p \boldsymbol{Y}_{t-p} = \boldsymbol{\omega}_1 \boldsymbol{u}_{t-1} + \cdots + \boldsymbol{\omega}_r \boldsymbol{u}_{t-r} + \boldsymbol{\varepsilon}_t, \tag{9.79}$$

where $\dim(\boldsymbol{Y}_t) = m$ and $\dim(\boldsymbol{u}_t) = s$. Furthermore $\boldsymbol{\phi}_1, \ldots, \boldsymbol{\phi}_p$ are $m \times m$ matrices containing the autoregressive parameters, $\boldsymbol{\omega}_1, \ldots, \boldsymbol{\omega}_r$ are $m \times s$ matrices, and $\boldsymbol{\varepsilon}_t$ ($m \times 1$) is multivariate white noise with covariance $\boldsymbol{\Sigma}$.

First we can write

$$\boldsymbol{Y}_t^T = -\sum_{i=1}^{p} \boldsymbol{Y}_{t-i}^T \boldsymbol{\phi}_i^T + \sum_{j=1}^{r} \boldsymbol{u}_{t-j}^T \boldsymbol{\omega}_j^T + \boldsymbol{\varepsilon}_t^T \tag{9.80}$$

or

$$\boldsymbol{Y}_t^T = \boldsymbol{X}_t^T \boldsymbol{\theta} + \boldsymbol{\varepsilon}_t^T \tag{9.81}$$

where

$$\boldsymbol{X}_t^T = [-\boldsymbol{Y}_{t-1}^T, \ldots, -\boldsymbol{Y}_{t-p}^T, \boldsymbol{u}_{t-1}^T, \ldots, \boldsymbol{u}_{t-r}^T] \tag{9.82}$$

$$\boldsymbol{\theta} = [\boldsymbol{\phi}_1^T, \ldots, \boldsymbol{\phi}_p^T, \boldsymbol{\omega}_1^T, \ldots, \boldsymbol{\omega}_r^T] \tag{9.83}$$

Equation (9.81) is a multivariate general linear model.

Given N observations $\boldsymbol{Y}_1, \ldots, \boldsymbol{Y}_N$ and the needed input variables, we have

$$\boldsymbol{Y} = \boldsymbol{X}\boldsymbol{\theta} + \boldsymbol{\varepsilon} \tag{9.84}$$

where

$$\boldsymbol{Y} = \begin{pmatrix} \boldsymbol{Y}_{p+1}^T \\ \vdots \\ \boldsymbol{Y}_N^T \end{pmatrix}, \quad \boldsymbol{X} = \begin{pmatrix} \boldsymbol{X}_{p+1}^T \\ \vdots \\ \boldsymbol{X}_N^T \end{pmatrix}, \quad \boldsymbol{\varepsilon} = \begin{pmatrix} \boldsymbol{\varepsilon}_{p+1}^T \\ \vdots \\ \boldsymbol{\varepsilon}_N^T \end{pmatrix} \tag{9.85}$$

Notice the similarity with (6.35) on page 160.

The LS estimator is given by

$$\widehat{\boldsymbol{\theta}} = (\boldsymbol{X}^T \boldsymbol{X})^{-1} \boldsymbol{X}^T \boldsymbol{Y} \tag{9.86}$$

As an estimate of $\boldsymbol{\Sigma}$ we can take

$$\widehat{\boldsymbol{\Sigma}} = \frac{(\boldsymbol{Y} - \boldsymbol{X}\widehat{\boldsymbol{\theta}})^T (\boldsymbol{Y} - \boldsymbol{X}\widehat{\boldsymbol{\theta}})}{(N-p)} = \sum_{p+1}^{N} \frac{\boldsymbol{\varepsilon}_t(\widehat{\boldsymbol{\theta}}) \boldsymbol{\varepsilon}_t^T(\widehat{\boldsymbol{\theta}})}{(N-p)} \tag{9.87}$$

The properties for the LS estimator in the multivariate non-dynamic GLM model where $\boldsymbol{\varepsilon}_t \sim \mathrm{N}(\mathbf{0}, \boldsymbol{\Sigma})$ are

1. $\widehat{\boldsymbol{\theta}}$ is joint multivariate normal with

2. $\mathrm{E}[\widehat{\boldsymbol{\theta}}] = \boldsymbol{\theta}$

3. The variance of the estimator is

$$\mathrm{Var}[\widehat{\boldsymbol{\theta}}] = \boldsymbol{\Sigma} \otimes (\boldsymbol{X}^T \boldsymbol{X})^{-1}, \tag{9.88}$$

where $\otimes$ denotes the Kronecker product.

In the case of a dynamical model such as the ARX model, the properties are valid asymptotically—cf. Theorem 6.6 on page 162.

In the case where $\boldsymbol{\varepsilon}_t \sim \mathrm{N}(\mathbf{0}, \boldsymbol{\Sigma})$ the estimators for $\boldsymbol{\theta}$ and $\boldsymbol{\Sigma}$ presented are also the maximum likelihood estimates.

▸ **Remark 9.3 (Nonzero mean values)**
It should be noted that nonzero mean values can be treated by extending (9.80) to

$$\boldsymbol{Y}_t^T = \boldsymbol{\alpha}^T - \sum_{i=1}^{p} \boldsymbol{Y}_{t-i}^T \boldsymbol{\phi}_i^T + \sum_{j=1}^{r} \boldsymbol{u}_{t-j}^T \boldsymbol{\omega}_j^T + \boldsymbol{\varepsilon}_t^T \tag{9.89}$$

and then $\boldsymbol{X}_t$ and $\boldsymbol{\theta}$ in (9.81) should be changed accordingly. It is noticed that (9.88) is the multivariate extension of (6.47) on page 162. ◂

9.7.2 An extended LS method for multivariate ARMAX models (the Spliid method)

As pointed out previously, the LS methods cannot be used in the case of MA parameters, because that the model in that case is nonlinear in the parameters. However, a method which circumvents this problem is suggested by Spliid (1983). The method is based on repeated use of multivariate regression.

The algorithm converges, but not to the ML estimates. Experience suggests that the estimates are reasonably close to the ML estimates.

The initial estimate of the parameters $\boldsymbol{\theta}$ could be computed by fitting a pure ARX model and taking $\boldsymbol{\theta}_1 = \cdots = \boldsymbol{\theta}_q = \mathbf{0}$ in the first reconstruction.

9.7.3 ML estimates

Let $\boldsymbol{\theta}$ denote the unknown parameters, and $\mathcal{Y}_{N^*} = (\boldsymbol{Y}_1, \boldsymbol{Y}_2, \ldots, \boldsymbol{Y}_{N^*})$ denote the available observations ($N^* \leq N$). For all time series data *the conditional likelihood function* (conditioned on $\boldsymbol{Y}_0$) can be written

$$\begin{aligned} L(\boldsymbol{\theta}; \mathcal{Y}_{N^*}) &= f(\mathcal{Y}_{N^*} | \boldsymbol{\theta}) \\ &= f(\boldsymbol{Y}_{N^*} | \mathcal{Y}_{N^*-1}, \boldsymbol{\theta}) \, f(\boldsymbol{Y}_{N^*-1} | \mathcal{Y}_{N^*-2}, \boldsymbol{\theta}) \cdots f(\boldsymbol{Y}_1 | \boldsymbol{Y}_0, \boldsymbol{\theta}). \end{aligned} \tag{9.90}$$

For the moment we just assume that the following conditional density is normal, i.e.,

$$f(\boldsymbol{Y}_{t+1} | \mathcal{Y}_t) = ((2\pi)^m \det \boldsymbol{R}_{t+1})^{-1/2} \exp\left[-\frac{1}{2} \widetilde{\boldsymbol{Y}}_{t+1}^T \boldsymbol{R}_{t+1}^{-1} \widetilde{\boldsymbol{Y}}_{t+1}\right] \tag{9.91}$$

where we have introduced the one-step prediction error (the innovation) and its variance:

$$\begin{aligned}\widetilde{\boldsymbol{Y}}_{t+1} &= \boldsymbol{Y}_{t+1} - \widehat{\boldsymbol{Y}}_{t+1|t}, \\ \boldsymbol{R}_{t+1} &= \mathrm{Var}\left[\widetilde{\boldsymbol{Y}}_{t+1}\right] = \mathrm{Var}\left[\boldsymbol{Y}_{t+1}|\mathcal{Y}_t\right] = \boldsymbol{\Sigma}^{yy}_{t+1|t}\end{aligned} \tag{9.92}$$

Assuming that the white noise in the model is normal, the above formulation is valid for all linear models, i.e., models which are linear in previous values of the noise term. This is due to the fact that any linear combination of normally distributed random variables is again normal.

By combining (9.90) and (9.91) we get

$$L(\boldsymbol{\theta}; \mathcal{Y}_{N^*}) = \prod_{i=1}^{N^*} ((2\pi)^m \det \boldsymbol{R}_i)^{-1/2} \exp\left[-\frac{1}{2}\widetilde{\boldsymbol{Y}}_i^T \boldsymbol{R}_i^{-1} \widetilde{\boldsymbol{Y}}_i\right], \tag{9.93}$$

where $\widetilde{\boldsymbol{Y}}_i$ and $\boldsymbol{R}_i$ are the innovation and the corresponding variance for the i'th observation, respectively.

Notice that in the case of missing observations the prediction in (9.92) and the associated variance must be calculated appropriately, e.g., as described in Section 10.5.

The maximization of the likelihood function is equivalent to maximization of

$$\log L(\boldsymbol{\theta}; \mathcal{Y}_{N^*}) = -\frac{1}{2}\sum_{i=1}^{N^*} (\log \det \boldsymbol{R}_i + \widetilde{\boldsymbol{Y}}_i^T \boldsymbol{R}_i^{-1} \widetilde{\boldsymbol{Y}}_i) + \text{const.} \tag{9.94}$$

and the ML estimate of $\boldsymbol{\theta}$ is the argument which maximizes this, i.e.,

$$\widehat{\boldsymbol{\theta}} = \arg\left\{\max_{\boldsymbol{\theta}} \log L(\boldsymbol{\theta}; \mathcal{Y}_{N^*})\right\}. \tag{9.95}$$

As an approximation of the variances of the parameter estimates we can use

$$\mathrm{Var}[\widehat{\boldsymbol{\theta}}] \simeq -\boldsymbol{H}^{-1}, \tag{9.96}$$

where

$$\{\boldsymbol{H}\}_{ij} = \left.\frac{\partial^2 \log L(\boldsymbol{\theta}; \mathcal{Y}_{N^*})}{\partial\theta_i \partial\theta_j}\right|_{\boldsymbol{\theta}=\widehat{\boldsymbol{\theta}}}. \tag{9.97}$$

For the maximization a numerical method, such as the Newton-Raphson procedure described in Chapter 3, must be used.

9.7.3.1 ML estimates under stationary conditions

It is clearly seen that in a stationary situation (e.g., no missing observations) the problem reduces to the following likelihood function

$$L(\boldsymbol{\theta}; \mathcal{Y}_N) = \prod_{t=1}^{N} ((2\pi)^m \det \boldsymbol{\Sigma})^{-1/2} \exp\left[-\frac{1}{2}\widetilde{\boldsymbol{Y}}_t^T \boldsymbol{\Sigma}^{-1} \widetilde{\boldsymbol{Y}}_t\right] \\ = [(2\pi)^m \det \boldsymbol{\Sigma}]^{-N/2} \exp\left[-\frac{1}{2}\sum_{t=1}^{N} \widetilde{\boldsymbol{Y}}_t^T \boldsymbol{\Sigma}^{-1} \widetilde{\boldsymbol{Y}}_t\right] \tag{9.98}$$

Compared with the notation above we now use the symbol $\boldsymbol{\Sigma}$ for the steady state variance of the innovation. Hence, the problem is simplified to calculate the one-step prediction errors $\widetilde{\boldsymbol{Y}}_t = \widetilde{\boldsymbol{Y}}_{t|t-1} = \boldsymbol{Y}_t - \widehat{\boldsymbol{Y}}_{t|t-1}$, which is easily done in, e.g., a multivariate ARMAX model.

The problem is most conveniently separated in the cases $\boldsymbol{\Sigma}$ known and unknown.

$\boldsymbol{\Sigma}$ known For this case, maximization of the likelihood function is equivalent to minimization of $S_1(\boldsymbol{\theta})$ where

$$S_1(\boldsymbol{\theta}) = \frac{1}{N}\sum_{t=1}^{N} \widetilde{\boldsymbol{Y}}_t^T \boldsymbol{\Sigma}^{-1} \widetilde{\boldsymbol{Y}}_t \tag{9.99}$$

Let us introduce the sample covariance of the predictions error, i.e.,

$$\boldsymbol{D}(\boldsymbol{\theta}) = \frac{1}{N}\sum_{t=1}^{N} \widetilde{\boldsymbol{Y}}_t \widetilde{\boldsymbol{Y}}_t^T \tag{9.100}$$

then the cost function S_1 is written

$$S_1(\boldsymbol{\theta}) = \operatorname{tr} \boldsymbol{\Sigma}^{-1} \boldsymbol{D}(\boldsymbol{\theta}) \tag{9.101}$$

However, if the distribution assumption is postponed, this method is called a *prediction error method*, and the cost function is hence S_1 in the case where $\boldsymbol{\Sigma}$ is known.

$\boldsymbol{\Sigma}$ unknown In this case the maximization of the likelihood function above is equivalent to minimization of

$$S(\boldsymbol{\theta}, \boldsymbol{\Sigma}) = \frac{1}{2}N \log\det \boldsymbol{\Sigma} + \frac{1}{2}\sum_{t=1}^{N} \widetilde{\boldsymbol{Y}}_t^T \boldsymbol{\Sigma}^{-1} \widetilde{\boldsymbol{Y}}_t \tag{9.102}$$

Differentiating with respect to $\boldsymbol{\Sigma}$ gives

$$\frac{\partial S}{\partial \boldsymbol{\Sigma}} = \frac{N}{2}\boldsymbol{\Sigma}^{-1} - \frac{1}{2}\boldsymbol{\Sigma}^{-1}\left(\sum_{t=1}^{N} \widetilde{\boldsymbol{Y}}_t \widetilde{\boldsymbol{Y}}_t^T\right)\boldsymbol{\Sigma}^{-1} \tag{9.103}$$

which equals zero for

$$\boldsymbol{\Sigma} = \frac{1}{N}\sum_{t=1}^{N} \widetilde{\boldsymbol{Y}}_t \widetilde{\boldsymbol{Y}}_t^T = \boldsymbol{D}(\boldsymbol{\theta}) \tag{9.104}$$

Therefore the problem can be reduced by imposing the constraint $\boldsymbol{\Sigma} = \boldsymbol{D}(\boldsymbol{\theta})$. Substituting this constraint into the above cost function gives

$$S_c(\boldsymbol{\theta}) = \frac{1}{2}\log\det \boldsymbol{D}(\boldsymbol{\theta}) + \frac{1}{2}\sum_{t=1}^{N} \widetilde{\boldsymbol{Y}}_t^T \boldsymbol{D}^{-1}(\boldsymbol{\theta})\widetilde{\boldsymbol{Y}}_t \tag{9.105}$$

or

$$\begin{aligned} S_c(\boldsymbol{\theta}) &= \frac{1}{2}\log\det \boldsymbol{D}(\boldsymbol{\theta}) + \frac{1}{2}\operatorname{tr}\left[\left(\sum_{t=1}^{N} \widetilde{\boldsymbol{Y}}_t \widetilde{\boldsymbol{Y}}_t^T\right)\boldsymbol{D}^{-1}(\boldsymbol{\theta})\right] \\ &= \frac{1}{2}\log\det \boldsymbol{D}(\boldsymbol{\theta}) + \frac{N}{2}\operatorname{tr}\boldsymbol{I}_m \\ &= \frac{1}{2}\log\det \boldsymbol{D}(\boldsymbol{\theta}) + \frac{Nm}{2} \end{aligned}$$

Since the last term is constant, the ML estimate is found by minimizing the cost function

$$S_2(\boldsymbol{\theta}) = \log\det \boldsymbol{D}(\boldsymbol{\theta}) \tag{9.106}$$

Again this is a *prediction error method* if the assumption about the distribution must be postponed, where, in the present case, the cost function is given by S_2.

9.8 Model checking

The tools for checking multivariate models are the same as for univariate models and transfer function models in Chapters 6 and 8. Hence for each sequence of residuals the tools from Chapter 6 should be used, whereas for each combination of two sequences of residuals the cross-correlation analysis as explained in Chapter 8 should be applied. If any problem is discovered an adjusted model structure might solve it, as illustrated in Figure 6.1 on page 145.

An important aspect of model checking is the analysis of the residuals. A fundamental and important part of the model checking is to plot the residuals and compare the individual residuals with some control limits obtained by using the variance of this sequence of residuals. If large residuals occur some actions must be taken. In some cases it might be useful to add an *intervention model* component to explain the large residuals.

However, whereas there is just one set of residuals to examine in the univariate analysis, in multivariate modeling there will be one set of residuals associated with each time series.

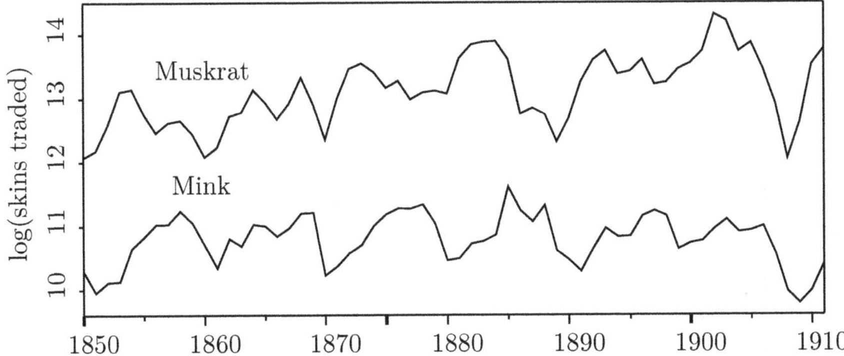

Figure 9.3: *Annually traded skins of muskrat ($Z_{1,t}$) and mink ($Z_{2,t}$) after logarithmic transformation.*

The goal is to verify that the multivariate white noise assumption is not contradicted by the observed sequence of residuals. However, while it is convenient for model checking to work with each series of residuals individually, for model checking it might be difficult to make independent checks due to a significant correlation between the individual signals. When large correlations occur it is worthwhile plotting and investigating a transformed set of residuals which are uncorrelated in addition to the original residual sequence. Suppose that the estimated covariance matrix for the sequence of multivariate residuals $\{\widehat{\varepsilon}_t\}$ is $\widehat{\boldsymbol{\Sigma}}$. Now suppose that $\boldsymbol{\Lambda}$ is the diagonal matrix of eigenvalues of $\widehat{\boldsymbol{\Sigma}}$, and $\boldsymbol{Q}$ the matrix of the eigenvectors, so that

$$\widehat{\boldsymbol{\Sigma}}\boldsymbol{Q} = \boldsymbol{Q}\boldsymbol{\Lambda}. \tag{9.107}$$

Then the transformation $\widehat{\varepsilon}_t^* = \boldsymbol{Q}\widehat{\varepsilon}_t$ will establish a sequence of uncorrelated residuals whose components $\widehat{\varepsilon}_{it}^*$ have the variance λ_i which is the i'th value on the diagonal of $\boldsymbol{\Lambda}$. Now, control limits of, say, $\pm 2\sqrt{\lambda_i}$ can be used for making judgments of the presence of large residuals in the transformed sequence of residuals.

Example 9.7 (Modeling multivariate time series: Muskrat and mink skins traded)

In this example, the model structure of a multivariate time series will be identified, and the parameters of the model will be estimated using the maximum likelihood method. The bivariate time series analysed consists of the number of muskrat ($Z_{1,t}$) and mink ($Z_{2,t}$) skins traded from Hudson's Bay Company from 1850–1911. The two series after logarithmic transformation are shown in Figure 9.3.

Because of the upward trend in the muskrat series it is decided to difference the series. The mink series will, however, be left undifferenced in the analysis.

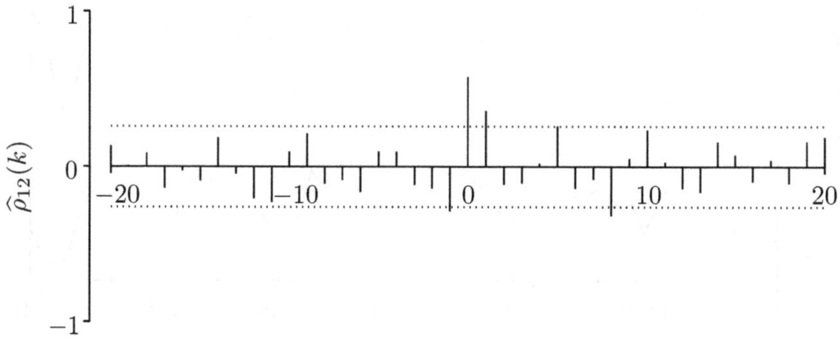

Figure 9.4: *Cross-correlation function of the residuals from the univariate models along with its ±2 standard error limits under the assumption that the series are uncorrelated.*

It has been shown in Jenkins (1975) that the two series are adequately represented by the univariate models:

$$(1 - 0.65B + 0.6B^2 - 0.23B^3 + 0.34B^4 + 0.06B^5 + 0.38B^6)\nabla \ln Z_{1,t} = (1 - 0.54)\varepsilon_{1,t} \tag{9.108}$$

$$(1 - 0.82B + 0.22B^2 + 0.00B^3 + 0.28B^4)(\ln Z_{2,t} - 10.79) = \varepsilon_{2,t} \tag{9.109}$$

with estimated residual variances $\widehat{\sigma}_1^2 = 0.0824$ and $\widehat{\sigma}_2^2 = 0.0681$.

The cross-correlation function between the residuals $\varepsilon_{1,t}$ and $\varepsilon_{2,t}$ resulting from the univariate models is shown in Figure 9.4. By inspecting the cross-correlation function the following can be noticed.

- The correlation in lag 0 and 1 is positive indicating that if the muskrat population increases in year t, the mink population increases in years t and $t + 1$.
- The correlation at lag -1 is negative indicating that if the mink population increases in year t, the muskrat population decreases in year $t + 1$.

Since the population of muskrats depends on the population of mink and vice versa, it is not possible to classify one of the series as input and the other one as output. Therefore, a bivariate model including both series is needed.

If the method based on pre-whitening as described in Section 9.6.1 is used, the observed cross-correlation function indicates that a bivariate MA(1) model should be used to link the two series. Using this approach a bivariate ARMA(6, 1) model is obtained.

Table 9.1: *Correlation matrices $\boldsymbol{R_k}$ and partial correlation matrices $\boldsymbol{S_k}$ based on $W_{1,t} = \nabla \ln Z_{1,t}$ and $W_{2,t} = \ln Z_{2,t}$.*

k	$\boldsymbol{R_k}$	$\boldsymbol{S_k}$
0	$\begin{pmatrix} 1.00 & -0.33 \\ -0.33 & 1.00 \end{pmatrix}$	
1	$\begin{pmatrix} 0.22 & 0.03 \\ -0.64 & 0.65 \end{pmatrix}$	$\begin{pmatrix} 0.01 & -0.60 \\ 0.30 & 0.74 \end{pmatrix}$
2	$\begin{pmatrix} -0.29 & 0.10 \\ -0.25 & 0.26 \end{pmatrix}$	$\begin{pmatrix} -0.21 & 0.30 \\ -0.01 & -0.13 \end{pmatrix}$
3	$\begin{pmatrix} -0.13 & 0.18 \\ 0.10 & -0.05 \end{pmatrix}$	$\begin{pmatrix} 0.09 & 0.09 \\ 0.14 & -0.36 \end{pmatrix}$
4	$\begin{pmatrix} -0.08 & 0.26 \\ 0.22 & -0.26 \end{pmatrix}$	$\begin{pmatrix} 0.00 & 0.10 \\ 0.10 & -0.12 \end{pmatrix}$
5	$\begin{pmatrix} -0.17 & 0.32 \\ 0.24 & -0.34 \end{pmatrix}$	$\begin{pmatrix} 0.00 & -0.02 \\ 0.21 & 0.02 \end{pmatrix}$
6	$\begin{pmatrix} -0.25 & 0.21 \\ 0.17 & -0.29 \end{pmatrix}$	$\begin{pmatrix} -0.14 & -0.27 \\ 0.11 & 0.15 \end{pmatrix}$

A more successful approach is to consider the method based on the correlation matrices as described in Section 9.6. Hence, in order to identify the model structure, the correlation and the partial correlation matrices of the transformed and differenced series are used. The matrices are shown in Table 9.1.

Approximated one standard error limits under the assumption that the series are unrelated are $\pm 1/\sqrt{N} = 1/\sqrt{62} = 0.13$, giving a 95% confidence interval of ± 0.26. By looking at the matrices it is noticed that the values of the partial correlation matrices are smaller than 0.26 after lag 3 indicating a bivariate autoregressive model with maximum order $p = 3$.

The parameters of the model are estimated using maximum likelihood as described in Section 9.7.3 resulting in the model

$$\begin{pmatrix} 1 - 0.232B + 0.175B^2 - 0.070B^3 & 0.733B - 0.265B^2 - 0.122B^3 \\ -0.436B + 0.254B^2 - 0.178B^3 & 1 - 0.747B - 0.215B^2 + 0.417B^3 \end{pmatrix}$$
$$\times \begin{pmatrix} \nabla \ln Z_{1,t} \\ \ln Z_{2,t} - 10.78 \end{pmatrix} = \begin{pmatrix} \varepsilon_{1,t} \\ \varepsilon_{2,t} \end{pmatrix}$$

and the covariance matrix

$$\widehat{\Sigma} = \begin{pmatrix} 0.0623 & 0.0147 \\ 0.0147 & 0.0638 \end{pmatrix}.$$

As mentioned previously univariate models fitted to each of the series lead to the models (9.108) and (9.109) on page 276, which are a sixth and fourth order model, respectively. By using (9.78) it is seen that univariate fitting of the above estimated third order model will lead to a sixth order model if univariate models are fitted to each of the series. This illustrates the mis-specification of the model obtained by fitting univariate models to each of the series.

9.9 Problems

Exercise 9.1
A bivariate process is given as

$$\begin{aligned} X_{1,t} &= \alpha_{11} X_{1,t-1} + \alpha_{12} X_{2,t-1} + \varepsilon_{1,t} \\ X_{2,t} &= \alpha_{21} X_{1,t-1} + \alpha_{22} X_{2,t-1} + \varepsilon_{2,t} \end{aligned}$$

where $\{\varepsilon_{1,t}\}$ and $\{\varepsilon_{2,t}\}$ are white noise processes with mean value 0 and variance σ_1^2 and σ_2^2, respectively. The covariance between $\varepsilon_{1,t}$ and $\varepsilon_{2,t}$ is σ_{12}^2.

Question 1 Find a suitable recursion formula for determining the covariance functions $\gamma_{12}(k)$, $\gamma_{11}(k)$, $\gamma_{21}(k)$, and $\gamma_{22}(k)$.

Question 2 Consider now the process

$$\begin{aligned} X_{1,t} &= 0.6 X_{1,t-1} - 0.5 X_{2,t-1} + \varepsilon_{1,t} \\ X_{2,t} &= 0.4 X_{1,t-1} + 0.5 X_{2,t-1} + \varepsilon_{2,t} \end{aligned}$$

where $\{\varepsilon_{1,t}\}$ and $\{\varepsilon_{2,t}\}$ are mutually uncorrelated white noise processes with mean value 0 and variance 1. Use the result found above to calculate and sketch the autocorrelation and cross-correlation functions for the process (only for $|k| \le 5$).

Exercise 9.2
Consider the bivariate AR(1) process

$$\begin{pmatrix} X_{1,t} \\ X_{2,t} \end{pmatrix} = \begin{pmatrix} -1.0 & -2.5 \\ 1.0 & 2.0 \end{pmatrix} \begin{pmatrix} X_{1,t-1} \\ X_{2,t-1} \end{pmatrix} + \begin{pmatrix} \varepsilon_{1,t} \\ \varepsilon_{2,t} \end{pmatrix}$$

where $\{\varepsilon_{1,t}\}$ and $\{\varepsilon_{2,t}\}$ are white noise processes both with variance 1. The correlation between $\varepsilon_{1,t}$ and $\varepsilon_{2,t}$ is 0.75.

Question 1 Establish recursions for determining the autocovariance and cross-covariance functions of the process.

Question 2 Calculate and sketch the autocorrelation and cross-correlation functions for the process.

Question 3 Calculate and sketch the impulse response function for the transfer from $\{\varepsilon_{1,t}\}$ to $\{X_{2,t}\}$.

Exercise 9.3

Let X_t denote number of houses sold in month t, and let Y_t denote the number of initiated new building projects in month t.

This exercise considers the relation between X_t and Y_t. It is expected that the number of initiated new building projects will depend on the number of sold houses in the most recent past, and hence we will initially consider X_t as an input process. The investigation is based on monthly observations from the period January 1965 to December 1974.

In the first phase an ARMA model is fitted for the time series for number of houses sold in month t, $\{X_t\}$. The time series of residuals resulting from this model is called $\{\alpha_t\}$, i.e., a pre-whitening of the input series has been carried out. The time series Y_t is filtered with the same ARMA model and the filtered time series is called $\{\beta_t\}$.

The following cross-correlation function between $\{\alpha_t\}$ and $\{\beta_t\}$ has been estimated.

k	0	1	2	3	4
$\widehat{\gamma}_{\alpha\beta}(k)$	0.262	0.274	0.103	0.079	–0.044

k	5	6	7	8	9	10
$\widehat{\gamma}_{\alpha\beta}(k)$	0.080	−0.043	0.054	−0.043	0.114	0.053

Finally an estimate for the variances of α_t and β_t is found as

$$\widehat{\sigma}_\alpha^2 = 45.60 \qquad \text{and} \qquad \widehat{\sigma}_\beta^2 = 13.27$$

Question 1 Estimate the impulse response function and make a sketch of it.

Question 2 Find a model to describe how the number of initiated new building projects depends on the number of sold houses.

Question 3 Let us assume a model of the form

$$Y_t = \frac{\omega_0 + \omega_1 B}{1 + \delta B} X_t$$

Find moment estimates of the parameters of this model.

Question 4 What can be done if it turns out that number of sold houses depends on the number of initiated new building projects during the past months?

Exercise 9.4

Question 1 A process can be described by the bivariate MA(1) process

$$\begin{pmatrix} X_{1,t} \\ X_{2,t} \end{pmatrix} = \begin{pmatrix} 1-\theta_{11}B & -\theta_{12}B \\ -\theta_{21}B & 1-\theta_{22}B \end{pmatrix} \begin{pmatrix} \varepsilon_{1,t} \\ \varepsilon_{2,t} \end{pmatrix}$$

which can be represented in matrix form

$$\boldsymbol{X}_t = \boldsymbol{\theta}(B)\boldsymbol{\varepsilon}_t$$

where

$$\text{Var}[\boldsymbol{\varepsilon}_t] = \begin{pmatrix} \sigma_{11} & \sigma_{12} \\ \sigma_{21} & \sigma_{22} \end{pmatrix}$$

Assume that only $X_{2,t}$ is measured, and let us introduce Y_t for describing the measured output, i.e.,

$$Y_t = \begin{pmatrix} 0 & 1 \end{pmatrix} \boldsymbol{X}_t$$

The process $\{Y_t\}$ is therefore a univariate process.

Show that $\{Y_t\}$ can be described by a univariate MA(1) process.

Question 2 Consider now another process described by the bivariate AR(1) process

$$\begin{pmatrix} 1-\phi_{11}B & -\phi_{12}B \\ -\phi_{21}B & 1-\phi_{22}B \end{pmatrix} \begin{pmatrix} X_{1,t} \\ X_{2,t} \end{pmatrix} = \begin{pmatrix} \varepsilon_{1,t} \\ \varepsilon_{2,t} \end{pmatrix}$$

Or in matrix form

$$\boldsymbol{\phi}(B)\boldsymbol{X}_t = \boldsymbol{\varepsilon}_t$$

Show that the variation of Y_t defined by

$$Y_t = \begin{pmatrix} 0 & 1 \end{pmatrix} \boldsymbol{X}_t$$

can be described by a univariate ARMA(2, 1) process.

Question 3 Derive equations for finding the parameters of the ARMA(2, 1) process as a function of the parameters of the bivariate process.

Exercise 9.5

Question 1 Consider the process

$$\begin{pmatrix} 1-2B & 2B \\ -1.25B & 1+B \end{pmatrix} \begin{pmatrix} X_{1,t} \\ X_{2,t} \end{pmatrix} = \begin{pmatrix} 1 & 0.5B \\ -0.5B & 1-B \end{pmatrix} \begin{pmatrix} \varepsilon_{1,t} \\ \varepsilon_{2,t} \end{pmatrix}$$

Is this process stationary? Invertible?

Question 2 Find the covariance matrix functions of the process

$$\begin{pmatrix} X_{1,t} \\ X_{2,t} \end{pmatrix} = \begin{pmatrix} 1 & 0.5B \\ -0.5B & 1-B \end{pmatrix} \begin{pmatrix} \varepsilon_{1,t} \\ \varepsilon_{2,t} \end{pmatrix}$$

where $\boldsymbol{\varepsilon}_t^T = (\varepsilon_{1,t}, \varepsilon_{2,t})$ is white noise with covariance matrix given by

$$\begin{pmatrix} 2 & -2 \\ -2 & 4 \end{pmatrix}$$

Draw the cross-correlation function $\rho_{12}(k)$.

Exercise 9.6

In order to calculate the k-step predictions in a multivariate ARMA(p,q) process, the first $k-1$ values of the sequence of $\boldsymbol{\psi}$ matrices are needed.

Show that the recurrence relationships for the $\boldsymbol{\psi}$ matrices for the multivariate ARMA(p,q) model are the following set of equations:

$$\begin{aligned}
\boldsymbol{\psi}_1 &= -\boldsymbol{\phi}_1 + \boldsymbol{\theta}_1 \\
\boldsymbol{\psi}_2 &= -\boldsymbol{\phi}_1 \boldsymbol{\psi}_1 - \boldsymbol{\phi}_2 + \boldsymbol{\theta}_2 \\
&\ \vdots \\
\boldsymbol{\psi}_j &= -\boldsymbol{\phi}_1 \boldsymbol{\psi}_{j-1} - \boldsymbol{\phi}_2 \boldsymbol{\psi}_{j-2} - \cdots - \boldsymbol{\phi}_j + \boldsymbol{\theta}_j
\end{aligned}$$

Exercise 9.7

Consider Example 9.4 on page 258. For this example the value of the coefficients of the AR part of the univariate model in (9.47) is readily seen. In this exercise you should derive equations for finding the parameter θ of the MA part of the model and the variance σ_ξ^2 of the combined white noise process $\{\xi_t\}$.

CHAPTER 10

State space models of dynamic systems

Dynamic systems are often described in terms of differential equations (in continuous time) or in terms of difference equations (in discrete time). The *state* of the system (or the *state vector*) is a set of numbers $x_1(t_0), x_2(t_0), \ldots, x_m(t_0)$, characterized by the fact that, together with possible input-signals and noise disturbances at time $t \geq t_0$, they uniquely determine the system at times $t \geq t_0$. For *deterministic systems* (i.e., systems without noise), the state contains the initial values necessary to determine the particular solution to the differential or difference equation. The complete solution and the set of initial conditions will determine the future evolution of the system without uncertainty. For *stochastic systems*, the state vector at a given time contains all information available for the future evaluation of the system. The state (or the state vector) is thus a (first order) *Markov process*. A system having m states is called an *m'th order system*.

In the previous chapters both univariate and multivariate input-output models have been presented. Such a model represents an *external description* of the system, since only the input to output relations are modelled. The description is said to provide a model on input-output form (also called external or transfer function form) of the system.

The state space model introduced in this chapter defines an *internal description* of the system. This is obtained by defining a state vector such that the dynamics of the system can be described by a first order Markov process. The Kalman filter will be introduced as a method for estimating and predicting the unmeasured state vector. In Section 1.1.3 the heat dynamics of a building were used to introduce the concept of state space models.

State space models are often used in modern control theory and some of the important classical books in this area are Åström (1970), Kailath (1980), Ljung (1987), and Söderström and Stoica (1989). In economics and applied statistics the book Harvey (1996) is often used as a reference. The Kalman filter was first introduced in Kalman (1960). Later on filtering were covered in more detail in the books by Jazwinski (1970), Maybeck (1982), and Meinhold and Singpurwalla (1983).

10.1 The linear stochastic state space model

In discrete time we define the *linear stochastic state space model* by a model containing two equations; namely, the *system equation*

$$\boldsymbol{X}_t = \boldsymbol{A}_t \boldsymbol{X}_{t-1} + \boldsymbol{B}_t \boldsymbol{u}_{t-1} + \boldsymbol{e}_{1,t}, \tag{10.1}$$

and the *observation equation*

$$\boldsymbol{Y}_t = \boldsymbol{C}_t \boldsymbol{X}_t + \boldsymbol{e}_{2,t}, \tag{10.2}$$

where $\boldsymbol{X}_t$ is an m-dimensional, latent (not directly observable), stochastic *state vector*. Here $\boldsymbol{u}_t$ is a deterministic *input vector*, $\boldsymbol{Y}_t$ is a vector of observable (measurable) stochastic outputs, and $\boldsymbol{A}_t$, $\boldsymbol{B}_t$, and $\boldsymbol{C}_t$ are known matrices of suitable dimensions. Finally, $\{\boldsymbol{e}_{1,t}\}$ and $\{\boldsymbol{e}_{2,t}\}$ are random vectors where

$$\mathrm{E}\left[\boldsymbol{e}_{1,t}\right] = \mathrm{E}\left[\boldsymbol{e}_{2,t}\right] = 0 \tag{10.3}$$

$$\mathrm{C}\left[\boldsymbol{e}_{1,t}, \boldsymbol{e}_{1,s}\right] = \begin{cases} \mathrm{Var}\left[\boldsymbol{e}_{1,t}\right] = \boldsymbol{\Sigma}_{1,t} & \text{for } s = t \\ 0 & \text{for } s \neq t \end{cases} \tag{10.4}$$

$$\mathrm{C}\left[\boldsymbol{e}_{2,t}, \boldsymbol{e}_{2,s}\right] = \begin{cases} \mathrm{Var}\left[\boldsymbol{e}_{2,t}\right] = \boldsymbol{\Sigma}_{2,t} & \text{for } s = t \\ 0 & \text{for } s \neq t \end{cases} \tag{10.5}$$

$$\mathrm{C}\left[\boldsymbol{e}_{1,t}, \boldsymbol{e}_{2,s}\right] = 0, \quad \text{for all } s, t \tag{10.6}$$

As the notation indicates, the matrices $\boldsymbol{A}_t$, $\boldsymbol{B}_t$, $\boldsymbol{C}_t$, $\boldsymbol{\Sigma}_{1,t}$, and $\boldsymbol{\Sigma}_{2,t}$ may depend on time, but for most applications they are constant.

In the state space model (10.1)–(10.2), the system equation describes the evolution of the system states whereas the observation equation describes what can be directly observed (or measured).

Example 10.1 (A state space model of an ARMA(2, 1) process)
Consider the process $\{Y_t\}$ given by

$$Y_t + \phi_1 Y_{t-1} + \phi_2 Y_{t-2} = \varepsilon_t + \theta_1 \varepsilon_{t-1}, \tag{10.7}$$

where $\{\varepsilon_t\}$ is white noise with variance σ_ε^2.

This ARMA(2, 1) process can be written in a state space form as

$$\begin{pmatrix} X_{1,t} \\ X_{2,t} \end{pmatrix} = \begin{pmatrix} -\phi_1 & 1 \\ -\phi_2 & 0 \end{pmatrix} \begin{pmatrix} X_{1,t-1} \\ X_{2,t-1} \end{pmatrix} + \begin{pmatrix} 1 \\ \theta_1 \end{pmatrix} \varepsilon_t, \tag{10.8}$$

$$Y_t = \begin{pmatrix} 1 & 0 \end{pmatrix} \begin{pmatrix} X_{1,t} \\ X_{2,t} \end{pmatrix}, \tag{10.9}$$

where $\mathrm{Var}[\varepsilon_t] = \sigma_\varepsilon^2$. It is seen that (10.8) and (10.9) correspond to (10.7) by eliminating $X_{1,t}$ and $X_{2,t}$.

The state space description for ARMA models in general is discussed in Section 10.4.

When the Markov property becomes essential there are advantages in writing ARMA models in state space form, e.g., in estimation problems of time series with missing observations. This is discussed in Section 10.5.

Often the state space formulation may be obtained directly from considering the physical problem which is being considered. This is illustrated in the following example.

Example 10.2 (A falling body)
(The example is taken from Olbjer, Holst, and Holst (2005).) We consider a body affected only by gravity, i.e., the position of the body, $z(t)$ may be described by the following differential equation

$$\frac{d^2 z}{dt^2} = -g. \tag{10.10}$$

In order to determine the position of the body $z(t)$ for $t \geq t_0$, we need two initial conditions. (An m'th order differential equation needs m initial conditions.) If we select the initial condition for the position $x_1(t_0) = z(t_0)$ and for the velocity $x_2(t_0) = z'(t_0)$, we can obtain the state vector

$$\boldsymbol{x}(t) = \begin{pmatrix} x_1(t) \\ x_2(t) \end{pmatrix} \tag{10.11}$$

Applying these two state variables, we obtain the following state space formulation

$$\begin{pmatrix} x_1'(t) \\ x_2'(t) \end{pmatrix} = \begin{pmatrix} 0 & 1 \\ 0 & 0 \end{pmatrix} \begin{pmatrix} x_1(t) \\ x_2(t) \end{pmatrix} + \begin{pmatrix} 0 \\ -1 \end{pmatrix} g, \tag{10.12}$$

$$z(t) = \begin{pmatrix} 1 & 0 \end{pmatrix} \boldsymbol{x}(t). \tag{10.13}$$

Applying the initial conditions, we get the solution to (10.10)

$$x_2(t) = \frac{dz(t)}{dt} = -g\,(t - t_0) + x_2\,(t_0) \tag{10.14}$$

$$x_1(t) = z(t) = -\frac{1}{2} g\,(t - t_0)^2 + x_2\,(t_0)\,(t - t_0) + x_1\,(t_0) \tag{10.15}$$

The system can now be formulated in discrete time by setting $t = kT$, $t_0 = (k-1)T$, and $T = 1$. In summary, $t = k$ and $t_0 = k - 1$, and we get

$$x_1(k) = x_1(k-1) + x_2(k-1) - \frac{1}{2} g, \tag{10.16}$$

$$x_2(k) = x_2(k)(k-1) - g. \tag{10.17}$$

If the body is affected by other disturbances besides gravity, e.g., by the wind, it is possible to add these disturbances to the equations (10.16) and (10.17). In this case we can write the system in a stochastic state space form (please note, that since $T = 1$, $x_i(k) \simeq x_{i,t}$):

$$\begin{pmatrix} X_{1,t} \\ X_{2,t} \end{pmatrix} = \begin{pmatrix} 1 & 1 \\ 0 & 1 \end{pmatrix} \begin{pmatrix} X_{1,t-1} \\ X_{2,t-1} \end{pmatrix} + \begin{pmatrix} -\frac{1}{2} \\ -1 \end{pmatrix} u_{t-1} + \begin{pmatrix} e_{11,t} \\ e_{12,t} \end{pmatrix}, \tag{10.18}$$

where $u_t = g$ is a deterministic input to the system and $(e_{11,t}, e_{12,t})^T$ are stochastic disturbances. Using $X_{1,t}$ and $X_{2,t}$ we can calculate the position and velocity of the body at time t. Compare (10.18) with (10.1) on page 284. The system is observed by the fact that the position of the body is measured. These observations are measured with error $e_{2,t}$, and hereby the observation equation becomes

$$Z_t = \begin{pmatrix} 1 & 0 \end{pmatrix} \begin{pmatrix} X_{1,t} \\ X_{2,t} \end{pmatrix} + e_{2,t}. \tag{10.19}$$

Compare with (10.2) on page 284.

The Kalman filter, which will be introduced in Section 10.3.1, makes it possible to reconstruct (or estimate) and predict the position and velocity of the body based on the measurements of the body's position (10.19).

10.2 Transfer function and state space formulations

We consider the *state space model*

$$\boldsymbol{X}_{t+1} = \boldsymbol{A}\boldsymbol{X}_t + \boldsymbol{B}\boldsymbol{u}_t + \boldsymbol{e}_{1,t}, \tag{10.20}$$

$$\boldsymbol{Y}_t = \boldsymbol{C}\boldsymbol{X}_t + \boldsymbol{e}_{2,t}, \tag{10.21}$$

where $\{\boldsymbol{e}_{1,t}\}$ and $\{\boldsymbol{e}_{2,t}\}$ are mutually independent white noise. The model (10.20) corresponds to (10.1) on page 284 with the following exceptions: (a) the noise term is entering at time t instead of the time $t+1$ and (b) the parameters are constant. It is seen that $\{\boldsymbol{X}_t\}$ and $\{\boldsymbol{Y}_t\}$ are stationary processes.

Applying the z-transform yields

$$z\boldsymbol{X}(z) = \boldsymbol{A}\boldsymbol{X}(z) + \boldsymbol{B}\boldsymbol{u}(z) + \boldsymbol{e}_1(z), \tag{10.22}$$

$$\boldsymbol{Y}(z) = \boldsymbol{C}\boldsymbol{X}(z) + \boldsymbol{e}_2(z). \tag{10.23}$$

By eliminating $\boldsymbol{X}(z)$ in (10.22)—(10.23) we get

$$\boldsymbol{Y}(z) = \boldsymbol{C}(z\boldsymbol{I} - \boldsymbol{A})^{-1}\boldsymbol{B}\boldsymbol{u}(z) + \boldsymbol{C}(z\boldsymbol{I} - \boldsymbol{A})^{-1}\boldsymbol{e}_1(z) + \boldsymbol{e}_2(z). \tag{10.24}$$

This form is called the *input-output* or *transfer function form.*

If $\{\boldsymbol{Y}_t\}$ is stationary ($\boldsymbol{A}$ stable) the noise terms in (10.24) can be aggregated into one single term (Goodwin and Payne 1977), yielding

$$\boldsymbol{Y}(z) = \boldsymbol{C}(z\boldsymbol{I} - \boldsymbol{A})^{-1}\boldsymbol{B}\boldsymbol{u}(z) + \left[\boldsymbol{C}(z\boldsymbol{I} - \boldsymbol{A})^{-1}\boldsymbol{K} + \boldsymbol{I}\right]\boldsymbol{\varepsilon}(z) \tag{10.25}$$

or

$$\boldsymbol{Y}(z) = \boldsymbol{H}_1(z)\boldsymbol{u}(z) + \boldsymbol{H}_2(z)\boldsymbol{\varepsilon}(z), \tag{10.26}$$

where $\boldsymbol{\varepsilon}(z)$ is white noise with covariance $\boldsymbol{\Sigma}$ and

$$\boldsymbol{H}_1(z) = \boldsymbol{C}(z\boldsymbol{I} - \boldsymbol{A})^{-1}\boldsymbol{B}, \tag{10.27}$$

$$\boldsymbol{H}_2(z) = \boldsymbol{C}(z\boldsymbol{I} - \boldsymbol{A})^{-1}\boldsymbol{K} + \boldsymbol{I}. \tag{10.28}$$

The matrix $\boldsymbol{K}$ (which we will later denote as the *stationary Kalman gain*—cf. Section 10.3.3) and $\boldsymbol{\Sigma}$ is determined based on $\boldsymbol{\Sigma}_1$, $\boldsymbol{\Sigma}_2$, $\boldsymbol{A}$, and $\boldsymbol{C}$, since

$$\boldsymbol{K} = \boldsymbol{A}\boldsymbol{P}\boldsymbol{C}^T(\boldsymbol{C}\boldsymbol{P}\boldsymbol{C}^T + \boldsymbol{\Sigma}_2)^{-1}, \tag{10.29}$$

$$\boldsymbol{\Sigma} = \boldsymbol{C}\boldsymbol{P}\boldsymbol{C}^T + \boldsymbol{\Sigma}_2, \tag{10.30}$$

where $\boldsymbol{P}$ is determined from the *Ricatti equation*

$$\boldsymbol{P} = \boldsymbol{A}\boldsymbol{P}\boldsymbol{A}^T + \boldsymbol{\Sigma}_1 - \boldsymbol{A}\boldsymbol{P}\boldsymbol{C}^T(\boldsymbol{C}\boldsymbol{P}\boldsymbol{C}^T + \boldsymbol{\Sigma}_2)\boldsymbol{C}\boldsymbol{P}\boldsymbol{A}^T \tag{10.31}$$

If we consider the state space model

$$\widehat{\boldsymbol{X}}_{t+1|t} = \boldsymbol{A}\widehat{\boldsymbol{X}}_{t|t-1} + \boldsymbol{B}\boldsymbol{u}_t + \boldsymbol{K}\boldsymbol{\varepsilon}_t \tag{10.32}$$

$$\boldsymbol{Y}_t = \boldsymbol{C}\widehat{\boldsymbol{X}}_{t|t-1} + \boldsymbol{\varepsilon}_t \tag{10.33}$$

we see that (10.25) is the corresponding transfer function form. It is obvious, that if $\widehat{\boldsymbol{X}}_{1|0}$ is known, then we can use the observation $\boldsymbol{Y}_1$ to calculate $\boldsymbol{\varepsilon}_1$ and then $\widehat{\boldsymbol{X}}_{2|1}$ can be calculated from (10.32). Subsequently we can calculate $\widehat{\boldsymbol{X}}_{3|2}$ based on $\boldsymbol{Y}_2$, etc. Thus, the sequence of $\widehat{\boldsymbol{X}}_{t|t-1}$ can be calculated from past observations $\boldsymbol{Y}_{t-1}, \boldsymbol{Y}_{t-2}, \ldots$. In the section on the Kalman filter, we will demonstrate that $\widehat{\boldsymbol{X}}_{t|t-1}$ is the conditional mean of $\boldsymbol{X}_t$ given $\boldsymbol{Y}_{t-1}, \boldsymbol{Y}_{t-2}, \ldots$. Correspondingly, $\boldsymbol{P}$ is the (stationary) conditional covariance of $\boldsymbol{X}_t$ given $\boldsymbol{Y}_{t-1}, \boldsymbol{Y}_{t-2}, \ldots$; hence, we will use the notation $\boldsymbol{\Sigma}^{xx}_{t|t-1}$.

Let us substitute the prediction in (10.32) and (10.33) by $\boldsymbol{Z}_t$. We hereby obtain the state space model

$$\boldsymbol{Z}_{t+1} = \boldsymbol{A}\boldsymbol{Z}_t + \boldsymbol{B}\boldsymbol{u}_t + \boldsymbol{K}\boldsymbol{\varepsilon}_t \tag{10.34}$$

$$\boldsymbol{Y}_t = \boldsymbol{C}\boldsymbol{Z}_t + \boldsymbol{\varepsilon}_t, \tag{10.35}$$

which is also called the *innovation form* of the system. The name is due to the fact that the noise process equals the *innovation* (one-step prediction error), i.e.,

$$\boldsymbol{\varepsilon}_t = \boldsymbol{Y}_t - \widehat{\boldsymbol{Y}}_{t|t-1} \tag{10.36}$$

When the initial condition $\boldsymbol{Z}_1$ is known, all future values can be reconstructed without uncertainty based on the observed output (and input, u_t) since

$$\boldsymbol{Z}_{t+1} = \boldsymbol{A}\boldsymbol{Z}_t + \boldsymbol{B}\boldsymbol{u}_t + \boldsymbol{K}(\boldsymbol{Y}_t - \boldsymbol{C}\boldsymbol{Z}_t), \quad t = 1, 2, \ldots \tag{10.37}$$

Thus, for a system written on the innovation form, the only unknown quantity is the initial condition.

As illustrated it is possible to find the corresponding transfer function form based on the state space form by eliminating the state vector. Correspondingly, it is necessary to select a state vector if one wishes to find a state space description based on a transfer function. In principle there are infinitely many possible ways to do this since the *state space form is not unique.* This is illustrated in the following example.

Example 10.3 (Two examples of state space forms)
The AR(2) model: $Y_t + \phi_1 Y_{t-1} + \phi_2 Y_{t-2} = \varepsilon_t$ can be written in the state space form

$$\begin{pmatrix} X_{1,t} \\ X_{2,t} \end{pmatrix} = \begin{pmatrix} -\phi_1 & 1 \\ -\phi_2 & 0 \end{pmatrix} \begin{pmatrix} X_{1,t-1} \\ X_{2,t-1} \end{pmatrix} + \begin{pmatrix} 1 \\ 0 \end{pmatrix} \varepsilon_t$$
$$Y_t = (1 \quad 0) \begin{pmatrix} X_{1,t} \\ X_{2,t} \end{pmatrix}, \tag{10.38}$$

or in the state space form

$$\begin{pmatrix} X'_{1,t} \\ X'_{2,t} \end{pmatrix} = \begin{pmatrix} -\phi_1 & -\phi_2 \\ 1 & 0 \end{pmatrix} \begin{pmatrix} X'_{1,t-1} \\ X'_{2,t-1} \end{pmatrix} + \begin{pmatrix} 1 \\ 0 \end{pmatrix} \varepsilon_t$$
$$Y_t = (1 \quad 0) \begin{pmatrix} X'_{1,t} \\ X'_{2,t} \end{pmatrix}. \tag{10.39}$$

There are infinitely many other state space descriptions, and thus it is more adequate to obtain a state space description by selecting a *canonical form*, characterized by the fact that the *transfer function uniquely determines the state space description.* The equations (10.38) and (10.39) are examples of canonical forms which each determine a state space description for the AR(2) model.

10.3 Interpolation, reconstruction, and prediction

In the introduction to Section 5.7 on page 135, we defined the terms *prediction* and *interpolation* in the case where, based on a stochastic process $\{\boldsymbol{Y}_t\}$, we sought to estimate future values of the process, e.g., $\boldsymbol{Y}_{t+k}$. This section concerns the same issues, namely, those where, based on an observed stochastic process $\{\boldsymbol{Y}_t\}$, we seek to estimate a thusly correlated process $\{\boldsymbol{X}_t\}$.

Let there be given a sequence of observations $\mathcal{Y}_t = \{\boldsymbol{Y}_s, s = t, t-1, \ldots\}$, and a process $\{\boldsymbol{X}_t\}$ which is correlated with $\{\boldsymbol{Y}_t\}$. We now consider the *discrete time estimation problem* that is concerned with the calculation of an estimate of $\boldsymbol{X}_{t+k}$ given $\mathcal{Y}_t$.

We now distinguish between

i) **Prediction** of a future value of $\{\boldsymbol{X}_t\}$, i.e., $k > 0$;

ii) **Reconstruction** (*filtering*), estimation of $\boldsymbol{X}_t$, i.e., $k = 0$; and

iii) **Interpolation** (*smoothing*), estimation of a past value of $\{\boldsymbol{X}_t\}$, i.e., $k < 0$.

In this section we will limit ourselves to the case where the processes $\{\boldsymbol{X}_t\}$ and $\{\boldsymbol{Y}_t\}$, as well as the relations between them, can be described by the linear stochastic state space model

$$\boldsymbol{X}_t = \boldsymbol{A}\boldsymbol{X}_{t-1} + \boldsymbol{B}\boldsymbol{u}_{t-1} + \boldsymbol{e}_{1,t}, \tag{10.40}$$

$$\boldsymbol{Y}_t = \boldsymbol{C}\boldsymbol{X}_t + \boldsymbol{e}_{2,t}, \tag{10.41}$$

which was introduced in Section 10.1. In order to make the notation simpler we consider the matrices $\boldsymbol{A}$, $\boldsymbol{B}$, and $\boldsymbol{C}$ in (10.40)–(10.41) as being constant matrices. Similarly, we consider $\mathrm{Var}[\boldsymbol{e}_{1,t}] = \boldsymbol{\Sigma}_1$ and $\mathrm{Var}[\boldsymbol{e}_{2,t}] = \boldsymbol{\Sigma}_2$ to be constant. The results presented in the following may, however, be generalized to time-varying systems.

Finally, we assume that the system in (10.40)—(10.41) is observable which ensures that all the states in $\boldsymbol{X}_t$ can be estimated. An m'th order system such as (10.40)–(10.41) having one output (Y_t scalar) is said to be *observable* if

$$\mathrm{rank}\left[\boldsymbol{C}^T \vdots (\boldsymbol{C}\boldsymbol{A})^T \vdots \cdots \vdots \left(\boldsymbol{C}\boldsymbol{A}^{m-1}\right)^T\right] = m. \tag{10.42}$$

For a system without any noise or any input, $\boldsymbol{u}_t$, (10.42) ensures that the following m equations

$$\begin{aligned} Y_t &= \boldsymbol{C}\boldsymbol{X}_t \\ Y_{t+1} &= \boldsymbol{C}\boldsymbol{X}_{t+1} = \boldsymbol{C}\boldsymbol{A}\boldsymbol{X}_t \\ &\vdots \\ Y_{t+m-1} &= \boldsymbol{C}\boldsymbol{X}_{t+m-1} = \boldsymbol{C}\boldsymbol{A}^{m-1}\boldsymbol{X}_t, \end{aligned}$$

can be solved with respect to $\boldsymbol{X}_t$. This means that, based on m observations of the output, we are able to calculate the m-dimensional state vector (which could not be measured). For stochastic systems equation (10.42) ensures that the state vector can be estimated.

10.3.1 The Kalman filter

The Kalman filter yields the optimal reconstruction and prediction of the state vector $\boldsymbol{X}_t$ given the observations of $\{\boldsymbol{u}_t\}$ and $\{Y_t\}$ in the state space model (10.40)–(10.41).

The foundation for the Kalman filter is given in Theorem 2.6 on page 24 or 2.8 on page 27 in the section on *linear projections.* Let there be two random vectors $\boldsymbol{X} = (X_1, \ldots, X_m)^T$ and $\boldsymbol{Y} = (Y_1, \ldots, Y_n)^T$, and let the $(m+n)$-dimensional vector $\binom{\boldsymbol{Y}}{\boldsymbol{X}}$ be normally distributed with mean and covariance

$$\begin{pmatrix} \boldsymbol{\mu}_Y \\ \boldsymbol{\mu}_X \end{pmatrix}, \qquad \begin{pmatrix} \boldsymbol{\Sigma}_{YY} & \boldsymbol{\Sigma}_{YX} \\ \boldsymbol{\Sigma}_{XY} & \boldsymbol{\Sigma}_{XX} \end{pmatrix}$$

respectively.

It then follows from Theorem 2.8 that $\boldsymbol{X}|\boldsymbol{Y}$ is normally distributed with mean value

$$\mathrm{E}[\boldsymbol{X}|\boldsymbol{Y}] = \boldsymbol{\mu}_X + \boldsymbol{\Sigma}_{XY}\boldsymbol{\Sigma}_{YY}^{-1}(\boldsymbol{Y} - \boldsymbol{\mu}_Y), \tag{10.43}$$

$$\mathrm{Var}[\boldsymbol{X}|\boldsymbol{Y}] = \boldsymbol{\Sigma}_{XX} - \boldsymbol{\Sigma}_{XY}\boldsymbol{\Sigma}_{YY}^{-1}\boldsymbol{\Sigma}_{XY}^T. \tag{10.44}$$

If the assumption of normality does not hold, then the assumption of $\boldsymbol{X}|\boldsymbol{Y}$ being normally distributed is no longer valid. However, (10.43) and (10.44) may still be applied, since the projection (10.43) can then be interpreted as the projection of $\boldsymbol{X}$ on $\boldsymbol{Y}$, which has the smallest variance among all linear projections (cf. Section 2.8 on page 24).

The theorems are also valid for the conditioned vector $\left(\binom{\boldsymbol{Y}}{\boldsymbol{X}}\middle|\boldsymbol{Z}\right)$ where (10.43)–(10.44) can be written

$$\mathrm{E}[\boldsymbol{X}|\boldsymbol{Y},\boldsymbol{Z}] = \mathrm{E}[\boldsymbol{X}|\boldsymbol{Z}] + \mathrm{C}[\boldsymbol{X},\boldsymbol{Y}|\boldsymbol{Z}]\,\mathrm{Var}^{-1}[\boldsymbol{Y}|\boldsymbol{Z}](\boldsymbol{Y} - \mathrm{E}[\boldsymbol{Y}|\boldsymbol{Z}]), \tag{10.45}$$

$$\mathrm{Var}[\boldsymbol{X}|\boldsymbol{Y},\boldsymbol{Z}] = \mathrm{Var}[\boldsymbol{X}|\boldsymbol{Z}] - \mathrm{C}[\boldsymbol{X},\boldsymbol{Y}|\boldsymbol{Z}]\,\mathrm{Var}^{-1}[\boldsymbol{Y}|\boldsymbol{Z}]\,\mathrm{C}^T[\boldsymbol{X},\boldsymbol{Y}|\boldsymbol{Z}]. \tag{10.46}$$

It is seen from (10.45) that the information gained from $\boldsymbol{Y}$ is small if one or more of the following issues hold.

1) The error $(\boldsymbol{Y} - \mathrm{E}[\boldsymbol{Y}|\boldsymbol{Z}])$ is small.

2) The conditional covariance $\mathrm{C}[\boldsymbol{X},\boldsymbol{Y}|\boldsymbol{Z}]$ is small compared to the conditional variance $\mathrm{Var}[\boldsymbol{Y}|\boldsymbol{Z}]$.

We now introduce

$$\mathcal{Y}_t^T = (\boldsymbol{Y}_1^T, \ldots, \boldsymbol{Y}_t^T), \tag{10.47}$$

which is a vector including all observations up to time t. Furthermore, we assume that we know $\boldsymbol{u}_1, \ldots, \boldsymbol{u}_t$.

In addition, we introduce

$$\widehat{\boldsymbol{X}}_{t+k|t} = \mathrm{E}\left[\boldsymbol{X}_{t+k}|\mathcal{Y}_t\right], \tag{10.48}$$

where $\widehat{\boldsymbol{X}}_{t|t}$ is called the *reconstruction* of $\boldsymbol{X}_t$, and $\widehat{\boldsymbol{X}}_{t+k|t}$ $(k \geq 1)$ is called the *prediction* of $\boldsymbol{X}_{t+k}$. We know from Theorem 3.9 that if we choose to minimize the expected value of the squared prediction error, the *optimal reconstruction and prediction are obtained by the conditional mean.*

We introduce the *prediction errors*

$$\widetilde{\boldsymbol{X}}_{t+k|t} = \boldsymbol{X}_{t+k} - \widehat{\boldsymbol{X}}_{t+k|t}, \tag{10.49}$$

$$\widetilde{\boldsymbol{Y}}_{t+k|t} = \boldsymbol{Y}_{t+k} - \widehat{\boldsymbol{Y}}_{t+k|t}. \tag{10.50}$$

Corresponding to (10.48) we get the covariance

$$\begin{aligned}\boldsymbol{\Sigma}^{xx}_{t+k|t} &= \operatorname{Var}\left[\boldsymbol{X}_{t+k}|\mathcal{Y}_t\right] \\ &= \operatorname{E}\left[\left(\boldsymbol{X}_{t+k} - \widehat{\boldsymbol{X}}_{t+k|t}\right)\left(\boldsymbol{X}_{t+k} - \widehat{\boldsymbol{X}}_{t+k|t}\right)^T \middle| \mathcal{Y}_t\right] \\ &= \operatorname{E}\left[\widetilde{\boldsymbol{X}}_{t+k|t}\widetilde{\boldsymbol{X}}^T_{t+k|t}|\mathcal{Y}_t\right] \\ &= \operatorname{Var}\left[\widetilde{\boldsymbol{X}}_{t+k|t}|\mathcal{Y}_t\right].\end{aligned} \tag{10.51}$$

Since the conditional means are calculated using linear projections, we have from Theorem 2.6 on page 24 that

$$\operatorname{C}\left[\boldsymbol{X}_{t+k} - \operatorname{E}\left[\boldsymbol{X}_{t+k}|\mathcal{Y}_t\right], \mathcal{Y}_t\right] = 0, \tag{10.52}$$

i.e., the prediction error $\widetilde{\boldsymbol{X}}_{t+k|t}$ and $\mathcal{Y}_t$ are uncorrelated (or orthogonal).

If we further assume that they are independent (equivalent to the assumption of normality) we get

$$\operatorname{Var}\left[\widetilde{\boldsymbol{X}}_{t+k|t}|\mathcal{Y}_t\right] = \operatorname{Var}\left[\widetilde{\boldsymbol{X}}_{t+k|t}\right], \tag{10.53}$$

which together with (10.51) implies

$$\boldsymbol{\Sigma}^{xx}_{t+k|t} = \operatorname{Var}\left[\boldsymbol{X}_{t+k}|\mathcal{Y}_t\right] = \operatorname{Var}\left[\widetilde{\boldsymbol{X}}_{t+k|t}\right]. \tag{10.54}$$

In a similar manner we can show that

$$\boldsymbol{\Sigma}^{yy}_{t+k|t} = \operatorname{Var}\left[\boldsymbol{Y}_{t+k}|\mathcal{Y}_t\right] = \operatorname{Var}\left[\widetilde{\boldsymbol{Y}}_{t+k|t}\right], \tag{10.55}$$

$$\boldsymbol{\Sigma}^{xy}_{t+k|t} = \operatorname{C}\left[\boldsymbol{X}_{t+k}, \boldsymbol{Y}_{t+k}|\mathcal{Y}_t\right] = \operatorname{C}\left[\widetilde{\boldsymbol{X}}_{t+k|t}, \widetilde{\boldsymbol{Y}}_{t+k|t}\right]. \tag{10.56}$$

Now the optimal reconstruction of $\boldsymbol{X}_{t|t}$ is found by using (10.45) through (10.47) directly, with $\boldsymbol{X} = \boldsymbol{X}_t$, $\boldsymbol{Y} = \boldsymbol{Y}_t$ and $\boldsymbol{Z} = \mathcal{Y}_{t-1}$.

THEOREM 10.1 (RECONSTRUCTION)
The optimal reconstruction *of $\boldsymbol{X}_{t|t}$ is obtained by*

$$\begin{aligned}\widehat{\boldsymbol{X}}_{t|t} &= \operatorname{E}\left[\boldsymbol{X}_t|\boldsymbol{Y}_t, \mathcal{Y}_{t-1}\right] \\ &= \operatorname{E}\left[\boldsymbol{X}_t|\mathcal{Y}_{t-1}\right] \\ &\quad + \operatorname{C}\left[\boldsymbol{X}_t, \boldsymbol{Y}_t|\mathcal{Y}_{t-1}\right]\operatorname{Var}^{-1}\left[\boldsymbol{Y}_t|\mathcal{Y}_{t-1}\right]\left(\boldsymbol{Y}_t - \operatorname{E}\left[\boldsymbol{Y}_t|\mathcal{Y}_{t-1}\right]\right),\end{aligned} \tag{10.57}$$

which can be written as

$$\widehat{\boldsymbol{X}}_{t|t} = \widehat{\boldsymbol{X}}_{t|t-1} + \boldsymbol{\Sigma}^{xy}_{t|t-1}\left(\boldsymbol{\Sigma}^{yy}_{t|t-1}\right)^{-1}\left(\boldsymbol{Y}_t - \widehat{\boldsymbol{Y}}_{t|t-1}\right). \tag{10.58}$$

The covariance of the reconstruction error *becomes*

$$\boldsymbol{\Sigma}^{xx}_{t|t} = \boldsymbol{\Sigma}^{xx}_{t|t-1} - \boldsymbol{\Sigma}^{xy}_{t|t-1}\left(\boldsymbol{\Sigma}^{yy}_{t|t-1}\right)^{-1}\left(\boldsymbol{\Sigma}^{xy}_{t|t-1}\right)^{T}. \tag{10.59}$$

Proof See above. ■

If we introduce *the Kalman gain*

$$\boldsymbol{K}_t = \boldsymbol{\Sigma}^{xy}_{t|t-1}\left(\boldsymbol{\Sigma}^{yy}_{t|t-1}\right)^{-1}, \tag{10.60}$$

the optimal reconstruction and the corresponding covariance can be written

$$\widehat{\boldsymbol{X}}_{t|t} = \widehat{\boldsymbol{X}}_{t|t-1} + \boldsymbol{K}_t\left(\boldsymbol{Y}_t - \widehat{\boldsymbol{Y}}_{t|t-1}\right), \tag{10.61}$$

$$\boldsymbol{\Sigma}^{xx}_{t|t} = \boldsymbol{\Sigma}^{xx}_{t|t-1} - \boldsymbol{K}_t\boldsymbol{\Sigma}^{yy}_{t|t-1}\boldsymbol{K}_t^T. \tag{10.62}$$

The formulas denote how information from past observations, i.e., $\mathcal{Y}_{t-1}$, can be combined with the new observation $\boldsymbol{Y}_t$ to improve the estimate of $\boldsymbol{X}$.

It is further clear from (10.60) through (10.62) that in order to calculate the reconstruction and the corresponding variance we need to calculate the one-step prediction of $\boldsymbol{X}_t$ and $\boldsymbol{Y}_t$ and its corresponding variances and covariances.

From (10.40) and (10.41) on page 289 it follows that *the one-step prediction* becomes

$$\widehat{\boldsymbol{X}}_{t+1|t} = \boldsymbol{A}\widehat{\boldsymbol{X}}_{t|t} + \boldsymbol{B}\boldsymbol{u}_t, \tag{10.63}$$

$$\widehat{\boldsymbol{Y}}_{t+1|t} = \boldsymbol{C}\widehat{\boldsymbol{X}}_{t+1|t}. \tag{10.64}$$

Thus, it holds that for the prediction error

$$\begin{aligned}\widetilde{\boldsymbol{X}}_{t+1|t} = \boldsymbol{X}_{t+1} - \widehat{\boldsymbol{X}}_{t+1|t} &= \boldsymbol{A}\left(\boldsymbol{X}_t - \widehat{\boldsymbol{X}}_{t|t}\right) + \boldsymbol{e}_{1,t+1} \\ &= \boldsymbol{A}\widetilde{\boldsymbol{X}}_{t|t} + \boldsymbol{e}_{1,t+1},\end{aligned} \tag{10.65}$$

$$\begin{aligned}\widetilde{\boldsymbol{Y}}_{t+1|t} = \boldsymbol{Y}_{t+1} - \widehat{\boldsymbol{Y}}_{t+1|t} &= \boldsymbol{C}\left(\boldsymbol{X}_{t+1} - \widehat{\boldsymbol{X}}_{t+1|t}\right) + \boldsymbol{e}_{2,t+1} \\ &= \boldsymbol{C}\widetilde{\boldsymbol{X}}_{t+1|t} + \boldsymbol{e}_{2,t+1}.\end{aligned} \tag{10.66}$$

From this it follows that

$$\operatorname{Var}\left[\widetilde{\boldsymbol{X}}_{t+1|t}\right] = \boldsymbol{A}\operatorname{Var}\left[\widetilde{\boldsymbol{X}}_{t|t}\right]\boldsymbol{A}^T + \boldsymbol{\Sigma}_1, \tag{10.67}$$

$$\operatorname{Var}\left[\widetilde{\boldsymbol{Y}}_{t+1|t}\right] = \boldsymbol{C}\operatorname{Var}\left[\widetilde{\boldsymbol{X}}_{t+1|t}\right]\boldsymbol{C}^T + \boldsymbol{\Sigma}_2, \tag{10.68}$$

$$\operatorname{C}\left[\widetilde{\boldsymbol{X}}_{t+1|t}, \widetilde{\boldsymbol{Y}}_{t+1|t}\right] = \operatorname{Var}\left[\widetilde{\boldsymbol{X}}_{t+1|t}\right]\boldsymbol{C}^T. \tag{10.69}$$

Applying (10.54)–(10.57) yields

$$\boldsymbol{\Sigma}^{xx}_{t+1|t} = \boldsymbol{A}\boldsymbol{\Sigma}^{xx}_{t|t}\boldsymbol{A}^T + \boldsymbol{\Sigma}_1, \tag{10.70}$$

$$\boldsymbol{\Sigma}^{yy}_{t+1|t} = \boldsymbol{C}\boldsymbol{\Sigma}^{xx}_{t+1|t}\boldsymbol{C}^T + \boldsymbol{\Sigma}_2, \tag{10.71}$$

$$\boldsymbol{\Sigma}^{xy}_{t+1|t} = \boldsymbol{\Sigma}^{xx}_{t+1|t}\boldsymbol{C}^T, \tag{10.72}$$

The equations (10.60)–(10.62), (10.63)–(10.64), and (10.67)–(10.69) can in turn be used to reconstruct and predict $\boldsymbol{X}_t$ and the corresponding covariance $\boldsymbol{\Sigma}^{xx}_{t+1|t}$. We have, thus, derived the Kalman filter in discrete time. The results are gathered in the following.

THEOREM 10.2 (THE KALMAN FILTER)
The optimal linear reconstruction $\widehat{\boldsymbol{X}}_{t|t}$ *and* prediction $\widehat{\boldsymbol{X}}_{t+1|t}$ *of the system* (10.40)–(10.41) *is obtained in terms of a* reconstruction *(updating)*

$$\widehat{\boldsymbol{X}}_{t|t} = \widehat{\boldsymbol{X}}_{t|t-1} + \boldsymbol{K}_t\left(\boldsymbol{Y}_t - \boldsymbol{C}\widehat{\boldsymbol{X}}_{t|t-1}\right), \tag{10.73}$$

$$\boldsymbol{\Sigma}^{xx}_{t|t} = \boldsymbol{\Sigma}^{xx}_{t|t-1} - \boldsymbol{K}_t\boldsymbol{\Sigma}^{yy}_{t|t-1}\boldsymbol{K}_t^T = \boldsymbol{\Sigma}^{xx}_{t|t-1} - \boldsymbol{K}_t\boldsymbol{C}\boldsymbol{\Sigma}^{xx}_{t|t-1}, \tag{10.74}$$

where the Kalman gain is

$$\boldsymbol{K}_t = \boldsymbol{\Sigma}^{xx}_{t|t-1}\boldsymbol{C}^T\left(\boldsymbol{\Sigma}^{yy}_{t|t-1}\right)^{-1}, \tag{10.75}$$

and the prediction *is*

$$\widehat{\boldsymbol{X}}_{t+1|t} = \boldsymbol{A}\widehat{\boldsymbol{X}}_{t|t} + \boldsymbol{B}\boldsymbol{u}_t, \tag{10.76}$$

$$\boldsymbol{\Sigma}^{xx}_{t+1|t} = \boldsymbol{A}\boldsymbol{\Sigma}^{xx}_{t|t}\boldsymbol{A}^T + \boldsymbol{\Sigma}_1, \tag{10.77}$$

$$\boldsymbol{\Sigma}^{yy}_{t+1|t} = \boldsymbol{C}\boldsymbol{\Sigma}^{xx}_{t+1|t}\boldsymbol{C}^T + \boldsymbol{\Sigma}_2. \tag{10.78}$$

With initial condition

$$\widehat{\boldsymbol{X}}_{1|0} = \mathrm{E}[\boldsymbol{X}_1] = \boldsymbol{\mu}_0, \tag{10.79}$$

$$\boldsymbol{\Sigma}^{xx}_{1|0} = \mathrm{Var}[\boldsymbol{X}_1] = \boldsymbol{V}_0. \tag{10.80}$$

Proof Follows directly. ■

For each recursion—or each new observation $\boldsymbol{Y}_t$—we first use $\widehat{\boldsymbol{X}}_{t|t-1}$ and $\boldsymbol{\Sigma}^{xx}_{t|t-1}$. Then we successively apply the recursion formulas (10.73)–(10.75) (after a calculation of $\boldsymbol{\Sigma}^{yy}_{t|t-1}$ via (10.78)) and the prediction formulas (10.76), (10.77), and (10.78). If we give this a Bayesian interpretation, then $\widehat{\boldsymbol{X}}_{t|t-1}$ and $\boldsymbol{\Sigma}^{xx}_{t|t-1}$ are the mean and covariance, respectively, in the prior distribution for $\boldsymbol{X}_t$. This will be further elaborated in the next section.

Example 10.4 (The Kalman filter and a falling body)
Let us consider the falling body in Example 10.2 on page 285. This system can be written in the state space form

$$\begin{pmatrix} X_{1,t} \\ X_{2,t} \end{pmatrix} = \begin{pmatrix} 1 & 1 \\ 0 & 1 \end{pmatrix} \begin{pmatrix} X_{1,t-1} \\ X_{2,t-1} \end{pmatrix} + \begin{pmatrix} -\frac{1}{2} \\ -1 \end{pmatrix} u_{t-1} + \begin{pmatrix} e_{11,t} \\ e_{12,t} \end{pmatrix}$$

where $X_{1,t}$ is the position and $X_{2,t}$ is the velocity. For the process noise we assume

$$\boldsymbol{\Sigma}_{1,t} = \begin{pmatrix} 0 & 0 \\ 0 & 0.1 \end{pmatrix}.$$

Let us assume that the position of the body is observed with high accuracy and that the velocity of the body is observed at every second sampling interval. The observation equation becomes

$$\begin{pmatrix} Y_{1,t} \\ Y_{2,t} \end{pmatrix} = \begin{pmatrix} 1 & 0 \\ 0 & 1 \end{pmatrix} \begin{pmatrix} X_{1,t} \\ X_{2,t} \end{pmatrix} + \begin{pmatrix} e_{21,t} \\ e_{22,t} \end{pmatrix}$$

with variance

$$\Sigma_{2,t} = \begin{pmatrix} 1 & 0 \\ 0 & \gamma_t \end{pmatrix}$$

where $\gamma_t = 2$ for $t = 2, 4, 6, \dots$ and $\gamma_t = \infty$ (in practice a large number) for $t = 1, 3, 5, \dots$.

As starting value we apply

$$\boldsymbol{\Sigma}^{xx}_{1|0} = \begin{pmatrix} 10 & 0 \\ 0 & 1 \end{pmatrix}$$

From (10.78) we get the covariance of $\boldsymbol{Y}_{1|0}$

$$\boldsymbol{\Sigma}^{yy}_{1|0} = \begin{pmatrix} 1 & 0 \\ 0 & 1 \end{pmatrix} \begin{pmatrix} 10 & 0 \\ 0 & 1 \end{pmatrix} \begin{pmatrix} 1 & 0 \\ 0 & 1 \end{pmatrix} + \begin{pmatrix} 1 & 0 \\ 0 & \infty \end{pmatrix} = \begin{pmatrix} 11 & 0 \\ 0 & \infty \end{pmatrix}$$

It is now possible to calculate the Kalman gain for $t = 1$, as well as the variances for the reconstruction and the prediction.

$$\boldsymbol{K}_1 = \begin{pmatrix} 10 & 0 \\ 0 & 1 \end{pmatrix} \begin{pmatrix} 1 & 0 \\ 0 & 1 \end{pmatrix} \begin{pmatrix} 1/11 & 0 \\ 0 & 1/\infty \end{pmatrix} = \begin{pmatrix} 10/11 & 0 \\ 0 & 1/\infty \end{pmatrix}$$

$$\boldsymbol{\Sigma}^{xx}_{1|1} = \begin{pmatrix} 10 & 0 \\ 0 & 1 \end{pmatrix} - \begin{pmatrix} 10/11 & 0 \\ 0 & 1/\infty \end{pmatrix} \begin{pmatrix} 11 & 0 \\ 0 & \infty \end{pmatrix} \begin{pmatrix} 10/11 & 0 \\ 0 & 1/\infty \end{pmatrix}$$

$$= \begin{pmatrix} 10/11 & 0 \\ 0 & 1 \end{pmatrix}$$

$$\boldsymbol{\Sigma}^{xx}_{2|1} = \begin{pmatrix} 1 & 1 \\ 0 & 1 \end{pmatrix} \begin{pmatrix} 10/11 & 0 \\ 0 & 1 \end{pmatrix} \begin{pmatrix} 1 & 0 \\ 1 & 1 \end{pmatrix} + \begin{pmatrix} 0 & 0 \\ 0 & 0.1 \end{pmatrix} = \begin{pmatrix} 21/11 & 1 \\ 1 & 1.1 \end{pmatrix}$$

We see that the Kalman gain ignores the velocity at $t = 1$.

The equivalent calculations for $t = 2$ yield

$$\boldsymbol{\Sigma}^{yy}_{2|1} = \begin{pmatrix} 2.909 & 1 \\ 1 & 3.1 \end{pmatrix} = \begin{pmatrix} 32/11 & 1 \\ 1 & 3.1 \end{pmatrix}$$

$$\boldsymbol{K}_2 = \begin{pmatrix} 0.631 & 0.125 \\ 0.249 & 0.274 \end{pmatrix}$$

$$\boldsymbol{\Sigma}^{xx}_{2|2} = \begin{pmatrix} 0.613 & 0.249 \\ 0.249 & .549 \end{pmatrix}$$

$$\boldsymbol{\Sigma}^{xx}_{3|2} = \begin{pmatrix} 1.661 & 0.798 \\ 0.798 & 0.649 \end{pmatrix}$$

The matrices obtain their stationary values relatively quickly. For example, the stationary value for the Kalman gain becomes

$$\boldsymbol{K}_{t \text{ odd}} = \begin{pmatrix} 0.526 & 0 \\ 0.194 & 0 \end{pmatrix}$$

$$\boldsymbol{K}_{t \text{ even}} = \begin{pmatrix} 0.519 & 0.091 \\ 0.181 & 0.114 \end{pmatrix}$$

Please note, that for t odd we are disregarding the velocity.

In the previous example the variance of the observations was time-varying. A simple example, where the $\boldsymbol{C}$ matrix is time-varying, is the following.

Example 10.5 (Estimation of sales numbers)
Let X_t denote the yearly sales of a given product and let Y_t denote the observed sales in month t modulus 12. From experience it is known how the sales are distributed during the months of the year. Let p_t denote the fraction of the yearly sales in month i.

A company is interested in successively estimating the yearly sales in order to track an increase or decrease in sales. From experience, the yearly sales are a random walk, i.e.,

$$X_t = X_{t-1} + e_{1,t}$$

The observation equation becomes

$$Y_t = p_t X_t + e_{2,t}$$

By applying a Kalman filter, the company can now successively estimate the yearly sales.

10.3.2 k-step predictions in state space models

Only one-step predictions are given by the Kalman filter.

The k-step prediction in the state space model (10.40) and (10.41) on page 289 can be calculated by applying the recursive formula

$$\widehat{\boldsymbol{X}}_{t+k+1|t} = \boldsymbol{A}\widehat{\boldsymbol{X}}_{t+k|t} + \boldsymbol{B}\boldsymbol{u}_{t+k}, \tag{10.81}$$

and the corresponding recursive formula for the *covariance*

$$\boldsymbol{\Sigma}^{xx}_{t+k+1|t} = \boldsymbol{A}\boldsymbol{\Sigma}^{xx}_{t+k|t}\boldsymbol{A}^T + \boldsymbol{\Sigma}_1. \tag{10.82}$$

The formulas (10.81) and (10.82) follow directly from the state space model.

10.3.3 Empirical Bayesian description of the Kalman filter

A Kalman filter is a recursive procedure for calculating an estimate for $\boldsymbol{X}_t$ based on the observations $\mathcal{Y}_t = (\boldsymbol{Y}_1, \ldots, \boldsymbol{Y}_t)$. Given these observations, inference about $\boldsymbol{X}_t$ can be made by applying Bayes' formula. We get

$$\mathrm{P}\{\boldsymbol{X}_t|\mathcal{Y}_t\} \propto \mathrm{P}\{\boldsymbol{Y}_t|\boldsymbol{X}_t, \mathcal{Y}_{t-1}\}\,\mathrm{P}\{\boldsymbol{X}_t|\mathcal{Y}_{t-1}\}, \tag{10.83}$$

where $\mathrm{P}\{A|B\}$ is the probability for the event A given the event B.

From a Bayesian point of view $\mathrm{P}\{\boldsymbol{X}_t|\mathcal{Y}_t\}$ is the *posterior distribution*, $\mathrm{P}\{\boldsymbol{Y}_t|\boldsymbol{X}_t, \mathcal{Y}_{t-1}\}$ is the *likelihood distribution*, and $\mathrm{P}\{\boldsymbol{X}_t|\mathcal{Y}_{t-1}\}$ is the *prior distribution*.

We assume that at times $t-1$, the information of $\boldsymbol{X}_{t-1}$ (assuming normality) is given by

$$(\boldsymbol{X}_{t-1}|\mathcal{Y}_{t-1}) \sim \mathrm{N}\left(\widehat{\boldsymbol{X}}_{t-1|t-1}, \boldsymbol{\Sigma}^{xx}_{t-1|t-1}\right), \tag{10.84}$$

i.e., the posterior distribution for $\boldsymbol{X}_{t-1}$.

We now consider the time t in two steps:

1) Before obtaining the observation $\boldsymbol{Y}_t$.

2) After obtaining the observation $\boldsymbol{Y}_t$.

Step 1 *Before obtaining the observation* $\boldsymbol{Y}_t$, the best estimate of $\boldsymbol{X}_t$ is obtained directly from the system equation

$$\boldsymbol{X}_t = \boldsymbol{A}\boldsymbol{X}_{t-1} + \boldsymbol{B}\boldsymbol{u}_{t-1} + \boldsymbol{e}_{1,t}. \tag{10.85}$$

And we get the *prior distribution*

$$(\boldsymbol{X}_t|\mathcal{Y}_{t-1}) \sim \mathrm{N}\left(\boldsymbol{A}\widehat{\boldsymbol{X}}_{t-1|t-1} + \boldsymbol{B}\boldsymbol{u}_{t-1}, \boldsymbol{A}\boldsymbol{\Sigma}^{xx}_{t-1|t-1}\boldsymbol{A}^T + \boldsymbol{\Sigma}_1\right) \tag{10.86}$$

Step 2 *After obtaining the observation* $\boldsymbol{Y}_t$, we can calculate the posterior distribution using Bayes' formula (10.83), but first we have to calculate the likelihood distribution $\mathrm{P}\{\boldsymbol{Y}_t|\boldsymbol{X}_t, \mathcal{Y}_{t-1}\}$. Let $\widetilde{\boldsymbol{Y}}_{t|t-1}$ be the prediction error given by $\mathcal{Y}_{t-1}$, i.e.,

$$\begin{aligned}\widetilde{\boldsymbol{Y}}_{t|t-1} &= \boldsymbol{Y}_t - \widehat{\boldsymbol{Y}}_{t|t-1} \\ &= \boldsymbol{Y}_t - \boldsymbol{C}\widehat{\boldsymbol{X}}_{t|t-1} \\ &= \boldsymbol{Y}_t - \boldsymbol{C}\boldsymbol{A}\widehat{\boldsymbol{X}}_{t-1|t-1} - \boldsymbol{C}\boldsymbol{B}\boldsymbol{u}_{t-1}.\end{aligned} \tag{10.87}$$

Since $\boldsymbol{C}$, $\boldsymbol{A}$, $\boldsymbol{B}$, $\widehat{\boldsymbol{X}}_{t-1|t-1}$ and $\boldsymbol{u}_{t-1}$ are all known, the observation $\boldsymbol{Y}_t$ is equivalent to observing $\boldsymbol{Y}_{t|t-1}$, i.e., (10.83) can be written

$$\begin{aligned}\mathrm{P}\left\{\boldsymbol{X}_t|\boldsymbol{Y}_t, \mathcal{Y}_{t-1}\right\} &= \mathrm{P}\left\{\boldsymbol{X}_t|\widetilde{\boldsymbol{Y}}_{t|t-1}, \mathcal{Y}_{t-1}\right\} \\ &\propto \mathrm{P}\left\{\widetilde{\boldsymbol{Y}}_{t|t-1}|\boldsymbol{X}_t, \mathcal{Y}_{t-1}\right\}\mathrm{P}\left\{\boldsymbol{X}_t|\mathcal{Y}_{t-1}\right\}.\end{aligned} \tag{10.88}$$

Since $\boldsymbol{Y}_t = \boldsymbol{C}\boldsymbol{X}_t + \boldsymbol{e}_{2,t}$ we may write $\widetilde{\boldsymbol{Y}}_{t|t-1}$ in (10.87) as

$$\widetilde{\boldsymbol{Y}}_{t|t-1} = \boldsymbol{C}\left(\boldsymbol{X}_t - \boldsymbol{A}\widehat{\boldsymbol{X}}_{t-1|t-1} - \boldsymbol{B}\boldsymbol{u}_{t-1}\right) + \boldsymbol{e}_{2,t}, \tag{10.89}$$

i.e.,

$$\mathrm{E}\left[\widetilde{\boldsymbol{Y}}_{t|t-1}|\boldsymbol{X}_t, \mathcal{Y}_{t-1}\right] = \boldsymbol{C}\left(\boldsymbol{X}_t - \boldsymbol{A}\widehat{\boldsymbol{X}}_{t-1|t-1} - \boldsymbol{B}\boldsymbol{u}_{t-1}\right), \tag{10.90}$$

$$\mathrm{Var}\left[\widetilde{\boldsymbol{Y}}_{t|t-1}|\boldsymbol{X}_t, \mathcal{Y}_{t-1}\right] = \boldsymbol{\Sigma}_2. \tag{10.91}$$

The *likelihood distribution* is thus

$$\begin{aligned}&\left(\widetilde{\boldsymbol{Y}}_{t|t-1}|\boldsymbol{X}_t, \mathcal{Y}_{t-1}\right) \\ &\qquad \sim \mathrm{N}\left(\boldsymbol{C}\left(\boldsymbol{X}_t - \boldsymbol{A}\widehat{\boldsymbol{X}}_{t-1|t-1} - \boldsymbol{B}\boldsymbol{u}_{t-1}\right), \boldsymbol{\Sigma}_2\right).\end{aligned} \tag{10.92}$$

In principle, we may apply Bayes' rule (10.88) and calculate the posterior distribution, but it is much simpler to apply the following procedure to calculate the conditional mean and conditional variance in the multivariate normal distribution (Theorem 2.8 on page 27). It holds that

$$\begin{pmatrix}\boldsymbol{X}_1 \\ \boldsymbol{X}_2\end{pmatrix} \sim \mathrm{N}\left[\begin{pmatrix}\boldsymbol{\mu}_1 \\ \boldsymbol{\mu}_2\end{pmatrix}, \begin{pmatrix}\boldsymbol{\Sigma}_{11} & \boldsymbol{\Sigma}_{12} \\ \boldsymbol{\Sigma}_{21} & \boldsymbol{\Sigma}_{22}\end{pmatrix}\right] \tag{10.93}$$

and it follows that

$$(\boldsymbol{X}_1|\boldsymbol{X}_2) \sim \mathrm{N}\left(\boldsymbol{\mu}_1 + \boldsymbol{\Sigma}_{12}\boldsymbol{\Sigma}_{22}^{-1}(\boldsymbol{X}_2 - \boldsymbol{\mu}_2), \boldsymbol{\Sigma}_{11} - \boldsymbol{\Sigma}_{12}\boldsymbol{\Sigma}_{22}^{-1}\boldsymbol{\Sigma}_{21}\right). \tag{10.94}$$

Similarly to (10.45) and (10.46) on page 290, we may condition on a third variable $\boldsymbol{X}_3$ in (10.93) and (10.94).

If we in (10.93) substitute $\boldsymbol{X}_1$, $\boldsymbol{X}_2$, $\boldsymbol{\mu}_2$, and $\boldsymbol{\Sigma}_{22}$ with $\widetilde{\boldsymbol{Y}}_{t|t-1}$, $\boldsymbol{X}_t$, $\boldsymbol{A}\widehat{\boldsymbol{X}}_{t-1|t-1} + \boldsymbol{B}\boldsymbol{u}_{t-1}$, and $\boldsymbol{R} = \boldsymbol{A}\boldsymbol{\Sigma}^{xx}_{t-1|t-1}\boldsymbol{A}^T + \boldsymbol{\Sigma}_1$ (the two latter are obtained from the prior distribution for $\boldsymbol{X}_t$—see (10.86)), we get from (10.94)

$$\mathrm{E}\left[\widetilde{\boldsymbol{Y}}_{t|t-1}|\boldsymbol{X}_t, \mathcal{Y}_{t-1}\right] = \boldsymbol{\mu}_1 + \boldsymbol{\Sigma}_{12}\boldsymbol{R}_t^{-1}\left(\boldsymbol{X}_t - \boldsymbol{A}\widehat{\boldsymbol{X}}_{t-1|t-1} - \boldsymbol{B}\boldsymbol{u}_{t-1}\right) \quad (10.95)$$

By comparing with (10.92) we see that

$$\boldsymbol{\mu}_1 = \boldsymbol{0} \quad \text{and} \quad \boldsymbol{\Sigma}_{12} = \boldsymbol{C}\boldsymbol{R}_t. \quad (10.96)$$

Similarly, we get from (10.94)

$$\mathrm{Var}\left[\widetilde{\boldsymbol{Y}}_{t|t-1}|\boldsymbol{X}_t, \mathcal{Y}_{t-1}\right] = \boldsymbol{\Sigma}_{11} - \boldsymbol{C}\boldsymbol{R}_t\boldsymbol{C}^T. \quad (10.97)$$

By comparing with (10.92) we get

$$\boldsymbol{\Sigma}_{11} = \boldsymbol{C}\boldsymbol{R}_t\boldsymbol{C}^T + \boldsymbol{\Sigma}_2. \quad (10.98)$$

Now, all the terms of (10.93) are identified

$$\begin{aligned}
&\left(\begin{matrix}\boldsymbol{X}_t \\ \widetilde{\boldsymbol{Y}}_{t|t-1}\end{matrix}\middle| \mathcal{Y}_{t-1}\right) \\
&\quad \sim \mathrm{N}\left[\begin{pmatrix}\boldsymbol{A}\widehat{\boldsymbol{X}}_{t-1|t-1} + \boldsymbol{B}\boldsymbol{u}_{t-1} \\ 0\end{pmatrix}, \begin{pmatrix}\boldsymbol{R}_t & \boldsymbol{R}_t\boldsymbol{C}^T \\ \boldsymbol{C}\boldsymbol{R}_t & \boldsymbol{\Sigma}_2 + \boldsymbol{C}\boldsymbol{R}_t\boldsymbol{C}^T\end{pmatrix}\right]
\end{aligned} \quad (10.99)$$

We can re-apply (10.93)–(10.94) by conditioning on $\widetilde{\boldsymbol{Y}}_{t|t-1}$. We get

$$\begin{aligned}
&\left(\boldsymbol{X}_t|\widetilde{\boldsymbol{Y}}_{t|t-1}, \mathcal{Y}_{t-1}\right) \\
&\quad \sim \mathrm{N}\Big[\boldsymbol{A}\widehat{\boldsymbol{X}}_{t-1|t-1} + \boldsymbol{B}\boldsymbol{u}_{t-1} \\
&\qquad + \boldsymbol{R}_t^T\boldsymbol{C}^T\left(\boldsymbol{\Sigma}_2 + \boldsymbol{C}\boldsymbol{R}_t\boldsymbol{C}^T\right)^{-1}\widetilde{\boldsymbol{Y}}_{t|t-1}, \\
&\qquad \boldsymbol{R}_t - \boldsymbol{R}_t\boldsymbol{C}^T\left(\boldsymbol{\Sigma}_2 + \boldsymbol{C}\boldsymbol{R}_t\boldsymbol{C}^T\right)^{-1}\boldsymbol{C}\boldsymbol{R}_t\Big],
\end{aligned} \quad (10.100)$$

which is the *posterior distribution*—see (10.88).

Summary: After time $t-1$, we have the posterior distribution for X_{t-1} having mean $\widehat{\boldsymbol{X}}_{t-1|t-1}$ and covariance $\boldsymbol{\Sigma}^{xx}_{t-1|t-1}$. Based on this, we get the prior distribution (10.86) for $\boldsymbol{X}_t$ having mean value $\boldsymbol{A}\widehat{\boldsymbol{X}}_{t-1|t-1} + \boldsymbol{B}\boldsymbol{u}_{t-1}$ and covariance $\boldsymbol{R} = \boldsymbol{A}\boldsymbol{\Sigma}^{xx}_{t-1|t-1}\boldsymbol{A}^T + \boldsymbol{\Sigma}_1$.

When the observation $\boldsymbol{Y}_t$ is available, we find the posterior distribution (10.88) with mean value

$$\widehat{\boldsymbol{X}}_{t|t} = \boldsymbol{A}\widehat{\boldsymbol{X}}_{t-1|t-1} + \boldsymbol{B}\boldsymbol{u}_{t-1} + \boldsymbol{R}_t\boldsymbol{C}^T\left(\boldsymbol{\Sigma}_2 + \boldsymbol{C}\boldsymbol{R}_t\boldsymbol{C}^T\right)^{-1}\widetilde{\boldsymbol{Y}}_{t|t-1} \quad (10.101)$$

and covariance

$$\Sigma_{t|t}^{xx} = R_t - R_t C^T \left(\Sigma_2 + C R_t C^T\right)^{-1} C R_t. \tag{10.102}$$

We can now repeat from the beginning using the calculated posterior distribution for $\boldsymbol{X}_t$.

Since

$$\begin{aligned}
\widetilde{\boldsymbol{Y}}_{t|t-1} &= \boldsymbol{Y}_t - \boldsymbol{C}\widehat{\boldsymbol{X}}_{t|t-1}, \qquad \text{(see (10.87))} \\
\Sigma_{t|t-1}^{xx} &= R_t = A\Sigma_{t-1|t-1}^{xx}A^T + \Sigma_1, \\
\Sigma_{t|t-1}^{yy} &= CR_tC^T + \Sigma_2, \\
K_t &= R_tC^T\left(\Sigma_2 + CR_tC^T\right)^{-1},
\end{aligned}$$

we see that (10.101) and (10.102) correspond to the Kalman filter in Theorem 10.2 on page 293.

10.4 Some common models in state space form

In this section we will exemplify how some of the previously defined models can be written in state space form. A state space formulation is useful in estimation of models based on time series with missing observations and when generalizing non-stationary models.

We begin by showing how an arbitrary ARMA(p, q) model (possibly multivariate) can be written in *state space form.*

Assume that $\{Y_t\}$ can be described by an ARMA(p, q) model, i.e.,

$$Y_t + \phi_1 Y_{t-1} + \cdots + \phi_p Y_{t-p} = \varepsilon_t + \theta_1 \varepsilon_{t-1} + \cdots + \theta_q \varepsilon_{t-q} \tag{10.103}$$

This process can be written in state space form

$$\boldsymbol{X}_t = \boldsymbol{A}\boldsymbol{X}_{t-1} + \boldsymbol{e}_{1,t}, \tag{10.104}$$

$$Y_t = \boldsymbol{C}\boldsymbol{X}_t, \tag{10.105}$$

where

$$\boldsymbol{X}_t = (X_{1,t}, X_{2,t}, \ldots, X_{d,t})^T, \quad d = \max(p, q+1) \tag{10.106}$$

$$\boldsymbol{A} = \begin{pmatrix} -\phi_1 & 1 & 0 & \cdots & 0 \\ -\phi_2 & 0 & 1 & \cdots & 0 \\ \vdots & \vdots & \vdots & \ddots & \vdots \\ -\phi_{d-1} & 0 & 0 & \cdots & 1 \\ -\phi_d & 0 & 0 & \cdots & 0 \end{pmatrix} \tag{10.107}$$

$$e_{1,t} = G\varepsilon_t = \begin{pmatrix} 1 \\ \theta_1 \\ \vdots \\ \theta_{d-1} \end{pmatrix} \varepsilon_t \tag{10.108}$$

$$C = \begin{pmatrix} 1 & 0 & \cdots & 0 \end{pmatrix}. \tag{10.109}$$

If we consider a multivariate ARMA process, then $\phi_1, \dots, \phi_d$ and $\theta_1, \dots, \theta_{d-1}$ and the 1's should be replaced by matrices of suitable dimensions.

This is an example of a canonical form, which thus yields a unique formulation of the state space form.

In Example 10.3 on page 288 we showed, using an AR(2) model, how the A matrix can be changed and thus how an alternative formulation can be achieved.

One should also note the change of the noise term in (10.108) due to the G matrix. Thus, when updating Σ^{xx}, we need to apply (10.138) on page 308 instead of the formerly used updating formula (10.77) on page 293.

Furthermore, one should be aware that if we include measurement noise in (10.105), there will be further problems with the uniqueness. This is illustrated in a simple example.

Example 10.6 (Random walk with measurement noise)
Consider the following state space model

$$\mu_t = \mu_{t-1} + \eta_t, \tag{10.110}$$
$$Y_t = \mu_t + \varepsilon_t, \tag{10.111}$$

where $\{\eta_t\}$ is white noise with variance σ_η^2 and $\{\varepsilon_t\}$ is white noise with variance σ_ε^2. Equation (10.110) gives the level of $\{Y_t\}$ which is a random walk, whereas (10.111) expresses that the level is affected by measurement noise.

By substituting we find

$$Y_t - Y_{t-1} = \mu_t - \mu_{t-1} + \varepsilon_t - \varepsilon_{t-1} = \eta_t + \varepsilon_t - \varepsilon_{t-1}. \tag{10.112}$$

The autocorrelation function for $Y_t - Y_{t-1}$ becomes

$$\rho(k) = \begin{cases} 1 & k = 0 \\ -\sigma_\varepsilon^2/(\sigma_\eta^2 + 2\sigma_\varepsilon^2) & |k| = 1 \\ 0 & |k| > 1 \end{cases} \tag{10.113}$$

which equals the autocorrelation function for an MA(1) process. Thus, we find

$$Y_t - Y_{t-1} = \xi_t + \theta_1 \xi_{t-1}, \quad (-1 \le \theta_1 \le 0) \tag{10.114}$$

where $\{\xi_t\}$ is white noise with variance σ_ξ^2. We see that $\{Y_t\}$ can be described by an *IMA*(1, 1) *model*. This model can be rewritten in a state space form using (10.104) on page 299 through (10.109).

The parameters in (10.114), i.e., θ_1 and σ_ξ, can be found by using the fact that the autocovariance function for (10.114) has to equal the autocovariance function for (10.112). Thus, the following equations should be solved

$$\left(1+\theta_1^2\right)\sigma_\xi^2 = \sigma_\eta^2 + 2\sigma_\varepsilon^2,$$
$$\theta_1\sigma_\xi^2 = -\sigma_\varepsilon^2,$$

with respect to θ_1 and σ_ξ^2.

From the example it is seen that the IMA(1, 1) model given by (10.114) can be written in the state space form (10.104)–(10.105), or alternatively on the state space form (10.110)–(10.111). The example illustrates how the parameters in an MA model can be written alternatively as variance parameters.

Also the classical variance component models (see Chapter 3) can be formulated as state space models. In Chapter 3 we considered the *global trend model*

$$Y_t = \alpha + \beta t + \varepsilon_t, \tag{10.115}$$

and it was illustrated how the *method of least squares* (or *general exponential smoothing*) was able to account for the variations in α and β in time.

A perhaps more rational method is a *state space model* where both the intercept and slope follow, e.g., a random walk:

$$\left.\begin{aligned}\mu_t &= \mu_{t-1} + \beta_{t-1} + \eta_t\\ \beta_t &= \qquad\quad\;\; \beta_{t-1} + \xi_t\end{aligned}\right\}\quad \text{system equation} \tag{10.116}$$

$$Y_t = \mu_t + \varepsilon_t \quad \text{observation equation} \tag{10.117}$$

The state vector here is $\boldsymbol{X}_t = (\mu_t, \beta_t)^T$. It should be noted that if $\text{Var}[\eta_t] = 0$ and $\text{Var}[\xi_t] = 0$, then (10.116)–(10.117) correspond to (10.115), i.e., the global trend model. For a further discussion on the classical variance components in state space form, see, e.g., Harvey and Pierse (1984).

10.4.1 Signal extraction

In Chapter 3 we considered the classical additive decomposition, where the variations are divided into components, i.e., a trend T_t, a seasonal component S_t and an error component ε_t. The use of a state space model is similar to such additive models and will be discussed in the following. These types of models are often used in, e.g., economics, since the model allows for a decomposition of the variation components into a trend, seasonal component, etc.

In this section we assume that Y_t can be described by the following *additive model*:

$$Y_t = T_t + S_t + V_t + \varepsilon_t, \tag{10.118}$$

where T_t, S_t, and V_t are trend, seasonal, and stochastic components, respectively.

The additive model (10.118) can be written in state space form as

$$\boldsymbol{X}_t = \begin{pmatrix} \boldsymbol{A}_1 & \boldsymbol{0} & \boldsymbol{0} \\ \boldsymbol{0} & \boldsymbol{A}_2 & \boldsymbol{0} \\ \boldsymbol{0} & \boldsymbol{0} & \boldsymbol{A}_3 \end{pmatrix} \boldsymbol{X}_{t-1} + \begin{pmatrix} \boldsymbol{G}_1 & \boldsymbol{0} & \boldsymbol{0} \\ \boldsymbol{0} & \boldsymbol{G}_2 & \boldsymbol{0} \\ \boldsymbol{0} & \boldsymbol{0} & \boldsymbol{G}_3 \end{pmatrix} \boldsymbol{w}_t, \tag{10.119}$$

$$Y_t = \begin{pmatrix} \boldsymbol{C}_1 & \boldsymbol{C}_2 & \boldsymbol{C}_3 \end{pmatrix} \boldsymbol{X}_t + \varepsilon_t, \tag{10.120}$$

where $\{\varepsilon_t\}$ is white noise with variance σ_ε^2 and the components are described by separate components: $(\boldsymbol{A}_j, \boldsymbol{G}_j, \boldsymbol{C}_j)$. The following example shows how the matrices as well as the noise components $\{\boldsymbol{w}_t\}$ can be specified.

1) Local polynomial trend $(\boldsymbol{A}_1, \boldsymbol{G}_1, \boldsymbol{C}_1)$ The polynomial trend component satisfies the k'th order difference equations with noise:

$$\nabla^k T_t = w_{1t}, \tag{10.121}$$

where $\{w_{1t}\}$ is white noise with variance σ_T^2. For example, we get for $k = 2$:

$$T_t = 2T_{t-1} - T_{t-2} + w_{1t},$$

which can be written in state space form by introducing

$$\boldsymbol{A}_1 = \begin{pmatrix} 2 & -1 \\ 1 & 0 \end{pmatrix}, \quad \boldsymbol{G}_1 = \begin{pmatrix} 1 \\ 0 \end{pmatrix}, \quad \boldsymbol{C}_1 = \begin{pmatrix} 1 \\ 0 \end{pmatrix}^T.$$

The model (10.121) is often called a *k'th order random walk*.

2) Local polynomial seasonal component $(\boldsymbol{A}_2, \boldsymbol{G}_2, \boldsymbol{C}_2)$ As discussed in Chapter 3 the following seasonal component model is often applied

$$\sum_{i=0}^{s-1} S_{t-i} = w_{2t}, \tag{10.122}$$

where s is the length of the season and $\{w_{2t}\}$ is white noise with variance σ_S^2. By rewriting we get

$$S_t = -\sum_{i=1}^{s-1} S_{t-i} + w_{2t} \tag{10.123}$$

and it is seen that the model can be written in state space form by introducing

$$\boldsymbol{A}_2 = \begin{pmatrix} -1 & -1 & \cdots & -1 & -1 \\ 1 & 0 & \cdots & 0 & 0 \\ 0 & 1 & \cdots & 0 & 0 \\ \vdots & \vdots & \ddots & \vdots & \vdots \\ 0 & 0 & \cdots & 1 & 0 \end{pmatrix}, \quad \boldsymbol{G}_2 = \begin{pmatrix} 1 \\ 0 \\ \vdots \\ 0 \end{pmatrix}, \quad \boldsymbol{C}_2 = \begin{pmatrix} 1 \\ 0 \\ \vdots \\ 0 \end{pmatrix}^T \tag{10.124}$$

The model (10.122) is often called a *seasonal walk.*

3) Stochastic trend component $(\boldsymbol{A}_3, \boldsymbol{G}_3, \boldsymbol{C}_3)$ The stochastic trend model is introduced as an AR(p) model, i.e.,

$$V_t = \phi_1 V_{t-1} + \cdots + \phi_p V_{t-p} + w_{3t}, \tag{10.125}$$

where $\{w_{3t}\}$ is white noise with variance σ_V^2. This model is usually written in state space form by introducing

$$\boldsymbol{A}_3 = \begin{pmatrix} \phi_1 & \phi_2 & \cdots & \phi_{p-1} & \phi_p \\ 1 & 0 & \cdots & 0 & 0 \\ 0 & 1 & \cdots & 0 & 0 \\ \vdots & \vdots & \ddots & \vdots & \vdots \\ 0 & 0 & \cdots & 1 & 0 \end{pmatrix}, \quad \boldsymbol{G}_3 = \begin{pmatrix} 1 \\ 0 \\ \vdots \\ 0 \end{pmatrix}, \quad \boldsymbol{C}_3 = \begin{pmatrix} 1 \\ 0 \\ \vdots \\ 0 \end{pmatrix}^T \tag{10.126}$$

The parameters in the additive model consist of $\phi_1, \ldots, \phi_p$ as well as the variances σ_T^2, σ_S^2, σ_V^2, and σ_ε^2. The first three variances denote the disturbance on the trend models, whereas σ_ε^2 is the observation noise. The ratio between $\sigma_T^2/\sigma_\varepsilon^2$, $\sigma_S^2/\sigma_\varepsilon^2$, and $\sigma_V^2/\sigma_\varepsilon^2$ is thus the *signal-to-noise ratio.* The parameters can be estimated by using a likelihood method. In Section 10.5 on page 307, we discuss how the likelihood function can be calculated for a general model in state space form.

When the model parameters are known, the Kalman filter can be applied to distinguish among the variations into a polynomial trend component, a seasonal component, and a stochastic component. Thereby we get the so-called *signal extraction* where each of the signals, e.g., the seasonal component, is smoothed according to the signal-to-noise ratio. This is illustrated in the following example.

Example 10.7 (Signal extraction)
The example, taken from Funch and Hansen (1987), concerns a simulated additive model

$$Y_t = T_t + S_t + V_t + \varepsilon_t,$$

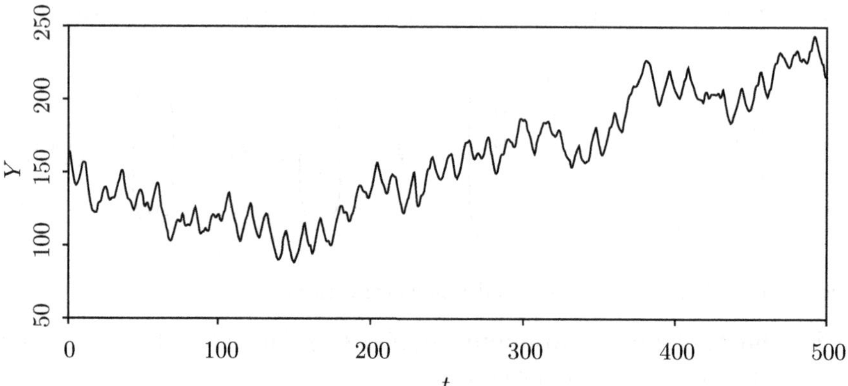

Figure 10.1: *A simulated time series consisting of the components: a random walk, a seasonal walk, AR(1), and white noise.*

Table 10.1: *ML estimates for the theoretical model based on the simulated values in Figure 10.1.*

	ϕ_1	σ_T^2	σ_S^2	σ_V^2
Theoretical	0.800	10.00	0.1000	5.00
Estimate	0.897	9.63	0.1010	5.49
Std. deviation	0.060	2.12	0.0165	2.21

where

$$T_t = T_{t-1} + w_{1t},$$

$$S_t = -\sum_{i=1}^{s-1} S_{t-i} + w_{2t},$$

$$V_t = \phi_1 V_{t-1} + w_{3t},$$

where $\text{Var}[\varepsilon_t] = \sigma_\varepsilon^2 = 1$, $\text{Var}[w_{1t}] = \sigma_T^2 = 10$, $\text{Var}[w_{2t}] = \sigma_S^2 = 0.1$, $\text{Var}[w_{3t}] = \sigma_V^2 = 5$, and $\phi_1 = 0.8$.

The simulated components are shown in Figure 10.2. The total time series is shown in Figure 10.1.

Applying an ML estimation procedure, corresponding to the one discussed in Section 10.5, we have found the estimates list in Table 10.1.

The corresponding one-step predictions for the decomposed time series are shown in Figure 10.3 on page 306, calculated by the Kalman filter in state space form. For the seasonal and AR components the standard deviations of the prediction errors are given. A comparison between Figures 10.2 and 10.3 indicates that the components are not equally well determined, which is probably due to an interaction between the estimates.

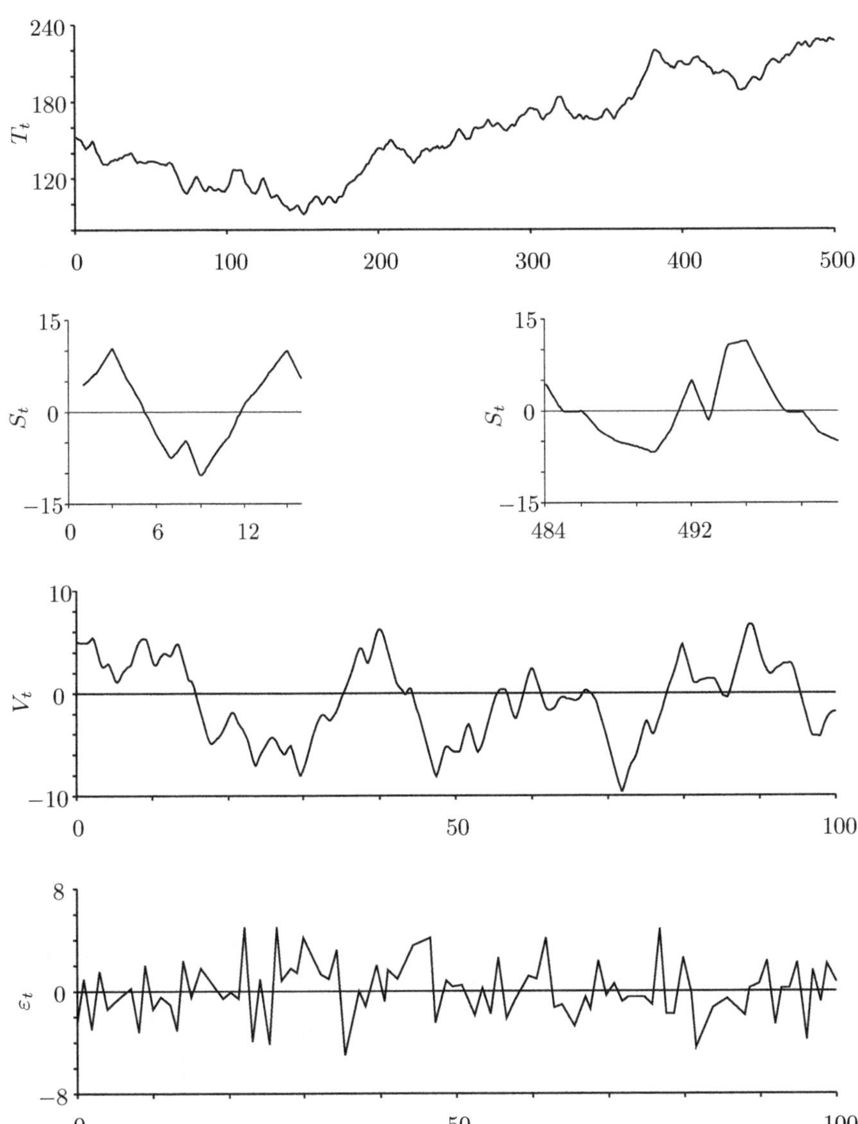

Figure 10.2: *Simulated components from the time series in Figure 10.1. From the top: trend, seasonal, AR(1), and white noise components.*

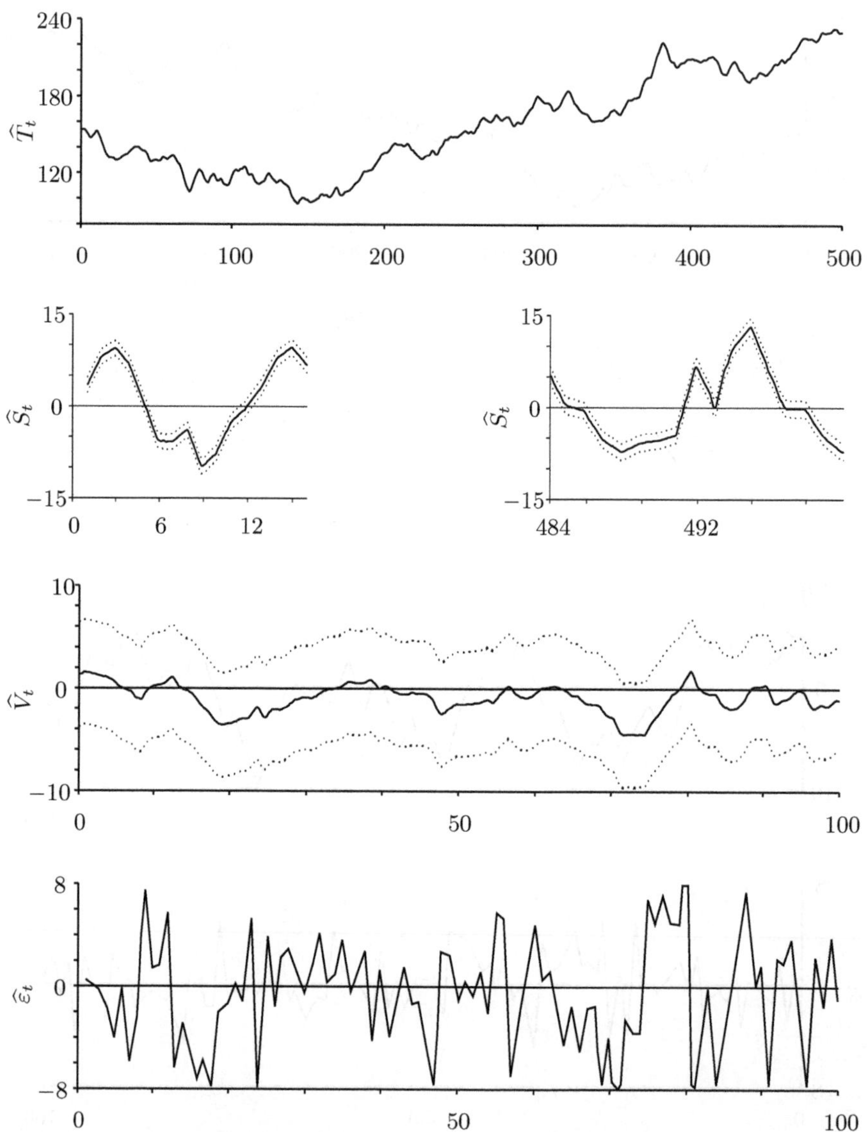

Figure 10.3: *Estimated components from the time series in Figure 10.1 on page 304. From the top: trend, seasonal, AR(1), and the white noise components.*

10.5 Time series with missing observations

10.5.1 Estimation of autocorrelation functions

In Dunsmuir and Robinson (1981) and Parzen (1963), methods for estimation of autocovariance functions on time series with missing observations are discussed and they introduce the following sequence:

$$a(t) = \begin{cases} 1 & \text{if } Y_t \text{ is observed.} \\ 0 & \text{if the observation of } Y_t \text{ is missing.} \end{cases} \tag{10.127}$$

Then we define

$$\mathrm{C}_a(k) = \frac{1}{N} \sum_{t=1}^{N-|k|} a(t)\, a(t+|k|) \tag{10.128}$$

$$\mathrm{C}^{\square}(k) = \frac{1}{N} \sum_{t=1}^{N-|k|} a(t)\, a(t+|k|) \left(Y_t - \overline{\mu}_y\right) \left(Y_{t+|k|} - \overline{\mu}_y\right) \tag{10.129}$$

$$\overline{\mu}_y = \frac{\sum_{t=1}^{N} a(t) Y_t}{\sum_{t=1}^{N} a(t)} \tag{10.130}$$

And the autocovariance function is estimated by

$$\mathrm{C}_{YY}(k) = \frac{\mathrm{C}^{\square}(k)}{\mathrm{C}_a(k)} \tag{10.131}$$

The autocorrelation and the partial autocorrelation are now estimated as discussed in Chapter 6. It follows from Dunsmuir and Robinson (1981) that if $\{Y_t\}$ is white noise, then $\widehat{\rho}(k)$ is asymptotically mutually independent and normally distributed with mean zero and variance

$$\mathrm{Var}\left[\widehat{\rho}(k)\right] \simeq \frac{1}{N\, \mathrm{C}_a(k)}. \tag{10.132}$$

It should be noted that if (10.128) is changed to

$$\mathrm{C}_a(k) = \frac{1}{N-|k|} \sum_{t=1}^{N-|k|} a(t)\, a(t+|k|),$$

and we have no missing observations, we get the same estimator for the autocovariance as the one proposed in Chapter 6.

10.6 ML estimates of state space models

In this section we will show how to find ML estimates of the parameters in a state space model. The method also allows for missing observations.

Let us consider the *state space model*

$$\boldsymbol{X}_t = \boldsymbol{A}\boldsymbol{X}_{t-1} + \boldsymbol{G}\boldsymbol{e}_{1,t}, \tag{10.133}$$
$$\boldsymbol{Y}_t = \boldsymbol{C}\boldsymbol{X}_t + \boldsymbol{e}_{2,t}, \tag{10.134}$$

where $\{\boldsymbol{e}_{1,t}\}$ and $\{\boldsymbol{e}_{2,t}\}$ are mutually uncorrelated normally distributed white noise with variances $\boldsymbol{\Sigma}_1$ and $\boldsymbol{\Sigma}_2$, respectively. The observation equation is assumed to be m-dimensional. Furthermore, we assume that the model may be identified so that the input-output relations uniquely determine the parameters in the state space formulation. This formulation includes, e.g., the ARMA models in state space form, and the method may thus be used to estimate the parameters in ARMA models where the observed time series has missing observations.

Let $\mathcal{Y}_t$ denote all the (non-missing) observations up to time t. Since both $\boldsymbol{e}_{1,t}$ and $\boldsymbol{e}_{2,t}$ are normally distributed, it implies that also $\boldsymbol{X}_t|\mathcal{Y}_t$ is normally distributed and characterized by mean and variance

$$\widehat{\boldsymbol{X}}_{t|t} = \mathrm{E}\left[\boldsymbol{X}_t|\mathcal{Y}_t\right], \tag{10.135}$$
$$\boldsymbol{\Sigma}^{xx}_{t|t} = \mathrm{Var}\left[\boldsymbol{X}_t|\mathcal{Y}_t\right]. \tag{10.136}$$

It follows directly from (10.134), that

$$\widehat{\boldsymbol{X}}_{t+1|t} = \boldsymbol{A}\widehat{\boldsymbol{X}}_{t|t} \tag{10.137}$$
$$\boldsymbol{\Sigma}^{xx}_{t+1|t} = \boldsymbol{A}\boldsymbol{\Sigma}^{xx}_{t|t}\boldsymbol{A}^T + \boldsymbol{G}\boldsymbol{\Sigma}_1\boldsymbol{G}^T \tag{10.138}$$
$$\widehat{\boldsymbol{Y}}_{t+1|t} = \boldsymbol{C}\boldsymbol{A}\widehat{\boldsymbol{X}}_{t|t}. \tag{10.139}$$

At time $t+1$ we have two possibilities

i) An observation $\boldsymbol{Y}_{t+1}$.

ii) No observation.

In the case i) we get from (10.73)–(10.75) on page 293 and (10.78) that

$$\begin{aligned}\widehat{\boldsymbol{X}}_{t+1|t+1} &= \mathrm{E}\left[\boldsymbol{X}_{t+1}|\mathcal{Y}_t, \boldsymbol{Y}_{t+1}\right] \\ &= \boldsymbol{A}\widehat{\boldsymbol{X}}_{t|t} + \boldsymbol{K}_{t+1}\left(\boldsymbol{Y}_{t+1} - \boldsymbol{C}\boldsymbol{A}\widehat{\boldsymbol{X}}_{t|t}\right),\end{aligned} \tag{10.140}$$

$$\boldsymbol{K}_{t+1} = \boldsymbol{\Sigma}^{xx}_{t+1|t}\boldsymbol{C}^T\left[\boldsymbol{C}\boldsymbol{\Sigma}^{xx}_{t+1|t}\boldsymbol{C}^T + \boldsymbol{\Sigma}_2\right]^{-1}, \tag{10.141}$$

$$\boldsymbol{\Sigma}^{xx}_{t+1|t+1} = \boldsymbol{\Sigma}^{xx}_{t+1|t} - \boldsymbol{K}_{t+1}\boldsymbol{C}\boldsymbol{\Sigma}^{xx}_{t+1|t}, \tag{10.142}$$

whereas in case ii) where we have no new information, we get

$$\widehat{\boldsymbol{X}}_{t+1|t+1} = \mathrm{E}\left[\boldsymbol{X}_{t+1}|\mathcal{Y}_{t+1}\right] = \mathrm{E}\left[\boldsymbol{X}_{t+1}|\mathcal{Y}_t\right] = \widehat{\boldsymbol{X}}_{t+1|t}, \tag{10.143}$$
$$\boldsymbol{\Sigma}^{xx}_{t+1|t+1} = \boldsymbol{\Sigma}^{xx}_{t+1|t}. \tag{10.144}$$

Finally, we introduce the one-step prediction error (the innovation) and the corresponding variance

$$\widetilde{\boldsymbol{Y}}_{t+1|t} = \boldsymbol{Y}_{t+1} - \widehat{\boldsymbol{Y}}_{t+1|t}, \tag{10.145}$$

$$\begin{aligned} \boldsymbol{R}_{t+1} &= \text{Var}\left[\widetilde{\boldsymbol{Y}}_{t+1|t}\right] \\ &= \text{Var}\left[\boldsymbol{Y}_{t+1}|\mathcal{Y}_t\right] \\ &= \boldsymbol{\Sigma}^{yy}_{t+1|t} \\ &= \boldsymbol{C}\boldsymbol{\Sigma}^{xx}_{t+1|t}\boldsymbol{C}^T + \boldsymbol{\Sigma}_2. \end{aligned} \tag{10.146}$$

The conditioned distribution becomes

$$f\left(\boldsymbol{Y}_{t+1}|\mathcal{Y}_t\right) = \left[(2\pi)^m \det \boldsymbol{R}_{t+1}\right]^{-1/2} \exp\left[-\frac{1}{2}\widetilde{\boldsymbol{Y}}^T_{t+1|t}\boldsymbol{R}^{-1}_{t+1}\widetilde{\boldsymbol{Y}}_{t+1|t}\right] \tag{10.147}$$

Let $\boldsymbol{\theta}$ denote the unknown parameters, $\mathcal{Y}_{N^*} = (\boldsymbol{Y}_1, \boldsymbol{Y}_2, \ldots, \boldsymbol{Y}_{N^*})$, and N^* denotes the available observations ($N^* \leq N$). The conditional likelihood function (conditioned on $\boldsymbol{Y}_0$) is

$$\begin{aligned} L\left(\boldsymbol{\theta}; \mathcal{Y}_{N^*}\right) &= f\left(\mathcal{Y}_{N^*}|\boldsymbol{\theta}\right) \\ &= f\left(\boldsymbol{Y}_{N^*}|\mathcal{Y}_{N^*-1}, \boldsymbol{\theta}\right) f\left(\boldsymbol{Y}_{N^*-1}|\mathcal{Y}_{N^*-2}, \boldsymbol{\theta}\right) \cdots f\left(\boldsymbol{Y}_1|\boldsymbol{Y}_0, \boldsymbol{\theta}\right). \end{aligned}$$

By applying (10.147) we get

$$L\left(\boldsymbol{\theta}; \mathcal{Y}_{N^*}\right) = \prod_{i=1}^{N^*} \left[(2\pi)^m \det \boldsymbol{R}(i)\right]^{-1/2} \exp\left[-\frac{1}{2}\widetilde{\boldsymbol{Y}}^T(i)\boldsymbol{R}^{-1}(i)\widetilde{\boldsymbol{Y}}(i)\right], \tag{10.148}$$

where $\widetilde{\boldsymbol{Y}}(i)$ and $\boldsymbol{R}(i)$ are the innovation and the corresponding covariance for the i'th observation. Please note that $\boldsymbol{R}(i)$ depends on both the initial condition and the structure of the missing observations (meaning the times when the observations are missing). If we, for example, obtain a long series with missing observations, then $\boldsymbol{R}(i)$ will increase from time to time, corresponding to (10.138) and (10.144) until we get a new observation. The starting values, i.e., $\widehat{\boldsymbol{X}}_{0|0}$ and $\boldsymbol{\Sigma}^{xx}_{0|0}$, may be found by back-forecasting or by setting $\widehat{\boldsymbol{X}}_{0|0} = \boldsymbol{0}$ and $\boldsymbol{\Sigma}^{xx}_{0|0} = \alpha\boldsymbol{I}$, where $\boldsymbol{I}$ is the identity matrix and α is a "large" constant.

The logarithm of the likelihood function becomes

$$\begin{aligned} \log L\left(\boldsymbol{\theta}; \mathcal{Y}_{N^*}\right) = &-\frac{1}{2}\sum_{i=1}^{N^*}\left[\log\det \boldsymbol{R}(i) + \widetilde{\boldsymbol{Y}}^T(i)\boldsymbol{R}^{-1}(i)\widetilde{\boldsymbol{Y}}(i)\right] \\ &+ \text{const.} \end{aligned} \tag{10.149}$$

The ML estimate of $\boldsymbol{\theta}$ is found as the argument which maximizes (10.149), i.e.,

$$\widehat{\boldsymbol{\theta}} = \arg\left\{\max_{\theta} \log L\left(\boldsymbol{\theta}; \mathcal{Y}_{N^*}\right)\right\}. \tag{10.150}$$

As an approximation of the variances of the parameter estimates we can apply

$$\text{Var}[\widehat{\boldsymbol{\theta}}] \simeq -\boldsymbol{H}^{-1}, \tag{10.151}$$

where

$$\{\boldsymbol{H}\}_{ij} = \left.\frac{\partial^2 \log L\left(\boldsymbol{\theta}; \mathcal{Y}_{N^*}\right)}{\partial \boldsymbol{\theta}_i \partial \boldsymbol{\theta}_j}\right|_{\boldsymbol{\theta}=\widehat{\boldsymbol{\theta}}}. \tag{10.152}$$

In the optimization one may apply e.g., a Newton-Raphson procedure, as described in Chapter 3.

10.7 Problems

Exercise 10.1

This exercise considers the heat dynamics of an ordinary single family house. By analyzing data for a suitable period, a stochastic model for the variation of the temperature states of the building has been established as

$$\begin{pmatrix} X_{i,t} \\ X_{m,t} \end{pmatrix} = \begin{pmatrix} 0.70 & 0.20 \\ 0.04 & 0.95 \end{pmatrix} \begin{pmatrix} X_{i,t-1} \\ X_{m,t-1} \end{pmatrix} + \begin{pmatrix} 0.8 \\ 0.02 \end{pmatrix} U_{t-1} + \begin{pmatrix} \varepsilon_{1,t} \\ \varepsilon_{2,t} \end{pmatrix}$$

where $X_{i,t}$ is the indoor air temperature at time t, and $X_{m,t}$ is the wall temperature. Finally, U_t is the heat input from the radiators.

$$\boldsymbol{\varepsilon}^T = (\varepsilon_{1,t}, \varepsilon_{2,t})$$

is white noise with the covariance

$$\boldsymbol{\Sigma} = \begin{pmatrix} 0.04 & 0 \\ 0 & 0.01 \end{pmatrix}$$

The temperatures X_i and X_m are the deviation from a reference point (21 °C).

Question 1 For a longer period the house has been without heating and the temperatures are now (at time t_0):

$$\begin{pmatrix} X_{i,t_0} \\ X_{m,t_0} \end{pmatrix} = \begin{pmatrix} -3.0 \\ -1.2 \end{pmatrix}$$

The maximum performance of the radiators is 2. Right after the above temperature measurement was conducted, the radiators were fully opened.

Find predictions of the indoor air temperature and the wall temperature for the time points $t_0 + 1$, $t_0 + 2$, and $t_0 + 3$.

What is the variance of the predictions?

Question 2 We assume that the radiators will remain fully open for infinitely long time, such that stationary temperatures varying around a new reference level will be observed.

Find the steady state solution, i.e., the new reference level.

Exercise 10.2
Weekly measurements of the oxygen content and the phosphate concentration in a stream have been conducted for a long time period and based on these data, a model is found

$$\begin{pmatrix} 1-0.8B & 0.5B \\ -0.4B & 1-0.7B \end{pmatrix} \begin{pmatrix} X_{1,t} \\ X_{2,t} \end{pmatrix} = \begin{pmatrix} 1 & 0 \\ 0 & 1-0.4B \end{pmatrix} \begin{pmatrix} \varepsilon_{1,t} \\ \varepsilon_{2,t} \end{pmatrix}$$

where $X_{1,t}$ and $X_{2,t}$ are the deviations from the long term mean for the oxygen content and the phosphate concentration, respectively.

$$\boldsymbol{\varepsilon}^T = (\varepsilon_{1,t}, \varepsilon_{2,t})$$

is white noise with the covariance

$$\boldsymbol{\Sigma} = \begin{pmatrix} 2 & -1 \\ -1 & 2 \end{pmatrix}$$

During the past six weeks the following measurements have been obtained.

Week	$t-5$	$t-4$	$t-3$	$t-2$	$t-1$	t
X_1	3.1	3.2	2.9	0.2	−2.3	−3.4
X_2	2.1	1.6	1.8	4.9	3.5	1.8

Question 1 Predict the oxygen content and phosphate concentration for the next two weeks.

Question 2 In week $t+1$ the measuring instrument for the oxygen broke down. As a consequence the oxygen concentration could not be measured, but the phosphate concentration was measured to -1.2. Using this knowledge give a revised estimate for the oxygen content and the corresponding variance in week $t+1$.

Exercise 10.3
A factory has established a quality control of the manufacturing process, and it is found that the relevant process can be described by

$$X_t = X_{t-1} + e_t \qquad \text{(System equation).}$$

However, it is only possible to measure X_t with a large uncertainty, i.e., the measured variable Y_t is given by

$$Y_t = X_t + v_t \qquad \text{(Observation equation)}$$

$\{e_t\}$ and $\{v_t\}$ are mutually uncorrelated white noise processes with variances $\sigma_e^2 = 1$ and $\sigma_v^2 = 2$,

Question 1 What is the name of the process describing the dynamics of the system?

Question 2 It is assumed that $\Sigma_{0|0}^{xx} = 1$. Use the Kalman filter to find the optimal prediction $\widehat{X}_{t+1|t}$.

Question 3 Show that the optimal forecast corresponds to exponential smoothing.

Exercise 10.4

Question 1 Write the process

$$(1 - \phi_1 B)(1 - \phi_4 B^4)X_t = (1 - \theta B)\varepsilon_t$$

in a state space form.

Question 2 Consider the following process in state space form

$$\begin{pmatrix} X_{1,t} \\ X_{2,t} \end{pmatrix} = \begin{pmatrix} -\phi_1 & 1 \\ -\phi_2 & 0 \end{pmatrix} \begin{pmatrix} X_{1,t-1} \\ X_{2,t-1} \end{pmatrix} + \begin{pmatrix} 1 \\ -\theta_1 \end{pmatrix} e_t$$
$$Y(t) = \begin{pmatrix} 1 & 0 \end{pmatrix} \begin{pmatrix} X_{1,t} \\ X_{2,t} \end{pmatrix}$$

where $\{e_t\}$ is white noise.

Show that $\{Y_t\}$ can be described by an ARMA process.

Question 3 Consider now the process $\{Y_t\}$ defined by

$$X_{1,t} = 0.8X_{1,t-1} + e_t$$
$$Y_t = X_{1,t} + v_t$$

where $\{e_t\}$ and $\{v_t\}$ are mutually uncorrelated white noise processes with variances σ_e^2 and σ_v^2.

Draw the autocorrelation function of $\{Y_t\}$ for the cases $\sigma_v^2/\sigma_e^2 = 0$, $\sigma_v^2/\sigma_e^2 = 1$, and $\sigma_v^2/\sigma_e^2 = \infty$.

CHAPTER 11

Recursive estimation

When implementing a model, (e.g., for prediction or control) it is often desirable to be able to update the parameter estimates as time passes and new information becomes available. Estimation methods, which in particular are suitable for recursive estimation for each new observation in time, are called *recursive estimation methods*. Other commonly used notations are *on-line*, *adaptive*, or *sequential parameter estimation*.

In Chapter 3 we considered the local trend model, where the updating of the parameter vector was introduced by adapting to the parameter vector as new observations became available. Also for dynamical models, it is often the case that the assumption (or adequacy) of using a fixed model is not valid as the dynamical characteristics change in time. A solution to this problem may be to apply *adaptive methods* which allow for on-line tuning of the parameters. By repeatedly applying the most recent parameter estimates in a prediction formula, a filter, or a controller, we have *adaptive prediction*, *adaptive filtering*, or *adaptive control*.

Often a systematic pattern is seen in the way a parameter estimate changes in time. For example, Madsen (1985) observes a yearly variation in the parameters of a transfer function model, which relates air temperature to solar radiation. When such a systematic pattern is seen, it can be included in the model and result in a *time-varying* model. Examples of such methods and models are given at the end of this chapter.

Methods for recursive and adaptive estimation have been introduced in both statistical journals and journals related to automatic control. Some references to classical work on recursive and adaptive estimation are Holt (1957), Brown (1963), Söderström (1973), Ljung (1976), Holst (1977), and Young (1984). In particular, the book by Ljung and Söderström (1983) was instrumental in bringing attention to the enormous flexibility of recursive and adaptive estimation.

11.1 Recursive LS

We consider the following *single-input/single-output* model

$$Y_t + \phi_1 Y_{t-1} + \cdots + \phi_p Y_{t-p} = \omega_1 U_{t-1} + \cdots + \omega_s U_{t-s} + \varepsilon_t \qquad (11.1)$$

where $\{\varepsilon_t\}$ is white noise and uncorrelated with $\{U_t\}$.

Please note, as discussed in Section 6.4.2 on page 159, that the LS method is related to models without MA parameters, see, e.g., (6.44) on page 162. The model in (11.1) can obviously be extended to any model in the class (6.44) and, thus, also for describing several input series.

If we introduce the regressor vector

$$\boldsymbol{X}_t^T = (-Y_{t-1}, \dots, -Y_{t-p}, U_{t-1}, \dots, U_{t-s}) \tag{11.2}$$

and parameter vector

$$\boldsymbol{\theta}^T = (\phi_1, \dots, \phi_p, \omega_1, \dots, \omega_s) \tag{11.3}$$

we can write the model (11.1) as

$$Y_t = \boldsymbol{X}_t^T \boldsymbol{\theta} + \varepsilon_t \tag{11.4}$$

Given N observations we get the *off-line LS estimate* as (cf. Definition 3.2 on page 34)

$$\widehat{\boldsymbol{\theta}} = \arg\min_{\boldsymbol{\theta}} S_N(\boldsymbol{\theta}) \tag{11.5}$$

where

$$S_N(\boldsymbol{\theta}) = \sum_{t=1}^{N} \varepsilon_t^2(\boldsymbol{\theta}) = \sum_{t=1}^{N} (Y_t - X_t^T \boldsymbol{\theta})^2 \tag{11.6}$$

By the *recursive (on-line) LS method*, we get the estimate at time t as

$$\widehat{\boldsymbol{\theta}}_t = \arg\min_{\boldsymbol{\theta}} S_t(\boldsymbol{\theta}) \tag{11.7}$$

where

$$S_t(\boldsymbol{\theta}) = \sum_{s=1}^{t} (Y_s - \boldsymbol{X}_s^T \boldsymbol{\theta})^2 \tag{11.8}$$

In Section 6.4.2 it was seen that the solution to (11.7) is given by

$$\widehat{\boldsymbol{\theta}}_t = \boldsymbol{R}_t^{-1} \boldsymbol{h}_t \tag{11.9}$$

where

$$\boldsymbol{R}_t = \sum_{s=1}^{t} \boldsymbol{X}_s \boldsymbol{X}_s^T \quad \text{and} \quad \boldsymbol{h}_t = \sum_{s=1}^{t} \boldsymbol{X}_s Y_s \tag{11.10}$$

and it is seen that the updating of $\boldsymbol{R}_t$ and $\boldsymbol{h}_t$ become

$$\boldsymbol{R}_t = \boldsymbol{R}_{t-1} + \boldsymbol{X}_t \boldsymbol{X}_t^T \tag{11.11}$$

and

$$h_t = h_{t-1} + X_t Y_t \tag{11.12}$$

Compare the results with Theorem 3.12 on page 55.

We now get

$$\begin{aligned}\widehat{\theta}_t &= R_t^{-1} h_t = R_t^{-1}\left[h_{t-1} + X_t Y_t\right]\\ &= R_t^{-1}\left[R_{t-1}\widehat{\theta}_{t-1} + X_t Y_t\right]\\ &= R_t^{-1}\left[R_t\widehat{\theta}_{t-1} - X_t X_t^T\widehat{\theta}_{t-1} + X_t Y_t\right]\\ &= \widehat{\theta}_{t-1} + R_t^{-1} X_t\left[Y_t - X_t^T\widehat{\theta}_{t-1}\right]\end{aligned}$$

Hereby, we have the *recursive least squares (RLS) method* for dynamical models.

The RLS algorithm

$$\widehat{\theta}_t = \widehat{\theta}_{t-1} + R_t^{-1} X_t\left[Y_t - X_t^T\widehat{\theta}_{t-1}\right] \tag{11.13a}$$

$$R_t = R_{t-1} + X_t X_t^T \tag{11.13b}$$

In order to avoid the inversion of R_t in each step we introduce

$$P_t = R_t^{-1} \tag{11.14}$$

and using the *matrix inversion rule*

$$[A + BCD]^{-1} = A^{-1} - A^{-1}B\left[DA^{-1}B + C^{-1}\right]^{-1} DA^{-1} \tag{11.15}$$

with $A = R_{t-1}$, $B = D^T = X_t$, and $C = I$, we get

$$P_t = P_{t-1} - \frac{P_{t-1}X_t X_t^T P_{t-1}}{1 + X_t^T P_{t-1} X_t} \tag{11.16}$$

Furthermore, we often apply

$$K_t = R_t^{-1}X_t = P_{t-1}X_t - \frac{P_{t-1}X_t X_t^T P_{t-1} X_t}{1 + X_t^T P_{t-1} X_t}$$

or

$$K_t = \frac{P_{t-1}X_t}{1 + X_t^T P_{t-1} X_t} \tag{11.17}$$

so that (11.13a) can be changed to

$$\widehat{\theta}_t = \widehat{\theta}_{t-1} + K_t\left[Y_t - X_t^T\widehat{\theta}_{t-1}\right] \tag{11.18}$$

Furthermore, we introduce the prediction error

$$\varepsilon_t(\widehat{\boldsymbol{\theta}}_{t-1}) = Y_t - \boldsymbol{X}_t^T \widehat{\boldsymbol{\theta}}_{t-1} \tag{11.19}$$

so that an alternative to (11.13a) is

$$\widehat{\boldsymbol{\theta}}_t = \widehat{\boldsymbol{\theta}}_{t-1} + \boldsymbol{K}_t \varepsilon_t(\widehat{\boldsymbol{\theta}}_{t-1}) \tag{11.20}$$

Please note that when using the starting value $\boldsymbol{R}_0 = 0$ (and an arbitrary value $\boldsymbol{\theta}_0$), the recursive method and the off-line method are identical. This starting value, however, cannot be transformed to a starting value for $\boldsymbol{P}_0$ due to the inversion in (11.14). In practical applications we often apply $\boldsymbol{P}_0 = \alpha \boldsymbol{I}$, where α is large.

11.1.1 Recursive LS with forgetting

In the previous section we considered the parameters as being constant in time. However, as in Section 3.4.4 on page 56, it may be adequate to allow the parameter vector, $\boldsymbol{\theta}$, to change in time. As in Section 3.4.4, we consider the *weighted least squares estimator*

$$\widehat{\boldsymbol{\theta}}_t = \arg\min_{\theta} S_t(\boldsymbol{\theta}) \tag{11.21}$$

where

$$S_t(\boldsymbol{\theta}_t) = \sum_{s=1}^{t} \beta(t,s)(Y_s - \boldsymbol{X}_s^T \boldsymbol{\theta}_t)^2 \tag{11.22}$$

which, apart from the weight $\beta(t,s)$, corresponds to (11.7)–(11.8) on page 314.

Furthermore, we assume that the *sequence of weights*, $\{\beta(t,s)\}$, can be expressed as

$$\beta(t,s) = \lambda(t)\beta(t-1,s) \qquad 1 \le s \le t-1 \tag{11.23a}$$

$$\beta(t,t) = 1 \tag{11.23b}$$

which implies that

$$\beta(t,s) = \prod_{j=s+1}^{t} \lambda(j) \tag{11.24}$$

If $\lambda(j) = \lambda = \text{constant}$, we get $\beta(t,s) = \lambda^{t-s}$, i.e., *exponential weights* (forgetting) as in Section 3.4.4. This is a generalization of the forgetting factor from Chapter 3.

The solution to the weighted least squares problem, (11.21)–(11.22), is found as

$$\widehat{\boldsymbol{\theta}}_t = \boldsymbol{R}_t^{-1} \boldsymbol{h}_t \tag{11.25}$$

where

$$\boldsymbol{R}_t = \sum_{s=1}^{t} \beta(t,s)\boldsymbol{X}_s\boldsymbol{X}_s^T, \quad \boldsymbol{h}_t = \sum_{s=1}^{t} \beta(t,s)\boldsymbol{X}_s Y_s \tag{11.26}$$

The updating of $\boldsymbol{R}_t$ and $\boldsymbol{h}_t$ become

$$\boldsymbol{R}_t = \lambda(t)\boldsymbol{R}_{t-1} + \boldsymbol{X}_t\boldsymbol{X}_t^T \tag{11.27}$$

$$\boldsymbol{h}_t = \lambda(t)\boldsymbol{h}_{t-1} + \boldsymbol{X}_t Y_t \tag{11.28}$$

In the case of weighted estimation we get

$$\begin{aligned}
\widehat{\boldsymbol{\theta}}_t &= \boldsymbol{R}_t^{-1}\boldsymbol{h}_t = \boldsymbol{R}_t^{-1}\left[\lambda(t)\boldsymbol{h}_{t-1} + \boldsymbol{X}_t Y_t\right] \\
&= \boldsymbol{R}_t^{-1}\left[\lambda(t)\boldsymbol{R}_{t-1}\widehat{\boldsymbol{\theta}}_{t-1} + \boldsymbol{X}_t Y_t\right] \\
&= \boldsymbol{R}_t^{-1}\left[\boldsymbol{R}_t\widehat{\boldsymbol{\theta}}_{t-1} - \boldsymbol{X}_t\boldsymbol{X}_t^T\widehat{\boldsymbol{\theta}}_{t-1} + \boldsymbol{X}_t Y_t\right] \\
&= \widehat{\boldsymbol{\theta}}_{t-1} + \boldsymbol{R}_t^{-1}\boldsymbol{X}_t\left[Y_t - \boldsymbol{X}_t^T\widehat{\boldsymbol{\theta}}_{t-1}\right]
\end{aligned} \tag{11.29}$$

We summarize as follows.

The RLS algorithm with forgetting

$$\widehat{\boldsymbol{\theta}}_t = \widehat{\boldsymbol{\theta}}_{t-1} + \boldsymbol{R}_t^{-1}\boldsymbol{X}_t\left[Y_t - \boldsymbol{X}_t^T\widehat{\boldsymbol{\theta}}_{t-1}\right] \tag{11.30a}$$

$$\boldsymbol{R}_t = \lambda(t)\boldsymbol{R}_{t-1} + \boldsymbol{X}_t\boldsymbol{X}_t^T \tag{11.30b}$$

If we have $\lambda(t) = \text{constant} = \lambda$, we denote λ the *forgetting factor*. Furthermore, we have the *memory*

$$T_0 = \frac{1}{1-\lambda} \tag{11.31}$$

Typical values for λ range from 0.90 to 0.995. The forgetting factor can be chosen based on assumptions of the dynamics or it can be a part of the global optimization as shown previously in Section 3.4.2.2 on page 52.

An alternative method, *variable forgetting*, is based on selecting the forgetting factor so that the loss function is constant, i.e.,

$$S_t(\widehat{\boldsymbol{\theta}}_t) = S_{t-1}(\widehat{\boldsymbol{\theta}}_{t-1}) = \cdots = S_0 \tag{11.32}$$

The forgetting factor which satisfies (11.32) is determined by

$$\lambda(t) \simeq 1 - \frac{\varepsilon_t^2}{S_0\left[1 + \boldsymbol{X}_t^T\boldsymbol{P}_{t-1}\boldsymbol{X}_t\right]} \tag{11.33}$$

For practical applications it may be necessary to apply a lower bound $\lambda_{\min}$, for $\lambda(t)$.

The approximation in (11.33) is good if $\varepsilon_t^2 \ll S_0$. It is seen that if the squared error, ε_t^2, at a given time t is large then $\lambda(t)$ becomes small, which implies that the information from past observations (contained in $\boldsymbol{R}_{t-1}$ in (11.30b)) is not given much weight. This criterion results in a method which is able to track relatively fast changes in the parameters.

If there is a sudden and total change in the dynamics then using a method with a forgetting factor is inappropriate because the matrix, $\boldsymbol{R}_t$, which contains information on the direction of the parameter changes, may be totally wrong following such a radical change as it has been determined based on observations obtained prior to the change. In such a case, model-based estimation methods, as discussed in Section 11.4, may be more appropriate.

Example 11.1 (RLS estimation)

In Madsen (1985), dynamic models for the variations in the air temperature (T) are presented. The presented model is a transfer function model using the net solar radiation (R) at the earth surface as input. The net solar radiation is the difference between the incoming and reflected radiation. During the day the net radiation is dominated by the incoming radiation, whereas the radiation at night is dominated by the reflected radiation (especially when there are no clouds).

The following transfer function has been applied

$$\phi(\mathrm{B})(T_t - \mu_T) = \omega(\mathrm{B})(R_t - \mu_R) + \varepsilon_t \tag{11.34}$$

where the order of the ϕ and ω polynomial are $p = 3$ and $s = 1$, respectively. This model can also be written

$$\phi(\mathrm{B})T_t = \omega(\mathrm{B})R_t + d + \varepsilon_t \tag{11.35}$$

where $d = \phi(1)\mu_T - \omega(1)\mu_R$.

Contrary to equation (11.34), the model (11.35) is linear in the parameters and can be written on the linear form

$$T_t = \boldsymbol{X}_t^T \boldsymbol{\theta} + \varepsilon_t \tag{11.36}$$

where

$$\begin{aligned} \boldsymbol{X}_t^T &= (-T_{t-1}, -T_{t-2}, -T_{t-3}, u_{t-1}, u_{t-2}, 1) \\ \boldsymbol{\theta}^T &= (\phi_1, \phi_2, \phi_3, \omega_0, \omega_1, d) \end{aligned}$$

In the study 69,384 hourly observations of the air temperature and net solar radiation from a climate station (Højbakkegård in Taastrup, Denmark) have been used. By using the RLS with forgetting factor of $\lambda = 0.999$ we get the pattern of the parameter estimate as shown in Figure 11.1.

The choice of forgetting factor implies that the memory is about 6 weeks. From the figure it is seen that ω_0 and ω_1 vary over the year. For the parameters of the ϕ polynomial, we cannot directly see this yearly variation, but further investigations of the three poles in $\phi(\mathrm{B})$ have shown that the pole closest to the unit circle shows a yearly seasonal effect.

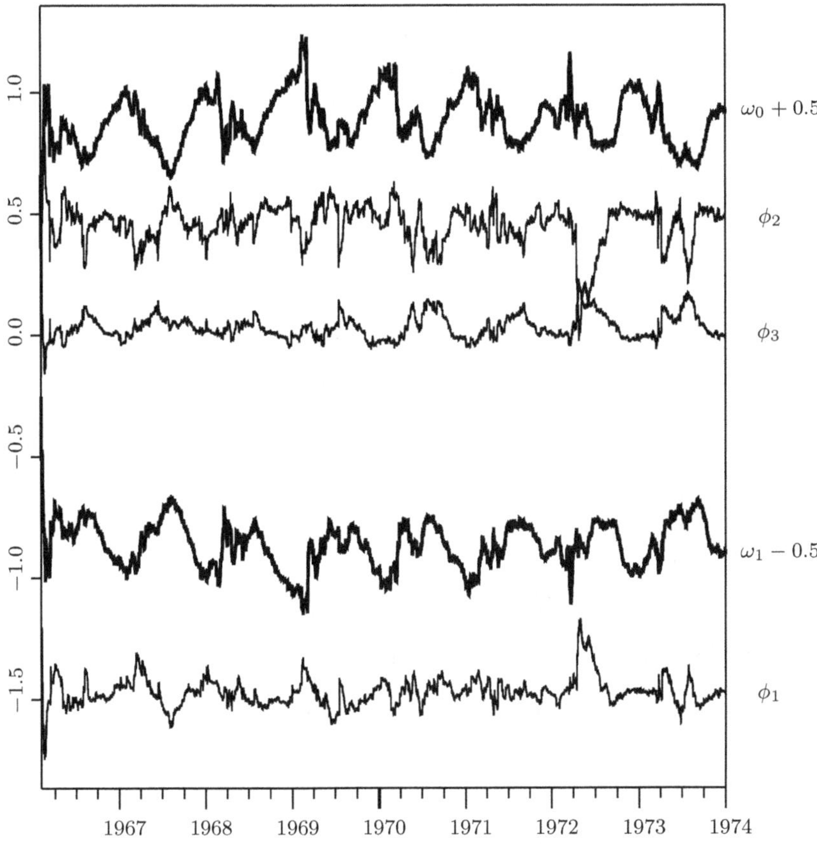

Figure 11.1: *Recursive estimation of the parameters in the transfer function model (11.34). The forgetting factor is* $\lambda = 0.999$.

11.2 Recursive pseudo-linear regression (RPLR)

Let us consider the model where the one step prediction can be written in the form

$$\widehat{Y}_{t|t-1}(\boldsymbol{\theta}) = \boldsymbol{X}_t^T(\boldsymbol{\theta})\boldsymbol{\theta} \tag{11.37}$$

The model (11.38) is a *pseudo-linear* representation of the prediction since $\boldsymbol{X}_t$ depends on $\boldsymbol{\theta}$. The model class includes the ARMA model—compare (11.37) with (6.52) on page 163.

At time t the parameters are estimated by

$$\widehat{\boldsymbol{\theta}}_t = \arg\min_{\theta} S_t(\boldsymbol{\theta}) \tag{11.38}$$

where (when we consider the weighted case)

$$\begin{aligned} S_t(\boldsymbol{\theta}_t) &= \sum_{s=1}^{t} \beta(t,s)\varepsilon_s^2(\boldsymbol{\theta}_t) = \sum_{s=1}^{t} \beta(t,s)\left(Y_s - \widehat{Y}_{s|s-1}(\boldsymbol{\theta}_t)\right)^2 \\ &= \sum_{s=1}^{t} \beta(t,s)\left(Y_s - \boldsymbol{X}_s^T(\boldsymbol{\theta}_t)\boldsymbol{\theta}_t\right)^2 \\ &= \lambda(t)S_{t-1}(\boldsymbol{\theta}_t) + \left(Y_t - \boldsymbol{X}_t^T(\boldsymbol{\theta}_t)\boldsymbol{\theta}_t\right)^2 \end{aligned} \tag{11.39}$$

If we assume $S_t(\boldsymbol{\theta})$ to be quadratic in $\boldsymbol{\theta}$, which is the case for $\boldsymbol{X}_t(\boldsymbol{\theta}) = \boldsymbol{X}_t$ (e.g., corresponding to no MA parameters in an ARMA model), the minimum will be obtained after one single Newton-Raphson iteration:

$$\widehat{\boldsymbol{\theta}}_t = \widehat{\boldsymbol{\theta}}_{t-1} - \left[\boldsymbol{H}_t(\widehat{\boldsymbol{\theta}}_{t-1})\right]^{-1} \nabla_\theta S_t(\widehat{\boldsymbol{\theta}}_{t-1}) \tag{11.40}$$

where $\boldsymbol{H}$ is the Hessian matrix and ∇S_t is the gradient.

Whenever S is quadratic in $\boldsymbol{\theta}$ then $\boldsymbol{H}$ is independent of $\boldsymbol{\theta}$. Thus, it follows from (11.39) that

$$\boldsymbol{H}_t = \lambda(t)\boldsymbol{H}_{t-1} + 2\boldsymbol{X}_t\boldsymbol{X}_t^T \tag{11.41}$$

and since $\nabla S_{t-1}(\widehat{\boldsymbol{\theta}}_{t-1}) = 0$, it follows that

$$\nabla S_t(\widehat{\boldsymbol{\theta}}_{t-1}) = -2\boldsymbol{X}_t\left[Y_t - \boldsymbol{X}_t^T\widehat{\boldsymbol{\theta}}_{t-1}\right] \tag{11.42}$$

If we set $\boldsymbol{R}_t = \frac{1}{2}\boldsymbol{H}_t$ and substitute (11.41)–(11.42) into (11.40), we obtain the RLS algorithm.

Even though we do not have a quadratic dependency of $\boldsymbol{\theta}$, we apply models where the one step predictions can be written on pseudo-linear form, (11.37), and we get the following.

The recursive pseudo-linear regression (RPLR)

$$\widehat{\boldsymbol{\theta}}_t = \widehat{\boldsymbol{\theta}}_{t-1} + \boldsymbol{R}_t^{-1}\boldsymbol{X}_t\left[Y_t - \boldsymbol{X}_t^T\widehat{\boldsymbol{\theta}}_{t-1}\right] \tag{11.43a}$$

$$\boldsymbol{R}_t = \lambda(t)\boldsymbol{R}_{t-1} + \boldsymbol{X}_t\boldsymbol{X}_t^T \tag{11.43b}$$

▸ **Remark 11.1**

It should be noted that the RPLR algorithm in terms of the programming implementation is equivalent to the RLS algorithm so that the same program may be used in both cases. ◂

For ARMAX models we have

$$\phi(\mathrm{B})Y_t = \omega(\mathrm{B})u_t + \theta(\mathrm{B})\varepsilon_t \tag{11.44}$$

and the procedure is also known as *extended least squares (ELS)*.

It can be shown (Ljung 1987) that a *sufficient condition* for the ELS estimator to converge toward the true values is

$$\Re\left[\frac{1}{\theta_0(e^{i\omega})}\right] \geq \frac{1}{2}, \quad \forall\omega \in [-\pi, \pi] \tag{11.45}$$

where $\theta_0(B)$ is the MA-polynomial in the true model of the system.

Example 11.2 (The ARMAX model on pseudo-linear form)
Let us consider the ARMAX model (11.44) and introduce the corresponding parameter vector

$$\boldsymbol{\theta}^T = (\phi_1, \ldots, \phi_p, \omega_1, \ldots, \omega_s, \theta_1, \ldots, \theta_q) \tag{11.46}$$

It is seen that the one step prediction can be written

$$\widehat{Y}_{t|t-1}(\boldsymbol{\theta}) = [1 - \phi(B)]Y_t + \omega(B)u_t + [\theta(B) - 1]\varepsilon_t(\boldsymbol{\theta}) \tag{11.47}$$

where the prediction errors

$$\varepsilon_t(\boldsymbol{\theta}) = Y_t - \widehat{Y}_{t|t-1}(\boldsymbol{\theta}) \tag{11.48}$$

depend on $\boldsymbol{\theta}$.

If we introduce the vector

$$\begin{aligned}\boldsymbol{X}_t(\boldsymbol{\theta}) = (-Y_{t-1}, \ldots, -Y_{t-p}, u_{t-1}, \ldots, u_{t-s},\\ \varepsilon_{t-1}(\boldsymbol{\theta}), \ldots, \varepsilon_{t-q}(\boldsymbol{\theta}))^T\end{aligned} \tag{11.49}$$

the one step prediction can be written on pseudo-linear form

$$\widehat{Y}_{t|t-1}(\boldsymbol{\theta}) = X_t^T(\boldsymbol{\theta})\boldsymbol{\theta} \tag{11.50}$$

The parameters can now be recursively estimated using the RPLR method (11.43b).

11.3 Recursive prediction error methods (RPEM)

For the recursive prediction error method we find the estimate as

$$\widehat{\boldsymbol{\theta}}_t = \arg\min S_t(\boldsymbol{\theta})$$

where the loss function is given by

$$S_t(\boldsymbol{\theta}_t) = \sum_{s=1}^{t} \beta(t,s)\varepsilon_s^2(\boldsymbol{\theta}_t) = \sum_{s=1}^{t} \beta(t,s)\left(Y_s - \widehat{Y}_{s|s-1}(\boldsymbol{\theta}_t)\right)^2 \tag{11.51}$$

The gradient with respect to $\boldsymbol{\theta}$ becomes

$$\nabla_\theta S_t(\boldsymbol{\theta}_t) = -2\sum_{s=1}^{t} \beta(t,s)\boldsymbol{\psi}_s(\boldsymbol{\theta}_t)\varepsilon_s(\boldsymbol{\theta}_t) \tag{11.52}$$

where

$$\boldsymbol{\psi}_s(\boldsymbol{\theta}_t) = \nabla_\theta \widehat{Y}_{s|s-1}(\boldsymbol{\theta}_t). \tag{11.53}$$

Correspondingly, for the Hessian matrix,

$$\boldsymbol{H}_t(\boldsymbol{\theta}) = 2\sum_{s=1}^{t} \beta(t,s)\boldsymbol{\psi}_s(\boldsymbol{\theta})\boldsymbol{\psi}_s^T(\boldsymbol{\theta}) - 2\sum_{s=1}^{t} \beta(t,s)\nabla_\theta\boldsymbol{\psi}_s(\boldsymbol{\theta})\varepsilon_s(\boldsymbol{\theta}) \tag{11.54}$$

In the region close to the true value, $\boldsymbol{\theta}_0$, the last term will be close to zero. It holds (Ljung and Söderström 1983) that $\mathrm{E}[\nabla_\theta\boldsymbol{\psi}_s(\boldsymbol{\theta})\varepsilon_s(\boldsymbol{\theta})] = 0$. Furthermore, neglecting the last term ensures that $\boldsymbol{H}_t(\boldsymbol{\theta})$ remains positive definite. Thus, disregarding the last term in (11.54), we get

$$\begin{aligned}\boldsymbol{H}_t(\boldsymbol{\theta}) &= 2\lambda(t)\sum_{s=1}^{t-1} \beta(t-1,s)\boldsymbol{\psi}_s(\boldsymbol{\theta})\boldsymbol{\psi}_s^T(\boldsymbol{\theta}) + 2\boldsymbol{\psi}_t(\boldsymbol{\theta})\boldsymbol{\psi}_t^T(\boldsymbol{\theta}) \\ &= \lambda(t)\boldsymbol{H}_{t-1}(\boldsymbol{\theta}) + 2\boldsymbol{\psi}_t(\boldsymbol{\theta})\boldsymbol{\psi}_t^T(\boldsymbol{\theta})\end{aligned} \tag{11.55}$$

Again we apply the Newton-Raphson algorithm

$$\widehat{\boldsymbol{\theta}}_t = \widehat{\boldsymbol{\theta}}_{t-1} - \left[\boldsymbol{H}_t(\widehat{\boldsymbol{\theta}}_{t-1})\right]^{-1} \nabla_\theta S_t(\widehat{\boldsymbol{\theta}}_{t-1}) \tag{11.56}$$

From (11.52) it follows

$$\begin{aligned}\nabla_\theta S_t(\widehat{\boldsymbol{\theta}}_t) &= -2\lambda(t)\sum_{s=1}^{t-1} \beta(t-1,s)\boldsymbol{\psi}_s(\boldsymbol{\theta}_t)\varepsilon_s(\boldsymbol{\theta}_t) - 2\boldsymbol{\psi}_t(\boldsymbol{\theta}_t)\varepsilon_t(\boldsymbol{\theta}_t) \\ &= -2\lambda(t)\nabla_\theta S_{t-1}(\boldsymbol{\theta}_t) - 2\boldsymbol{\psi}_t(\boldsymbol{\theta}_t)\varepsilon_t(\boldsymbol{\theta}_t)\end{aligned} \tag{11.57}$$

Let us assume that $\widehat{\boldsymbol{\theta}}_{t-1}$ minimizes $S_{t-1}(\boldsymbol{\theta})$. We get

$$\nabla_\theta S_t(\widehat{\boldsymbol{\theta}}_{t-1}) = -2\boldsymbol{\psi}_t(\widehat{\boldsymbol{\theta}}_{t-1})\varepsilon_t(\widehat{\boldsymbol{\theta}}_{t-1}) \tag{11.58}$$

If we set $\boldsymbol{R}_t = \frac{1}{2}\boldsymbol{H}_t$, we get the algorithm

$$\varepsilon_t(\widehat{\boldsymbol{\theta}}_{t-1}) = Y_t - \widehat{Y}_{t|t-1}(\widehat{\boldsymbol{\theta}}_{t-1}) \tag{11.59}$$

$$\widehat{\boldsymbol{\theta}}_t = \widehat{\boldsymbol{\theta}}_{t-1} + \boldsymbol{R}_t^{-1}(\widehat{\boldsymbol{\theta}}_{t-1})\boldsymbol{\psi}_t(\widehat{\boldsymbol{\theta}}_{t-1})\varepsilon_t(\widehat{\boldsymbol{\theta}}_{t-1}) \tag{11.60}$$

$$\boldsymbol{R}_t(\widehat{\boldsymbol{\theta}}_{t-1}) = \lambda(t)\boldsymbol{R}_{t-1}(\widehat{\boldsymbol{\theta}}_{t-1}) + \boldsymbol{\psi}_t(\widehat{\boldsymbol{\theta}}_{t-1})\boldsymbol{\psi}_t^T(\widehat{\boldsymbol{\theta}}_{t-1}) \tag{11.61}$$

For practical calculations we apply the following procedure. In the recursion for determining $\boldsymbol{\psi}_t(\boldsymbol{\theta})$ and $\widehat{Y}_{t|t-1}(\boldsymbol{\theta})$ for a given $\boldsymbol{\theta}$ at time k, we substitute the parameter $\boldsymbol{\theta}$ with the current estimate of $\widehat{\boldsymbol{\theta}}$. The obtained approximations to $\boldsymbol{\psi}_t(\boldsymbol{\theta}_{t-1})$ and $\widehat{Y}_{t|t-1}(\widehat{\boldsymbol{\theta}}_{t-1})$ are denoted $\boldsymbol{\psi}_t$ and $\widehat{Y}_{t|t-1}$, respectively. We have the following.

The recursive prediction error method (RPEM)

$$\varepsilon_t = Y_t - Y_{t|t-1} \tag{11.62a}$$

$$\widehat{\boldsymbol{\theta}}_t = \widehat{\boldsymbol{\theta}}_{t-1} + \boldsymbol{R}_t^{-1}\boldsymbol{\psi}_t\varepsilon_t \tag{11.62b}$$

$$\boldsymbol{R}_t = \lambda(t)\boldsymbol{R}_{t-1} + \boldsymbol{\psi}_t\boldsymbol{\psi}_t^T \tag{11.62c}$$

In order to apply the previous formulas, we need to calculate the *gradient* $\boldsymbol{\psi}_t(\boldsymbol{\theta})$. If we, for example, consider the model

$$\phi(\mathrm{B})Y_t = \omega(\mathrm{B})u_t + \theta(\mathrm{B})\varepsilon_t \tag{11.63}$$

or

$$Y_t = \frac{\omega(\mathrm{B})}{\phi(\mathrm{B})}u_t + \frac{\theta(\mathrm{B})}{\phi(\mathrm{B})}\varepsilon_t = H_1(\mathrm{B})u_t + H_2(\mathrm{B})\varepsilon_t$$

i.e.,

$$\begin{aligned}\widehat{Y}_{t|t-1} &= H_1(\mathrm{B})u_t + [H_2(\mathrm{B}) - 1]\,\varepsilon_t \\ &= H_1(\mathrm{B})u_t + \left[1 - H_2^{-1}(\mathrm{B})\right] H_2(\mathrm{B})\varepsilon_t \\ &= H_1(\mathrm{B})u_t + \left[1 - H_2^{-1}(\mathrm{B})\right][Y_t - H_1(\mathrm{B})u_t] \\ &= H_2^{-1}H_1(\mathrm{B})u_t + \left[1 - H_2^{-1}(\mathrm{B})\right] Y_t\end{aligned} \tag{11.64}$$

then

$$\widehat{Y}_{t|t-1} = \frac{\omega(\mathrm{B})}{\theta(\mathrm{B})}u_t + \left[1 - \frac{\phi(\mathrm{B})}{\theta(\mathrm{B})}\right] Y_t \tag{11.65}$$

or

$$\theta(\mathrm{B})\widehat{Y}_{t|t-1} = \omega(\mathrm{B})u_t + [\theta(\mathrm{B}) - \phi(\mathrm{B})]Y_t \tag{11.66}$$

Differentiation of (11.66) with respect to ϕ_k, ω_k, and θ_k, respectively, yields

$$\theta(\mathrm{B})\frac{\partial \widehat{Y}_{t|t-1}}{\partial \phi_k} = -\mathrm{B}^k Y_t \tag{11.67}$$

$$\theta(\mathrm{B})\frac{\partial \widehat{Y}_{t|t-1}}{\partial \omega_k} = \mathrm{B}^k u_t \tag{11.68}$$

$$\mathrm{B}^k\widehat{Y}_{t|t-1} + \theta(\mathrm{B})\frac{\partial \widehat{Y}_{t|t-1}}{\partial \theta_k} = \mathrm{B}^k Y_t \tag{11.69}$$

Since $Y_t - \widehat{Y}_{t|t-1} = \varepsilon_t$, this can be summarized in

$$\theta(\mathrm{B})\boldsymbol{\psi}_t(\boldsymbol{\theta}) = \boldsymbol{X}_t(\boldsymbol{\theta}) \tag{11.70}$$

where

$$\boldsymbol{X}_t(\boldsymbol{\theta}) = (-Y_{t-1}, \ldots, -Y_{t-p}, u_{t-1}, \ldots, u_{t-s}, \varepsilon_{t-1}(\boldsymbol{\theta}), \ldots, \varepsilon_{t-q}(\boldsymbol{\theta}))^T$$

and

$$\psi_t(\boldsymbol{\theta}) = \nabla \widehat{Y}_{t|t-1}(\boldsymbol{\theta})$$

From (11.70) it is seen that the gradient is obtained by filtering $\boldsymbol{X}_t$ using $1/\theta(\mathrm{B})$. The RPEM method used on ARMAX models is also known as *Recursive Maximum Likelihood (RML)*, (Söderström 1973).

11.4 Model-based adaptive estimation

In Section 10.4 on page 299, it was shown how, e.g., a trend model $Y_t = \mu_t + \beta_t + \varepsilon_t$ can be written in state space form. Subsequently, the variations of the parameters in time could be tracked using a Kalman filter. The method in this section is similarly based on a state space description where the *Kalman filter is applied to the adaptive parameter estimation.*

We assume that the parameters are contained in the parameter vector and that the dynamic model is assumed to be linear in the parameters. Furthermore, we assume that the variations in the parameters can be described by a random walk. We get the following state space model

$$\boldsymbol{\theta}_{t+1} = \boldsymbol{\theta}_t + \mathbf{e}_{1,t}\,, \quad \mathrm{Var}\left[\mathbf{e}_{1,t}\right] = \boldsymbol{\Sigma}_1 \tag{11.71}$$

$$Y_t = \boldsymbol{X}_t^T \boldsymbol{\theta}_t + \mathbf{e}_{2,t}\,, \quad \mathrm{Var}\left[\mathbf{e}_{2,t}\right] = \boldsymbol{\Sigma}_2 \tag{11.72}$$

The white noise processes $\{\mathbf{e}_{1,t}\}$ and $\{\mathbf{e}_{2,t}\}$ are assumed to be mutually uncorrelated.

The Kalman filter applied to (11.71)–(11.72) yields the conditional mean and conditional variance of $\boldsymbol{\theta}_{t-1}$ given $Y_t, Y_{t-1}, \ldots$. Applying Theorem 10.2 on page 293 we get the following.

Model-based adaptive estimation

$$\widehat{\boldsymbol{\theta}}_{t+1|t} = \widehat{\boldsymbol{\theta}}_{t|t-1} + \boldsymbol{K}_t \left(Y_t - \boldsymbol{X}_t^T \widehat{\boldsymbol{\theta}}_{t|t-1} \right) \tag{11.73a}$$

$$\boldsymbol{K}_t = \frac{\boldsymbol{\Sigma}_{t|t-1} \boldsymbol{X}_t}{(\boldsymbol{X}_t^T \boldsymbol{\Sigma}_{t|t-1} \boldsymbol{X}_t + \boldsymbol{\Sigma}_2)} \tag{11.73b}$$

$$\boldsymbol{\Sigma}_{t+1|t} = \boldsymbol{\Sigma}_{t|t-1} + \boldsymbol{\Sigma}_1 - \frac{\boldsymbol{\Sigma}_{t|t-1} \boldsymbol{X}_t \boldsymbol{X}_t^T \boldsymbol{\Sigma}_{t|t-1}}{(\boldsymbol{X}_t^T \boldsymbol{\Sigma}_{t|t-1} \boldsymbol{X}_t + \boldsymbol{\Sigma}_2)} \tag{11.73c}$$

The initial values $\widehat{\boldsymbol{\theta}}_{1|0}$ and $\boldsymbol{\Sigma}_{1|0}$ can be selected according to the prior knowledge about the parameters. If no prior knowledge is available then $\widehat{\boldsymbol{\theta}}_{1|0} = 0$ and $\boldsymbol{\Sigma}_{1|0} = \alpha \boldsymbol{I}$, where α is large, can be used as initial values.

11.5 Models with time-varying parameters

This section concerns *regression models* and *dynamic models* with *time-varying parameters.*

11.5.1 The regression model with time-varying parameters

We defined the general linear model in Section 3.2 on page 33. This model with time-varying parameters can be written

$$Y_t = \sum_{i+1}^{k} x_{it}\theta_{it} + \varepsilon_t$$

or

$$Y_t = \boldsymbol{x}_t^T \boldsymbol{\theta}_t + \varepsilon_t \tag{11.74}$$

where $\boldsymbol{x}_t = (x_{1t}, \ldots, x_{kt})^T$ is a known vector, $\boldsymbol{\theta} = (\theta_{1t}, \ldots, \theta_{kt})^T$ are the model parameters, and $\{\varepsilon_t\}$ is white noise with variance σ_ε^2.

We now let $\{\theta_{it}\}$ be described by the ARMA process

$$\phi_i(\mathrm{B})(\theta_{it} - \bar{\theta}) = \theta_i(\mathrm{B})w_{it}\,, \quad i = 1, \ldots, k \tag{11.75}$$

where $\{w_{it}\}$ is white noise and $\phi_i(\mathrm{B})$ and $\theta_i(\mathrm{B})$ are polynomials of order p_i and q_i, respectively. In the following we assume, without loss of generality, that $\bar{\theta} = 0$.

In order to estimate the model parameters, we write the model $\boldsymbol{\theta} = (\theta_{1t}, \ldots, \theta_{kt})^T$ in state space form:

$$\boldsymbol{S}_t = \boldsymbol{A}\boldsymbol{S}_{t-1} + \boldsymbol{G}\boldsymbol{e}_{1,t} \tag{11.76}$$

$$\boldsymbol{\theta}_t = \boldsymbol{C}\boldsymbol{S}_t \tag{11.77}$$

where the state vector is of dimension $(n \times 1)$ and $n = \sum^k m_i$ and $m_i = \max\{p_i, q_i + 1\}$. Furthermore, we have

$$\boldsymbol{A} = \begin{pmatrix} \boldsymbol{A}_{11} & \boldsymbol{0} & \cdots & \boldsymbol{0} \\ \boldsymbol{0} & \boldsymbol{A}_{22} & \cdots & \boldsymbol{0} \\ \vdots & \vdots & \ddots & \vdots \\ \boldsymbol{0} & \boldsymbol{0} & \cdots & \boldsymbol{A}_{kk} \end{pmatrix}, \quad \boldsymbol{G} = \begin{pmatrix} \boldsymbol{G}_{11} & \boldsymbol{0} & \cdots & \boldsymbol{0} \\ \boldsymbol{0} & \boldsymbol{G}_{22} & \cdots & \boldsymbol{0} \\ \vdots & \vdots & \ddots & \vdots \\ \boldsymbol{0} & \boldsymbol{0} & \cdots & \boldsymbol{G}_{kk} \end{pmatrix}$$

For the blocks it holds that $\boldsymbol{A}_{ii}$ is an $(m_i \times m_i)$-matrix and $\boldsymbol{G}_{ii}$ is an $(m_i \times 1)$-matrix. Furthermore,

$$\boldsymbol{C} = \begin{pmatrix} \boldsymbol{C}_{11} & \boldsymbol{0} & \cdots & \boldsymbol{0} \\ \boldsymbol{0} & \boldsymbol{C}_{22} & \cdots & \boldsymbol{0} \\ \vdots & \vdots & \ddots & \vdots \\ \boldsymbol{0} & \boldsymbol{0} & \cdots & \boldsymbol{C}_{kk} \end{pmatrix}$$

where each $\boldsymbol{C}_{ii}$ is a $1 \times m_i$-matrix.

Each triple $(\boldsymbol{A}_{ii}, \boldsymbol{G}_{ii}, \boldsymbol{C}_{ii})$ has the purpose of providing a state space description of the ARMA model for θ_{it}, so the structure of the blocks and $\{e_{1,t}\}$ are given by (10.104) on page 299 through (10.109) on page 300. The description may naturally be extended to include the cross-correlation between θ_i and θ_j.

By combining (11.74), (11.76), and (11.77) we get the total *state space form for the linear model with time-varying parameters*

$$\boldsymbol{S}_{t+1} = \boldsymbol{A}\boldsymbol{S}_t + \boldsymbol{G}\boldsymbol{e}_{1,t} \tag{11.78}$$

$$Y_t = \boldsymbol{H}_t\boldsymbol{S}_t + \varepsilon_t \tag{11.79}$$

where $\boldsymbol{H}_t = \boldsymbol{x}_t^T\boldsymbol{C}$.

Wall (1987) has shown that the model parameters can be identified. The maximum likelihood estimated can be found as shown in Section 10.5 on page 307.

Example 11.3 (Regression with time-varying parameters)
Consider the model

$$Y_t = x_t\theta_t + \varepsilon_t$$

where

$$\theta_t + \phi_1\theta_{t-1} + \phi_2\theta_{t-2} = w_t$$

The state space description (11.76)–(11.77) on the preceding page becomes

$$\begin{pmatrix} S_{1,t+1} \\ S_{2,t+1} \end{pmatrix} = \begin{pmatrix} -\phi_1 & 1 \\ -\phi_2 & 0 \end{pmatrix} \begin{pmatrix} S_{1,t} \\ S_{2,t} \end{pmatrix} + \begin{pmatrix} 1 \\ 0 \end{pmatrix} w_t \tag{11.80}$$

$$Y_t = \begin{pmatrix} x_t & 0 \end{pmatrix} \begin{pmatrix} S_{1,t} \\ S_{2,t} \end{pmatrix} + \varepsilon_t \tag{11.81}$$

The theory from Chapter 10 may now be applied to estimate the model parameters and to make predictions of Y_t.

11.5.2 Dynamic models with time-varying parameters

Let us consider the following *time-varying ARMA(p, q) model*

$$Y_t + \phi_{1t}Y_{t-1} + \cdots + \phi_{pt}Y_{t-p} = \varepsilon_t + \theta_{1t}\varepsilon_{t-1} + \cdots + \theta_{qt}\varepsilon_{t-q} \tag{11.82}$$

where $\{\varepsilon_t\}$ is white noise with variance σ_ε^2. The model in (11.82) is denoted the overall model whereas the model for the parameters is called the *latent model.* The latent model may be a deterministic model or a stochastic model (e.g., an ARMA model). It is obvious that there exists a large number of possible models and that the models in the following are simple examples of these.

11.5.2.1 Deterministic latent model

Example 11.4 (Deterministic latent model)
In the example we will go back to Example 11.1 on page 318 where we looked at a transfer function model for the variations in air temperature of the form

$$\phi(\mathrm{B})T_t = \omega(\mathrm{B})R_{t-1} + d + \varepsilon_t$$

where the order of the ϕ and ω polynomials are $p = 3$ and $s = 1$, respectively. We found that some of the parameters were subjected to annual variations. Therefore, it seems reasonable to describe these variations by applying harmonic functions. By investigating the trace of the parameter estimates, Figure 11.1 on page 319 indicates that the trajectory for ω_1 is very similar to that of ω_0 but of opposite sign. This may indicate that the zero in the ω polynomial is constant. The variation in time of the poles of the ϕ polynomial was also investigated, showing a very clear annual variation for the pole nearest the unit circle in the complex plane. The other poles did not show a clear annual variation. Based on these observations we apply the following parametrization of the model

$$(1 - p_t \mathrm{B})\left(1 + \phi_1' \mathrm{B} + \phi_2' \mathrm{B}^2\right)(T_t - \gamma_t) = \omega_t \left(1 + \omega_1'\right) R_t + \varepsilon_t \tag{11.83}$$

where

$$\gamma_t = \frac{d}{\phi(1)} = \mu_T - \frac{\omega(1)}{\phi(1)}\mu_R = \mu_T - H(1)R$$

and where $H(z) = \omega(z^{-1})/\phi(z^{-1})$ is the transfer function from R to T.

In (11.83) the variations of the parameters are limited to p, γ, and ω. It seems reasonable to describe the yearly variations using harmonic functions obtained from a Fourier expansion. As an example, Figure 11.2 on the next page shows the observed mean of the yearly variation of p_t and a suitable Fourier approximation.

In the case where the underlying parameter variation shows only slow variations, it is possible, as shown in the example, to identify the time-varying effects by using forgetting factor methods. However, it is clear that the low-pass filtering being performed when applying the forgetting factor method will imply that the fast variations are disregarded. In such a case, e.g., model-based estimation methods may be adequate. A more detailed description of this example is given in Madsen and Holst (1989).

11.5.2.2 Stochastic latent model

Let us consider the simple case where both the overall and the latent model are AR(1) models, i.e.,

$$Y_t = \phi_t Y_{t-1} + \varepsilon_t \tag{11.84}$$

$$\phi_t - \mu_\phi = \varphi\left(\phi_{t-1} - \mu_\phi\right) + \xi_t \tag{11.85}$$

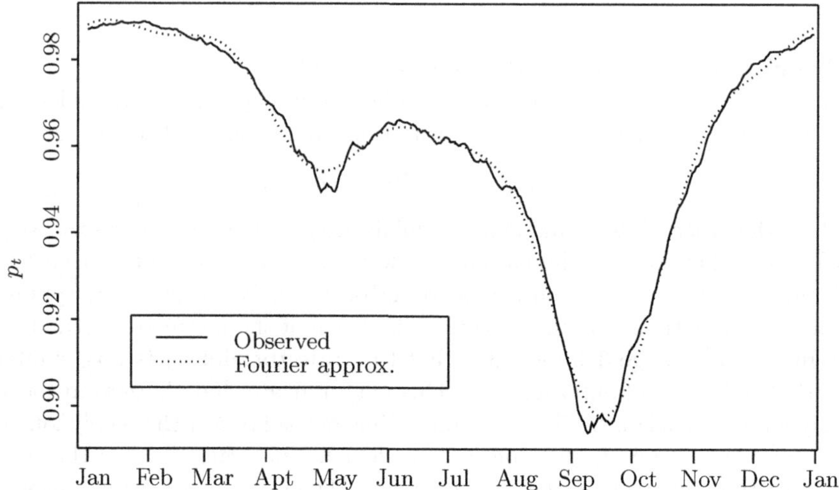

Figure 11.2: *Mean yearly variation of the time-varying pole, p_t, in* (11.83).

where $\{\varepsilon_t\}$ and $\{\xi_t\}$ are white noise with variances σ_ε^2 and σ_ξ^2. This is an example on the so-called *double stochastic model.*

The process $\{Y_t\}$ is obviously a non-linear process. In order to estimate the model parameters $(\varphi, \mu_\phi, \sigma_\varepsilon^2, \sigma_\xi^2)$, we write the model in state space form although (11.85) is initially written in a linear form

$$\phi_t = \varphi\phi_{t-1} + \mu_\phi(1-\varphi) + \xi_t.$$

If we put $\delta = \mu_\phi(1-\varphi)$, the state space form becomes

$$\begin{pmatrix}\phi_t\\ \delta_t\end{pmatrix} = \begin{pmatrix}\varphi & 1\\ 0 & 1\end{pmatrix}\begin{pmatrix}\phi_{t-1}\\ \delta_{t-1}\end{pmatrix} + \begin{pmatrix}1\\ 0\end{pmatrix}\xi_t \tag{11.86}$$

$$Y_t = \begin{pmatrix}Y_{t-1} & 0\end{pmatrix}\begin{pmatrix}\phi_t\\ \delta_t\end{pmatrix} + \varepsilon_t \tag{11.87}$$

where δ has the initial value δ_0.

Given the initial values, the Kalman filter can be applied to calculate the conditional mean and variance of Y_t given $Y_{t-1}, Y_{t-2}, \ldots$. Thereby, the maximum likelihood estimates can be found as shown in Section 10.5. Models of this type are further discussed in Porsholt (1989) and Madsen, Holst, and Lindström (2007).

Example 11.5 (An AR(1) model with time-varying parameter)
In this example we consider the process

$$Y_t = \phi_t Y_{t-1} + \varepsilon_t$$
$$(\phi_t - 0.8) = 0.85\,(\phi_{t-1} - 0.8) + \xi_t$$

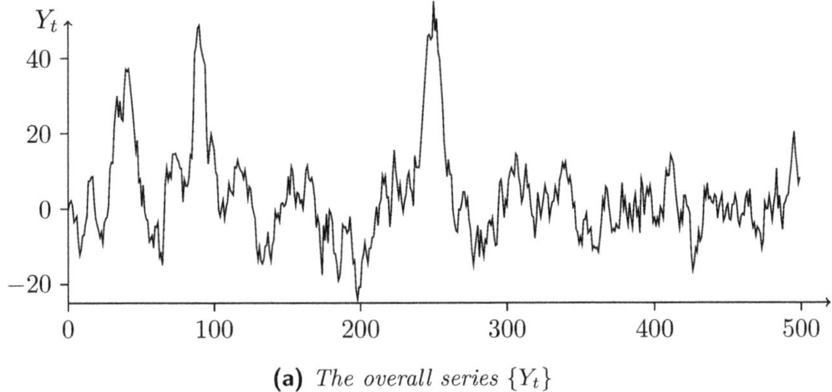

(a) *The overall series* $\{Y_t\}$

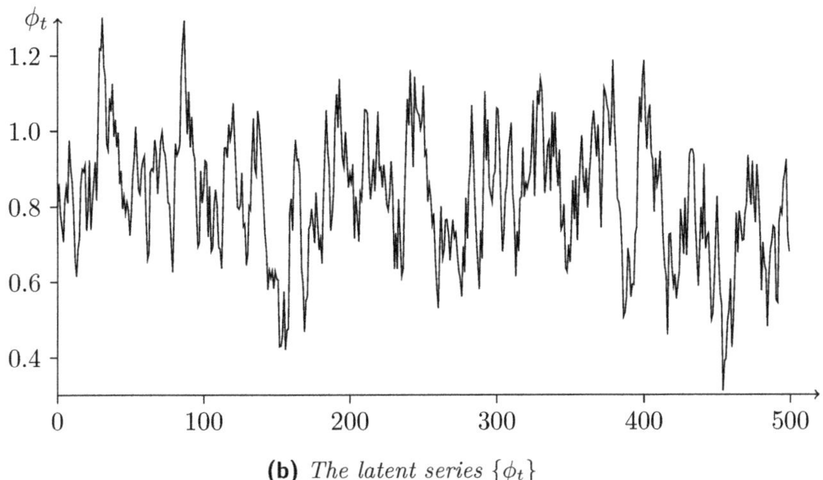

(b) *The latent series* $\{\phi_t\}$

Figure 11.3: *(a) The overall series* $\{Y_t\}$ *and (b) the latent series* $\{\phi_t\}$.

where $\sigma_\varepsilon^2 = 4.0^2$ and $\sigma_\xi^2 = 0.10^2$.

Figure 11.3 shows 500 simulated values of the overall series $\{Y_t\}$ and the latent series $\{\phi_t\}$. It is seen that when ϕ_t is greater than 1 there is a large probability that Y_t is either increasing or decreasing rapidly. This is in accordance with the fact that the AR(1) process is non-stationary for $|\phi| \geq 1$, where the influence from past values does not vanish with time.

Applying the Kalman filter in the calculation of the conditional mean and variance and applying the ML method on the series in Figure 11.3, the following estimates have been obtained

$$\widehat{\varphi} = 0.850^{\pm 0.004} \qquad \widehat{\mu}_\phi = 0.800^{\pm 0.033}$$
$$\widehat{\sigma}_\varepsilon^2 = \left(0.135^2\right)^{\pm 0.007} \qquad \widehat{\sigma}_\zeta^2 = \left(4.078^2\right)^{\pm 1.585}$$

The numbers in parentheses are estimates of the standard deviation of the estimator. By comparing the estimates with the true values we see a reasonable agreement.

CHAPTER 12

Real life inspired problems

This chapter contains a number of problems inspired by practical applications of time series analysis. The problems are more lengthly than those found at the end of each chapter throughout the book, and also, in general, a solution to the problems requires a theoretical background which cannot be provided by a single chapter.

The problems are aimed at illustrating how time series analysis can be used to solve a large number of real life problems. However, the problems are still kept in a form so that they can be solved without the need of using a computer.

The problems are related mostly to science and engineering. A few problems from econometrics and finance are also found; for more examples in these areas we refer to Madsen et al. (2004).

This chapter considers the following problems:

1) Prediction of wind power In order to operate a power system with a large penetration of wind power, such as in Denmark and Spain, on-line predictions of the wind power production are needed. This is essential in order to estimate the remaining power needed to fulfill the demand. The additional power is produced at conventional power plants. For more information about the methods used in practice see Madsen, Nielsen, and Nielsen (2005), Nielsen et al. (1999) or a large number of references at `http://www.enfor.eu`. Prediction of wind power is very interesting from a time series point of view since the uncertainty of the forecast is not just a function of the prediction horizon; a 6-hour prediction might be more uncertain than an 18-hour prediction. See also Nielsen, Madsen, and Nielsen (2006).

2) Optimal import of medicine A company which is responsible for importing a large amount of medicine to Denmark is seeking a model for predicting the expected sale of medicine in order to purchase what is needed and not much more.

3) Effect of chewing gum Chewing gum is most often used to improve the flavor of the exhaled air. Using some experiments on test persons a company producing chewing gums wants to derive models for describing how fast the flavor is dissolved when a person is chewing the gum. For more information of the methods used in practice see Haahr et al. (2003).

4) Prediction of stock price A model for the variation of an industrial stock is considered. The purpose is to predict whether it is profitable or not to invest in this stock. Such problems are studied in detail in Madsen et al. (2004).

5) Wastewater treatment using root zone plants The use of root zone plants for wastewater treatment is increasing. In this problem we will consider a root zone plant-based system used as wastewater treatment for a larger domestic area. Based on test experiments the purpose is to formulate a model describing the efficiency and dynamical characteristics of the system.

6) Scheduling system for oil delivery A large oil company wants a system for predicting when oil should be delivered to their customers. The company has a system which ensures an automatic schedule of the delivery of oil for heating. The purpose of this problem is to formulate a model which can be used for minimizing the number of oil deliveries, without taking the risk of running dry.

7) Warning system for slippery roads A county in Denmark wants a model which can be used for warnings of slippery roads. If it is possible to send out warnings in due time, some accidents can be avoided. On the other hand, it is possible to reduce the amount of salt used to prevent ice on roads since the model would allow salt to be used more efficiently.

8) Dynamical quality control Statistical quality control of a manufacturing process is most often based on the assumption that the individual test results are independent. However, a variation in time of some of the process parameters is very often seen, and in cases where some limits for the product have been verified, the level of inspection can be adapted to the predicted level of the underlying process. This enables methods for reducing the number of samples for quality insurance. See Thyregod and Madsen (1992) for more details.

9) Modeling and control of wastewater treatment plant For treatment of wastewater, biochemical processes are used still more frequently. On a pilot plant at Lundtofte, Denmark, biochemical processes are used for removal of phosphorus, nitrogen (mostly as ammonia), and organic material. In this problem the purpose is to establish a model for the variations of nitrate. This model will be used primarily to formulate an improved control of the shift from anoxic conditions to aerobic conditions. See Bechmann et al. (2002) for more information.

10) Sales numbers of a newspaper For this problem the purpose is to establish the model for the sales of a newspaper. Secondly, the problem focuses on modeling the effect of bankruptcy of a rival newspaper, and the long term increase in the sales due to the closing of the rival newspaper is estimated.

11) Modeling and prediction of stock prices Here a model for the variation of the stock prices for a large company is formulated. The purpose of the model is to be able to predict the stock prices. More information can be found in Nielsen, Vestergaard, and Madsen (2000) and Baadsgaard, Nielsen, and Madsen (2000).

12) Adaptive models for the interest rates This problem considers the observed interest rates for a longer period. Using conventional time series models, it is very difficult to predict the interest rates. However, adaptive models can be very useful, and this problem considers some adaptive models in order to provide on-line predictions of the interest rate. See Madsen et al. (2005) for more information.

12.1 Prediction of wind power production

In countries such as Spain, Germany, and Denmark, windmills contribute considerably to the power production. For Denmark about 20% of the power originates from windmills, and under windy conditions the windmills contribute more than 50% of the power production. In order to plan the production on the conventional power plants, it is thus very important to be able to predict the wind power production some hours in advance.

Based on a time series of 200 hourly observations of the wind power production (Y_t) in a larger area, the autocovariance function and the partial autocovariance function are estimated. The results are listed in Table 12.1 on the next page.

The mean value of the wind power production is estimated; the result is $\widehat{\mu} = 114$.

Question 1 Estimate the autocorrelation function for the wind power production.

Question 2 Formulate a reasonable model for the wind power production. The choice of model must be elaborated.

Question 3 Estimate the parameters of the model.

For the last hours the wind power production has been:

Time	6	7	8	9	10
Wind power production (MWh/h):	93.6	80.3	95.2	106.9	121.9

Question 4 Predict the wind power production for the next three hours. State the variance of these predictions.

For wind power prediction the performance of a prediction model is often compared with the performance of the *persistent predictor*, which is related to the model:

$$\nabla Y_t = \varepsilon_t$$

where ε_t is white noise.

Table 12.1: *Sample values of the autocovariance and partial autocorrelation functions for the wind power production.*

Lag k	$C(k)$	$\widehat{\phi}_{kk}$	Lag k	$C(k)$	$\widehat{\phi}_{kk}$
0	1621.9815	1.0000	13	421.9977	0.0960
1	1391.6113	0.8580	14	450.2248	0.0296
2	1157.5002	−0.0852	15	493.9226	0.0809
3	958.4999	−0.0015	16	526.1848	0.0267
4	792.5158	−0.0016	17	532.5273	0.0101
5	704.8090	0.1171	18	564.1671	0.0992
6	610.3871	−0.0564	19	574.0390	0.0090
7	564.0285	0.0951	20	573.9824	0.0135
8	524.0550	−0.0011	21	552.6435	−0.0215
9	469.2479	−0.0178	22	551.2704	0.0930
10	427.1069	0.0126	23	538.9900	−0.0294
11	398.1650	0.0468	24	509.5179	0.0031
12	392.0617	0.0411	25	482.8544	0.0102

Question 5 Find a prediction for the next 3 hours using the above mentioned persistent predictor.

Another often used model for predictions is the predictor

$$\widehat{Y}_{t+k|t} = \widehat{\mu}$$

where $\widehat{\mu}$ is as stated previously.

Question 6 Show that the variance of the prediction error for the persistent predictor in the case of a large prediction horizon (k large) is twice the variance of the prediction error for the mean value prediction, i.e., the predictor given by $\widehat{Y}_{t+k|t} = \mu$.

12.2 Prediction of the consumption of medicine

A company, Medicos, takes care of the import of nearly all medicine to Denmark. The company makes the purchase of the various medical products on a monthly basis.

In order to import a reasonable amount of various products, the company wants a model established for each product with the purpose of predicting the consumption of medicine on a monthly basis.

Based on the consumption of one product, Jodaex, which has been sold for several years, data for the monthly consumption is established for the past 15 years. Using these data, the autocovariance, the autocorrelation, and the partial autocorrelation functions are estimated. The results are listed in Table 12.2.

Table 12.2: *Sample values of the autocovariance, autocorrelation, and partial autocorrelation functions based on sales data for Jodaex.*

Lag k	$C(k)$	$\widehat{\rho}(k)$	$\widehat{\phi}_{kk}$
0	1900.727284	1.0000	1.0000
1	666.832609	0.3508	0.3508
2	−92.677043	−0.0488	•
3	69.184107	0.0364	0.1490
4	166.189602	0.0874	0.0076
5	39.808724	0.0209	−0.0054
6	−178.418743	−0.0939	−0.1016
7	−2.309981	−0.0012	0.0863
8	194.270778	0.1022	0.0474
9	128.448640	0.0676	0.0283
10	−102.621630	−0.0540	−0.0763
11	474.668896	0.2497	0.3792
12	1513.937469	0.7965	0.7195
13	561.688665	0.2955	−0.2916
14	−76.937014	−0.0405	0.1343
15	26.833374	0.0141	−0.0562
16	114.829297	0.0604	−0.0207
17	93.472536	0.0492	0.0707
18	−166.230375	−0.0875	−0.0583
19	−84.088404	−0.0442	−0.0523
20	118.786776	0.0625	−0.0786
21	90.895574	0.0478	−0.0472
22	−106.413088	−0.0560	0.0456
23	355.997246	0.1873	0.0160
24	1184.045173	0.6229	−0.0541
25	475.920187	0.2504	0.0513
26	−24.418636	−0.0128	0.0219
27	9.135132	0.0048	−0.0049
28	61.540287	0.0324	0.0085
29	89.484935	0.0471	−0.0457
30	−165.346009	−0.0870	0.0082

Question 1 Due to an error the partial autocorrelation function in lag 2 is not reported. Estimate the partial autocorrelation function in lag 2.

Question 2 Is there any seasonal variation in the monthly consumption of Jodaex? The answer should be elaborated.

Question 3 Formulate a reasonable model for the monthly consumption of Jodaex.

Question 4 Estimate the parameters of the model.

In the following it is assumed that the monthly consumption of Jodaex, Y_t, can be described by the following model:

$$(1 - 0.9\text{B})(1 - 0.6\text{B}^{12})(Y_t - 288) = \varepsilon_t$$

where $\{\varepsilon_t\}$ is white noise with the variance $\sigma_\varepsilon^2 = 22^2$.

Question 5 Find the poles of the transfer function from ε_t to Y_t.

During the recent months the consumption of Jodaex has been

...	383	411	389	348	371
443	420	424	414	422	451
380	362	387	372	323	382
380	412	393	385	373	357
281	258	290	318	265	365
399	436	449	414	359	362
275	238	246	287	233	303

The consumption for the most recent month of 303 units (as mentioned in the list above) has just been observed.

Question 6 Predict the monthly consumption of Jodaex for the coming two months. State an estimate for the uncertainty of the predictions.

12.3 Effect of chewing gum

A company, which manufactures chewing gum, wants to develop a method or a model in order to describe how the flavor is dissolving when a person is chewing the gum. The methods or models are going to be applied in the development of new products. One of the key flavor additives is menthol in the exhaled air.

During an experiment, where a test person was given the gum and started chewing at time 2.00 sec. (time index $t = 8$, the sampling time is 0.25 sec.), the concentration of menthol in the exhaled air was measured and is given in Table 12.3.

It is assumed that the concentration of menthol will increase towards a stationary value Y_s (after a longer period of chewing the concentration will naturally start to decrease, but this will be neglected in the following). Due to technical and economical reasons the experiment stopped at the time 3.50 sec.

Based on a preliminary analysis it is found that the concentration of menthol Y_t can be described by the following model

$$(Y_t - \mu) = \frac{\omega\text{B}}{(1 + \delta\text{B})(1 - \text{B})} I_t + N_t \tag{12.1}$$

Table 12.3: *Concentration of menthol in the exhaled air.*

t	0	1	2	3	4	5	6	7
Time (s)	0.00	0.25	0.50	0.75	1.00	1.25	1.50	1.75
Menthol Conc.	0.527	0.650	0.336	0.403	0.265	0.219	0.203	0.351
t	8	9	10	11	12	13	14	
Time (s)	2.00	2.25	2.50	2.75	3.00	3.25	3.50	
Menthol Conc.	0.348	1.239	1.896	2.297	3.081	3.118	3.414	

where

$$I_t = \begin{cases} 1 & \text{for } t = 8 \\ 0 & \text{otherwise} \end{cases} \tag{12.2}$$

and N_t is an ARMA process with mean value zero.

Question 1 What is the previously mentioned model called?

Question 2 Calculate estimates for μ, ω, and δ.

In the following it will be assumed that $\mu = 0.28$, $\omega = 0.81$, and $\delta = -0.78$.

Question 3 Calculate an estimate of the stationary value Y_s of the menthol concentration in the exhaled air.

The dynamics of a system is often characterized by its time constants. Corresponding to a non-negative real pole p_i ($|p_i| < 1$) in a transfer function, the time constant is defined as

$$\tau_i = -\frac{1}{\ln(p_i)} \tag{12.3}$$

Question 4 Calculate the time constant for the transfer function from I_t to Y_t.

Now, a description of the ARMA process N_t should be determined. Following a long time period, in which the test person has not been chewing gum, a time series of 110 values has been obtained. The sampling time remains 0.25 sec. The estimated autocovariance function and the estimated partial autocorrelation function for Y_t (which under the given circumstances correspond to N_t) are given in Table 12.4 on the following page.

Question 5 Calculate the autocorrelation function for N_t.

Question 6 Formulate a model for N_t.

Question 7 Estimate the parameters of the model for N_t, and provide the total model for Y_t.

Table 12.4: *Sample autocovariance and partial autocorrelation values for the contentration of menthol in the exhaled air.*

	Y_t			$(1-\mathrm{B})Y_t$	
Lag k	$C(k)$	$\widehat{\phi}_{kk}$	Lag k	$C(k)$	$\widehat{\phi}_{kk}$
0	0.0136839	1.0000	0	0.0377271	1.0000
1	−0.0055638	−0.4066	1	−0.0227523	−0.6031
2	−0.0006502	−0.2550	2	0.0036194	−0.4208
3	0.0000545	−0.1588	3	0.0007165	−0.2914
4	0.0003856	−0.0659	4	0.0007066	−0.1546
5	0.0001712	−0.0084	5	−0.0012965	−0.1095
6	0.0000313	0.0178	6	0.0007439	−0.0673
7	0.0003338	0.0571	7	0.0014838	0.0531
8	−0.0011036	−0.0487	8	−0.0034761	−0.2670
9	0.0008571	−0.0112	9	0.0028978	−0.0026
10	−0.0001773	−0.0008	10	−0.0003067	0.0421
11	−0.0007002	−0.0679	11	−0.0028361	−0.0795
12	0.0009251	0.0180	12	0.0020649	−0.1117
13	0.0011024	0.1427	13	0.0031495	0.0850
14	−0.0017087	−0.0043	14	−0.0041960	0.9030
15	−0.0000933	−0.0438	15	0.0002893	−0.0175

In the following we will assume the following model for N_t:

$$(1+\phi\mathrm{B})N_t = \varepsilon_t$$

where $\phi = -0.58$, and ε_t is white noise having the variance $\sigma_\varepsilon^2 = (0.082)^2$.

Question 8 Predict the concentration of menthol in the exhaled air at time 4 sec., conditioned that the test person is still chewing gum.

12.4 Prediction of stock prices

A model for the variations of an industrial stock should be formulated. Based on a time series consisting of 108 monthly observations of the stock price, Y_t, (this series does not contain any extraordinary behavior) the autocorrelation function as well as the partial autocorrelation function have been calculated for both Y_t, ∇Y_t, and $\nabla^2 Y_t$. These estimates are given in Tables 12.5 to 12.7.

Question 1 Based on the above information, a suitable model for the stock price should be formulated. The motivation for selecting the model should be explained briefly.

Question 2 Estimate the model parameters.

Table 12.5: *Autocorrelation and partial autocorrelation for the stock price, Y_t ($\widehat{\sigma}^2_{Y_t} = 99.7^2$).*

Lag k	r_k	$\widehat{\phi}_{kk}$	Lag k	r_k	$\widehat{\phi}_{kk}$
1	0.98	0.98	9	0.74	0.01
2	0.95	−0.20	10	0.71	−0.07
3	0.92	−0.06	11	0.68	−0.10
4	0.89	−0.02	12	0.65	−0.05
5	0.86	−0.02	13	0.61	−0.19
6	0.83	0.04	14	0.56	0.00
7	0.80	0.00	15	0.52	0.03
8	0.77	0.00	16	0.48	0.09

Table 12.6: *As in Table 12.5 but for $(1-\mathrm{B})Y_t$ ($\widehat{\sigma}^2_{(1-\mathrm{B})Y_t} = 9.78^2$).*

Lag k	r_k	$\widehat{\phi}_{kk}$	Lag k	r_k	$\widehat{\phi}_{kk}$
1	0.38	0.38	9	−0.07	0.02
2	0.20	0.05	10	0.12	0.18
3	0.09	0.00	11	0.02	−0.09
4	0.01	−0.04	12	0.07	−0.06
5	−0.06	−0.07	13	0.12	−0.07
6	−0.06	−0.02	14	0.06	−0.03
7	−0.11	−0.08	15	−0.13	−0.08
8	−0.12	−0.05	16	−0.10	−0.11

Table 12.7: *As in Table 12.5 but for $(1-\mathrm{B})^2Y_t$ ($\widehat{\sigma}^2_{(1-\mathrm{B})^2Y_t} = 10.5^2$).*

Lag k	r_k	$\widehat{\phi}_{kk}$	Lag k	r_k	$\widehat{\phi}_{kk}$
1	−0.39	−0.39	9	−0.10	−0.28
2	−0.04	−0.23	10	0.23	0.01
3	0.00	−0.13	11	−0.15	−0.13
4	0.00	−0.09	12	0.03	−0.12
5	−0.06	−0.07	13	0.09	−0.01
6	0.04	−0.07	14	0.01	0.03
7	−0.02	−0.07	15	0.00	0.06
8	−0.06	−0.15	16	−0.02	0.00

Based on similar observations of the stock price for a different time period, a model has been formulated as

$$(1 - \mathrm{B})(1 + \phi \mathrm{B})Y_t = \eta + \varepsilon_t$$

where $\phi = -0.6$, $\eta = 0.11$, and ε_t is white noise with variance $\sigma_\varepsilon^2 = (2.7)^2$.

Question 3 The model has been used to predict two months ahead. Calculate the variance of the prediction.

Question 4 Calculate the mean value of the monthly increase (or decrease) in the stock price.

An investor wants to invest 10.000 DKK in the stock and has decided to keep it for 2 months. The investment will be made at a time when the stock price is 783 DKK.

Question 5 Based on the information given previously, a 95% confidence interval for the expected profit (in DKK) should be calculated.

Question 6 If information about the stock price prior to the time of the investment was available, would this give reason to modify the estimate of the expected profit?

12.5 Wastewater treatment: Using root zone plants

Root zone plants are increasingly being used for wastewater treatment, but it can be difficult to estimate the effect of the purification. Using data from an existing plant, we will use time series analysis to investigate some conditions in connection with the use of root zone plants for removal of nitrogen.

A root zone plant in ordinary operation is considered since wastewater is constantly being supplied from a larger domestic area. The water flow can be assumed to be constant. At present, continuous recordings of the nitrogen concentration discharge from the root zone plant are being carried out, but not of the inlet to the plant. In preparation for determining the root zone plant dynamical characteristic one experiment has been carried out. On the 24th of April at 12 o'clock (and in the following 3-hour period) wastewater with a very high concentration (900 mg/l) is added. This inlet concentration is assumed to be higher than the normal inlet concentration. Measurements of the discharge concentration, which have been obtained through a longer period, where there have not been any experiments carried out, have shown that the average nitrogen concentration in the discharge is 38 mg/l.

In Table 12.8 the measured nitrogen concentration is given during the experimental period starting at midnight on the night to the 24th of April 1997.

Question 1 Determine the time of stay (defined at the time delay from inlet to discharge) in the root zone plant.

Table 12.8: *Measurements* (mg/l) *of the discharge of the root zone plant.*

Time (24th of April)	00	3	6	9	12	15	18	21
t	1	2	3	4	5	6	7	8
Y_t (mg/l)	44	42	35	34	39	41	35	97
Time (25th of April)	00	3	6	9	12	15	18	21
t	9	10	11	12	13	14	15	16
Y_t (mg/l)	84	83	79	72	73	64	60	54

Question 2 Identify a suitable transfer function component to describe the relation between inlet and discharge concentration.

Question 3 State the estimates of parameters of the transfer function components.

Question 4 Determine the efficiency of the root zone plant on the basis of the model (efficiency is defined as the proportion of nitrogen which is removed).

Changes of the inlet to the root zone plant are planned since the usual inlet is cut off. On the other hand plans are made to connect a new area where it is assumed that the daily variation in conducted nitrogen concentration can be described by a single harmonic function with a mean value 80 mg/l and an amplitude of 50 mg/l. The maximum is expected to occur at 3 p.m.

Question 5 Calculate the average discharge concentration with the new inlet. State an estimate of the expected average of the daily variation maximum discharge concentration together with an estimate of the time in the diurnal cycle where this maximum concentration is expected.

As mentioned earlier the inlet concentration was not measured and this gave rise to the mentioned experiment, which was conducted in order to determine the relation between inlet and discharge concentration.

Question 6 Discuss the assumptions which should hold in order to use the sketched procedure. Discuss (briefly) how the modeling could have been carried out as soon as the inlet concentration had been measured.

12.6 Scheduling system for oil delivery

One of the largest oil companies in the country wishes to develop a system to predict when oil deliveries are necessary. The background is a desire to minimize the number of oil deliveries without increasing the risk of running dry. Today interval deliveries are used, i.e., the supply follows a certain time interval. For most customers the oil consumption depends on the so-called degree day number. The degree day number for 24 hours is $\max(0, 17\,°\mathrm{C} - T_{\text{mean}})$, where T_{mean} is the 24-hour mean temperature.

In this assignment a customer with a 3500 l tank is considered. At present the oil is delivered every 7th day. The weekly consumption in liters is called Y_t and the degree day number for the same period is denoted X_t and is stated in °C.

A statistical adviser has found that the oil consumption can be described by the following model

$$Y_t - \alpha - \beta X_t = Z_t \tag{12.4}$$

$$\phi(\mathrm{B})Z_t = \theta(\mathrm{B})\varepsilon_t \tag{12.5}$$

where $\{\varepsilon_t\}$ is white noise with variance σ_ε^2. $\phi(\mathrm{B})$ and $\theta(\mathrm{B})$ are polynomials in the backward shift operator B. For the relevant customer $\alpha = 1204\,\mathrm{l}$ and $\beta = 5\,\mathrm{l}/°\mathrm{C}$.

On the basis of a number of deliveries to the customer, an analysis of the corrected consumption, Z_t, has been carried out. The results of this analysis are shown in Table 12.9.

Question 1 Specify a suitable model for Z_t. The choice of model must be explained.

Question 2 Estimate the parameters in the model.

The last deliveries and degree day numbers have been as follows:

Date		21/1	28/1	4/2	11/2	18/2
Delivery no.		181	182	183	184	185
Degree day no.	^{o}C	122.4	90.3	101.0	86.7	115.2
Delivery amount	l	2098	1873	1910	1957	2160

Question 3 It is assumed that the degree day number in the two following seven-day periods is 90 °C. State predictions of the two following oil deliveries. The uncertainty of the predictions must also be stated.

An analysis of the weekly degree day number shows that the following model can be used (provided that we are in a winter period):

$$(1 - 0.8\mathrm{B})(X_t - 80) = e_t$$

where $\{e_t\}$ white noise with the variance $\sigma_e^2 = (24.0\,°\mathrm{C})^2$. $\{e_t\}$ and $\{\varepsilon_t\}$ are mutually uncorrelated.

Question 4 Predict the two following oil deliveries. State the corresponding variances.

In the period from the 1st of May until the 1st of October the degree day number is either zero or so small that it is negligible.

Table 12.9: *Oil consumption data. The first column is the autocorrelation, the second column is the autocovariance, and third column is the partial autocorrelation. The data set consists of $N = 104$ observations.*

Lag k	$\widehat{\rho}(k)$	$\widehat{\gamma}(k)$	$\widehat{\phi}_{kk}$
1	1.000000000	150323.2344	1.000000000
2	0.959337115	144210.6562	0.959337115
3	0.888225853	133520.9844	−0.402924687
4	0.815250695	122551.1172	0.095117733
5	0.745057106	111999.3906	−0.046410114
6	0.669762373	100680.8438	−0.135191366
7	0.600345254	90245.8359	0.121635161
8	0.536711216	80680.1641	−0.069181748
9	0.475697219	71508.3438	−0.029990815
10	0.414694697	62338.2500	−0.026623955
11	0.355233639	53399.8711	−0.047014248
12	0.298067123	44806.4141	−0.011921338
13	0.247249082	37167.2812	0.035445973
14	0.200208023	30095.9180	−0.052443247
15	0.151253909	22736.9766	−0.082733408
16	0.102474608	15404.3145	0.004436771
17	0.058732785	8828.9023	−0.005888888
18	0.025725840	3867.1914	0.081523851
19	−0.001258631	−189.2015	−0.033358071
20	−0.021297373	−3201.4900	0.041945897
21	−0.036097620	−5426.3110	−0.014095530
22	−0.049451921	−7433.7725	−0.055693060

Question 5 The oil company is considering changing the delivery intervals for the concerned customer in the period from the 1st of May until the 1st of October. Calculate the probability of the tank running dry given that the delivery is every fortnight.

Another customer has always had the oil delivered on Tuesday, but not always every week. In the summer period it can be every second week or even, in rare cases, every third week. For this customer a model of the type (12.4)–(12.5) is wanted.

Question 6 State a model on state space form for the oil deliveries for the new customer which contains the same parameters as the model (12.4)–(12.5).

Question 7 Describe briefly (in words) a method to estimate the parameters in model for the above mentioned customer.

12.7 Warning system for slippery roads

A county has consulted some experts in time series analysis to establish models for warnings for slippery (icy) roads. If it is possible to warn of a slippery situation in due time, some accidents could be avoided. It would also be possible to reduce salt consumption since the quantity needed is at its minimum if the salt is spread on the roads before glaciation.

As an introduction, a model which connects the variation in the road temperature to the variation in the air temperature is wanted. Some time delay is expected between the air temperature and the road temperature due to the heat capacity of the road. By relating the temperature measurements to the humidity of the road, it can be decided whether the necessary condition for glaciation is fulfilled.

It is assumed that the simultaneous measurements of air temperature T_t and road temperature V_t are obtained hourly at 00, 01, 02, etc. Hourly measurements for a period of 14 days are available for modeling. In this period temperatures below zero degrees are observed several times. Based on the data, an impulse response function has been estimated, and a "first guess" for a model is

$$V_t = 0.9V_{t-1} + 0.25T_{t-1} - 0.15T_{t-2} + N_t$$

where $\{N_t\}$ is the sequence of residuals with a variance of $(0.4\,°\mathrm{C})^2$.

For this period the mean temperature of the air and the road temperature are both $0\,°\mathrm{C}$.

Question 1 Find the time delay, the poles, and the zeros in the transfer function from air temperature to road temperature. Sketch the impulse response function.

Question 2 Find the amplitude function corresponding to the transfer function from air temperature to road temperature. State the change of amplitude in the diurnal variation of the road temperature compared with the diurnal variation of the air temperature.

Based on the time series of residuals $\{N_t\}$ the autocorrelation function and partial autocorrelation function are estimated. The results are shown in Table 12.10.

Question 3 Find a model for $\{N_t\}$ and estimate the parameters of the model. Write the final model.

Question 4 Find a parametric estimate for the spectral density of $\{N_t\}$ and sketch the spectral density.

We now assume that the variations in the road temperature can be described by following model:

$$V_t = \frac{0.25(1-0.6\mathrm{B})}{1-0.9\mathrm{B}}T_{t-1} + \frac{1}{(1-0.9\mathrm{B})(1-0.8\mathrm{B})}\varepsilon_t$$

Table 12.10: *Autocorrelation function and partial autocorrelation function estimated on the basis of the time series of residuals $\{N_t\}$.*

Lag k	$\widehat{\rho}(k)$	$\widehat{\phi}_{kk}$	Lag k	$\widehat{\rho}(k)$	$\widehat{\phi}_{kk}$
1	0.810	0.810	9	0.028	−0.021
2	0.434	−0.643	10	0.053	−0.017
3	0.058	0.007	11	0.032	−0.019
4	−0.211	−0.058	12	−0.005	0.006
5	−0.322	0.055	13	−0.037	−0.026
6	−0.293	−0.019	14	−0.053	−0.007
7	−0.178	0.026	15	−0.048	−0.011
8	−0.054	−0.066	16	−0.030	0.004

where $\{\varepsilon_t\}$ is a white noise process with the variance $\sigma_\varepsilon^2 = (0.24\,°\text{C})^2$.

It is also assumed that the air temperature can be described by the model

$$T_t = 0.9T_{t-1} + \xi_t$$

where $\{\xi_t\}$ is a white noise process with the variance $\sigma_\xi^2 = (1.0\,°\text{C})^2$. Finally the noise processes $\{\varepsilon_t\}$ and $\{\xi_t\}$ are assumed to be uncorrelated.

The following measurements are available.

Time	12	13	14	15	16	17
V_t [°C]	0.37	0.56	0.81	0.48	0.12	0.11
T_t [°C]	2.31	2.71	1.30	−0.49	−0.22	−1.40

Question 5 Predict the road temperature at time 18 and 19. Find the corresponding estimate of the probability for the road temperature being less than −0.5 °C.

12.8 Statistical quality control

A company is producing a product with the promised level of quality parameter of at least 100. Due to variation in time of the raw material used for the production, the final quality of the product will vary in time.

Today the company is using traditional statistical quality control where 10 samples are taking every hour. If the average of these 10 samples is less than 102, then the production for the past hour has failed meaning that the products cannot be sold.

A further study of the production has shown that the variation in the quality parameter, Z_t, can be described as a stochastic process. In order to formulate the process model, it is useful to introduce $X_t = Z_t - 105$. Using time series analysis, it is found that the variations of $\{X_t\}$ can be described by

$$X_t = aX_{t-1} + v_t,$$

where $\{v_t\}$ is white noise with variance $\sigma_v^2 = 1$.

The quality of the product cannot be directly observed. More precisely a measurement of the quality parameter of the product at time t is given as

$$Y_t = X_t + e_t,$$

where $\{e_t\}$ is white noise with $\sigma_e^2 = 4$. $\{v_t\}$ and $\{e_t\}$ are mutually independent.

Question 1 Characterize the process $\{X_t\}$.

Question 2 Take $a = 0.8$. Find a process of the ARIMA type that describes the variation of $\{Y_t\}$.

Question 3 Determine the spectral density for $\{X_t\}$, and state for which values of a the spectrum exists.

The company wants to improve the quality control and wants to take advantage of the fact that the quality parameter can be described as a stochastic process.

Now it is assumed that an estimate of the quality parameter a is available at time t, and furthermore, the variance of the estimate is given. In the following we will use $a = 0.8$.

Question 4 Describe a procedure for predicting the value of the quality parameter and the associated uncertainty for $t + 1$.

As mention previously several measurements can be obtained at the same point in time. The uncertainty of the individual measured values at a given point in time is assumed independent.

Based on cost calculations the company has decided that the testing must be conducted such that the probability for rejection is less that 1%, and the criterion for rejecting a sample (and, hence, the related production) is that the average of the sample is less than 102.

Just after having obtained the measurements for a given time point t, the estimate of the quality parameter is 1.2, and the associated variance is 0.8.

Question 5 Determine the number of samples needed at the next point in time $(t + 1)$ in order to fullfil the above mention criterion.

Now it is assumed, that at time $t + 1$ the number of measurements is m.

Question 6 Provide an expression for updating the estimate of the quality parameter.

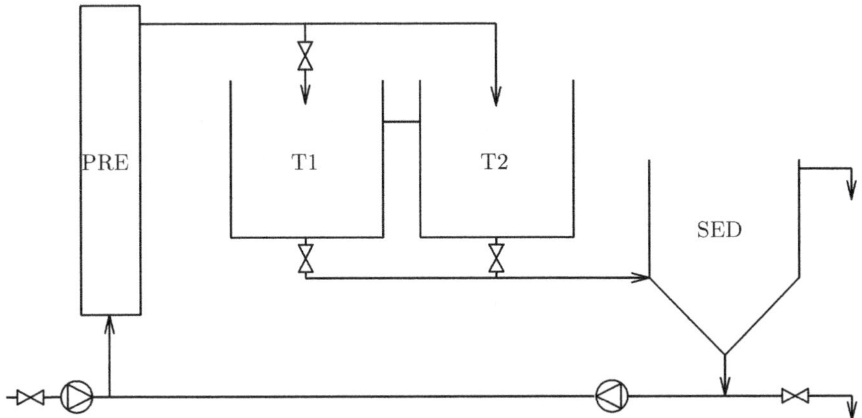

Figure 12.1: *Schematic diagram of the wastewater treatment plant at Lundtofte.*

12.9 Wastewater treatment: Modeling and control

For the treatment of wastewater, biochemical processes are used ever more frequently. On the pilot plant at Lundtofte, which is sketched in Figure 12.1, biochemical processes are used for removal of phosphorus, nitrogen (mostly as ammonia), and organic material.

The pilot plant consists of a pretreatment tank (PRE), two aeration tanks (T1 and T2), and a sedimentation tank (SED). The connection from SED to PRE illustrates, that a part of the so-called activated sludge from the sedimentation tank is pumped back to the pretreatment tank. This "feedback" concerns only the sludge and not the actual wastewater.

The aeration tanks alternate between aerobe and anoxic conditions. The biological removal of nitrogen takes place during the following two alternating processes (or phases): Nitrification during aerobic conditions and denitrification during anoxic conditions. During the nitrification process, ammonia, when supplying oxygen, is transformed into nitrate, which again, during the denitrification process, is transformed into free nitrogen, carbon dioxide, and water.

Today the wastewater plant is controlled by using a fixed length of the phase both for the nitrification process and for the denitrification process. This is far from optimal, since the length of the phase is roughly estimated in such a way that the process, independently of the nitrogen concentration of the wastewater, with a large probability "runs out." Furthermore, it is not optimal to use the actual measurements to cut off the phase, due to a large time delay from the time of sampling to the time where the process can be alternated.

The purpose of this exercise is to establish a model for the variations of nitrate and to use this model to formulate a more optimal control of the shift from anoxic conditions to aerobic conditions.

During a period of 24 hours, values of the nitrate concentration, Y_t, are sampled every ten minutes, and at the same time it is recorded, whether oxygen has been supplied. Let X_t take the value 1 during supply of oxygen, and otherwise set X_t equal to 0. In the following, one time unit corresponds to 10 minutes.

We want to identify a model of the form

$$(Y_t - \mu_0) = \frac{\omega(\mathrm{B})}{\delta(\mathrm{B})} X_{t-b} + N_t \tag{12.6}$$

where $\{N_t\}$ is a sequence of residuals and μ_0 denotes the mean value of Y_t, when no oxygen has been supplied for a long period of time.

Question 1 Explain why the usual procedure of pre-whitening is not well suited for the present case.

A minor part of the collected measurements is the following.

t	84	85	86	87	88	89	90	91	92
X_t	0	0	0	0	1	1	1	1	1
Y_t	3.52	3.41	3.56	3.98	3.58	4.81	6.58	7.60	8.73
t	93	94	95	96	97	98	99	100	101
X_t	1	1	1	1	1	1	1	1	1
Y_t	9.43	9.90	10.76	11.35	11.89	12.10	12.20	12.61	12.79

Question 2 Use the measurements shown to formulate a model of the form (12.6). State the reasons for your choice of model.

Question 3 Estimate the parameters in the transfer function component.

Based on all the measurements, the parameters in the model (12.6) are estimated. For the sequence of residuals, $\{\widehat{N}_t\}$, an analysis is carried out. The results are shown in Table 12.11.

Question 4 Formulate a reasonable model for N_t.

Question 5 Estimate the parameters in the formulated model for N_t.

A closer study of the measuring procedure shows that the nitrate censor is used for collecting samples at 4 different locations during a cycle of 8 measurements, with every second sample taken from T2. The time lag between two subsequent samples from T2 is still equal to 10 minutes. The remaining four measurements are taken in succession from the incoming stream, the PRE tank, the SED tank, and in a standard solution used for calibrating the censor.

This measuring procedure implies some carry over from measurement to measurement. After some studies an adviser in time series analysis finds that

Table 12.11: *Results of analysis of the sequence of residuals,* $\{\widehat{N}_t\}$.

Lag k	$\widehat{\rho}(k)$	$\widehat{\gamma}(k)$	$\widehat{\phi}_{kk}$
1	$1{,}000\,000 \cdot 10^{+00}$	$2{,}258\,578 \cdot 10^{-02}$	1,000 000 000
2	$-1{,}032\,155 \cdot 10^{-01}$	$-2{,}331\,202 \cdot 10^{-03}$	−0,103 215 471
3	$8{,}499\,245 \cdot 10^{-02}$	$1{,}919\,621 \cdot 10^{-03}$	0,075 139 515
4	$-5{,}727\,757 \cdot 10^{-02}$	$-1{,}293\,659 \cdot 10^{-03}$	−0,042 092 182
5	$6{,}739\,989 \cdot 10^{-01}$	$1{,}522\,279 \cdot 10^{-02}$	0,670 377 851
6	$-4{,}894\,063 \cdot 10^{-02}$	$-1{,}105\,363 \cdot 10^{-03}$	0,116 708 972
7	$6{,}289\,380 \cdot 10^{-02}$	$1{,}420\,506 \cdot 10^{-03}$	−0,005 449 079
8	$-6{,}347\,424 \cdot 10^{-02}$	$-1{,}433\,615 \cdot 10^{-03}$	−0,068 408 191
9	$4{,}764\,538 \cdot 10^{-01}$	$1{,}076\,108 \cdot 10^{-02}$	0,022 054 018
10	$-7{,}386\,434 \cdot 10^{-02}$	$-1{,}668\,284 \cdot 10^{-03}$	−0,089 624 122
11	$3{,}381\,935 \cdot 10^{-02}$	$7{,}638\,365 \cdot 10^{-04}$	−0,049 508 061
12	$-1{,}338\,409 \cdot 10^{-01}$	$-3{,}022\,902 \cdot 10^{-03}$	−0,148 202 196
13	$3{,}728\,714 \cdot 10^{-01}$	$8{,}421\,592 \cdot 10^{-03}$	0,044 511 799
14	$-1{,}042\,808 \cdot 10^{-01}$	$-2{,}355\,264 \cdot 10^{-03}$	−0,025 525 160
15	$8{,}665\,514 \cdot 10^{-04}$	$1{,}957\,174 \cdot 10^{-05}$	−0,022 004 291
16	$-1{,}606\,558 \cdot 10^{-01}$	$-3{,}628\,537 \cdot 10^{-03}$	−0,024 482 219
17	$2{,}760\,227 \cdot 10^{-01}$	$6{,}234\,189 \cdot 10^{-03}$	0,005 985 899
18	$-1{,}519\,213 \cdot 10^{-01}$	$-3{,}431\,261 \cdot 10^{-03}$	−0,085 066 617
19	$5{,}714\,627 \cdot 10^{-05}$	$1{,}290\,693 \cdot 10^{-06}$	0,011 653 523
20	$-1{,}162\,849 \cdot 10^{-01}$	$-2{,}626\,385 \cdot 10^{-03}$	0,090 905 897
21	$1{,}588\,861 \cdot 10^{-01}$	$3{,}588\,568 \cdot 10^{-03}$	−0,056 183 327

the variation in the measurements of nitrate in T2 (Y_t) can be described by the following model:

$$
\begin{aligned}
Y_t &= m_t + \nu_t \\
(1+\phi \mathrm{B})m_t &= \omega_o \mathrm{B} X_t + d + \varepsilon_t\,, \quad \varepsilon_t \in N(0,\sigma_\varepsilon^2) \\
(1+\psi \mathrm{B})\nu_t &= s_i + \zeta_{i,t}\,, \quad \zeta_{i,t} \in N(0,\sigma_{\zeta,i}^2) \\
i &= (t \text{ modulus } 4) + 1
\end{aligned}
$$

where m_t is the true concentration of nitrate in the tank T2. The quantity d corrects for the mean value of m_t. $\{\zeta_{1,t}\}$, $\{\zeta_{2,t}\}$, $\{\zeta_{3,t}\}$, $\{\zeta_{4,t}\}$, and $\{\varepsilon_t\}$ are sequences of mutually uncorrelated white noise. $s_i, i = 1, \cdots, 4$, are offset values, due to the fact that the measurement Y_t depends on from which location the previous sample in the measurement cycle was taken. All the parameters are assumed known.

Question 6 Write the above model in state space form.

Question 7 Describe how the true concentration of nitrate in T2 can be estimated.

In the following, take $\phi = -0.85$, $\omega_0 = 2.1\,\text{mg/l}$, and $d = 0.36\,\text{mg/l}$.

Question 8 Find the stationary mean value of the true concentration of nitrate in T2 during the nitrification phase and the denitrification phase.

Let us assume, that an optimal control of the wastewater plant is obtained if a shift from denitrification to nitrification occurs when the concentration of nitrate becomes less than $5\,\text{mg/l}$.

Also assume that the methods outlined in the solution to question 7 give an estimated true nitrate concentration in T2 equal to $\widehat{m}_{t|t} = 8.3\,\text{mg/l}$ during a denitrification phase.

Question 9 Given that an optimal control is wanted, calculate the points in time where it is expected that a shift from denitrification to nitrification should take place.

12.10 Sales numbers

Throughout a period of 150 days, where no unusual events have been observed, the actual number of sold copies of a certain newspaper The News has been registered on a daily basis. With this registration as the starting point an analysis is carried out and the results of this analysis are shown in Tables 12.12 and 12.13. The average sales number for the period was 28,600.

Question 1 On the basis of the provided results please argue that the model

$$\nabla Y_t = (1 + \theta B)\varepsilon_t$$

where $\{\varepsilon_t\}$ is white noise is a suitable model for the daily number of sold copies of The News.

Question 2 Please estimate the parameters in the above mentioned model.

The latest sales numbers are as follows.

Day No.	142	143	144	145	146
Sales number	28527	28306	28508	28618	28347

Question 3 Please state the predictions of the sales numbers for day 147 and day 148. Furthermore, please provide the corresponding uncertainties.

Now, a rival newspaper The Gossip goes bankrupt and closes down on day 147. The sales numbers for The News develop in the period for the closing down of The Gossip as follows.

Day No.	147	148	149	150	151	152	153
Sales number	29408	30034	30775	31400	31916	32360	32688

Table 12.12: *Original sales numbers.*

Lag k	$\widehat{\gamma}(k)$	$\widehat{\rho}(k)$	$\widehat{\phi}_{kk}$
1	534480.2	1.0000000	1.000000000
2	505630.8	0.9460235	0.946023583
3	490598.1	0.9178977	0.218367308
4	475583.5	0.8898057	0.044907372
5	464613.6	0.8692812	0.076172344
6	457213.1	0.8554350	0.097488314
7	442275.8	0.8274877	−0.099867120
8	429304.6	0.8032188	−0.021898558
9	415725.2	0.7778122	−0.015334969
10	408466.8	0.7642319	0.090475991
11	399977.3	0.7483482	0.008081225
12	391122.1	0.7317804	0.006759814
13	381890.9	0.7145090	0.001274695
14	372147.8	0.6962798	−0.004595928
15	360221.2	0.6739655	−0.078793243
16	352248.9	0.6590495	0.038385969

Table 12.13: *Differenced sales numbers.*

Lag k	$\widehat{\gamma}(k)$	$\widehat{\rho}(k)$	$\widehat{\phi}_{kk}$
1	44600.9258	1.000000000	1.000000000
2	−15834.5977	−0.355028450	−0.355028450
3	−2100.1267	−0.047087066	−0.198102057
4	−3954.8474	−0.088671871	−0.213221252
5	−851.2113	−0.019085059	−0.186961174
6	5306.2202	0.118971080	−0.002124049
7	−3882.7693	−0.087055802	−0.089882560
8	4201.1274	0.094193727	0.043004863
9	−6954.8032	−0.155934051	−0.125857756
10	2733.2632	0.061282657	−0.052672770
11	−241.4575	−0.005413732	−0.057255361
12	2698.5959	0.060505379	0.027490426
13	−1645.4272	−0.036892220	−0.034031499
14	1949.5852	0.043711767	0.078628004
15	−2958.1316	−0.066324443	−0.039179947
16	−1076.9310	−0.024145935	−0.044417605

Question 4 Please formulate a suitable model for the sales numbers of The News. The model must include a description of how the sales numbers develop in connection with the closing down of The Gossip.

Question 5 Please state an estimate of the parameters in the above mentioned model.

Question 6 Please state an estimate of the long term increased sales numbers of The News originating from the closing down of The Gossip.

12.11 Modeling and prediction of stock prices

A model of the variation of the stock price for stocks in a larger Danish company is wanted. Sometimes, however, the stock price is missing, e.g., due to a too low turnover of the stocks. Hence, a model of the following form is wanted.

$$\begin{pmatrix} X_{1,t} \\ X_{2,t} \\ \vdots \\ X_{m-1,t} \\ X_{m,t} \end{pmatrix} = \begin{pmatrix} -\phi_1 & 1 & 0 & \cdots & 0 \\ -\phi_2 & 0 & 1 & \cdots & 0 \\ \vdots & \vdots & \vdots & \ddots & \vdots \\ -\phi_{m-1} & 0 & 0 & \cdots & 1 \\ -\phi_m & 0 & 0 & \cdots & 0 \end{pmatrix} \begin{pmatrix} X_{1,t-1} \\ X_{2,t-1} \\ \vdots \\ X_{m-1,t-1} \\ X_{m,t-1} \end{pmatrix} + \begin{pmatrix} 1 \\ 0 \\ \vdots \\ 0 \\ 0 \end{pmatrix} \varepsilon_t \qquad (12.7)$$

where $\{\varepsilon_t\}$ is white noise with variance σ_ε^2 and $X_{1,t}$ is the stock price. The stocks are registered on a weekly basis, and therefore, one time interval in this problem corresponds to one week.

Question 1 Find the ARMA(p, q) model which is able to describe the variations of the stock price.

Based on 150 weekly registered stock prices, a preliminary analysis is done, and the results are given in Table 12.14. The average stock price was 220.0.

Question 2 Use the given information to formulate a reasonable model for the stock price in the class of models given by (12.7). Estimate the parameters of the model.

During a longer period the stock price has been registered every week, and the most recent registrations of the stock prices are the following.

Week No.	186	187	188	189
Price	220.3	221.8	220.7	219.9

Question 3 Predict the stock price for week No. 190, 191, and 192. Find the corresponding variances.

Question 4 Find the covariance between the 1-step prediction error and the 2-step prediction error.

Table 12.14: *Preliminary analysis of the weekly stock prices.*

Lag k	$\widehat{\gamma}(k)$	$\widehat{\rho}(k)$	$\widehat{\phi}_{kk}$
1	9.122377396	1.0000000000	1.000000000
2	8.336626053	0.9138655066	0.913865507
3	7.005969048	0.7679981589	−0.407352269
4	5.705332756	0.6254217029	0.072300062
5	4.642655849	0.5089304447	0.031614624
6	3.748092413	0.4108679295	−0.050892115
7	2.838299274	0.3111359179	−0.118939050
8	2.120466471	0.2324467003	0.129603818
9	1.583324432	0.1735648960	−0.040883355
10	1.177471876	0.1290751100	−0.012481641
11	0.874695539	0.0958846062	0.020636933
12	0.583921790	0.0640098229	−0.038749840
13	0.346806437	0.0380171128	−0.001615727
14	0.085030705	0.0093211122	−0.060234960
15	−0.317513943	−0.0348060504	−0.119047306
16	−0.743235350	−0.0814738646	0.014269204
17	−1.065813065	−0.1168350130	0.027178686
18	−1.028637171	−0.1127597690	0.162397087
19	−0.725705326	−0.0795522183	0.026990114
20	−0.319356680	−0.0350080542	0.048596863
21	0.004159517	0.0004559685	−0.063043706

It now turns out that the registration of the stock price is missing in week No. 190. For week No. 191 the registered price is 217.7.

Question 5 Use all the available observations to predict the stock price in week No. 192.

Question 6 Find the variance belonging to the prediction in Question 5.

12.12 Adaptive modeling of interest rates

Time series methods for analyzing the monthly observations Y_t of the interest rates R_t for a period from 1952 to 1982 have indicated that the observations can be described by the following model

$$\nabla Y_t = (1 - 0.678B)\varepsilon_t$$

where $\{\varepsilon_t\}$ is white noise with variance $\sigma_\varepsilon^2 = 14.5$.

Question 1 Is the process $\{Y_t\}$ stationary?

State whether $\{Y_t\}$ belongs to the class of ARIMA(p, d, q) processes, and if the answer is positive, which model?

Question 2 Find the autocorrelation function for ∇Y_t.

Instead, the interest rates are now described by the following model:

$$R_t = R_{t-1} + v_t$$

where $\{v_t\}$ is white noise with variance σ_v^2. The observations of the interest rates is

$$Y_t = R_t + e_t$$

where $\{e_t\}$ is white noise with σ_e^2. Finally, $\{e_t\}$ and $\{v_t\}$ are mutually independent.

Question 3 Find the parameters of the above formulated state space model.

Question 4 Provide a method for on-line (recursive) estimation of the interest rate.

It is now assumed that this procedure has been used for a long time.

Question 5 Show that the on-line estimate of the interest rate at time t can be found as

$$\widehat{R}_{t|t-1} = \frac{K}{1 - (1 - K)\,\mathrm{B}} Y_t$$

State a method for estimating the constant K.

APPENDIX A

The solution to difference equations

A set of equations on the form

$$X_{t+p} + \alpha_1 X_{t+p-1} + \cdots + \alpha_p X_t = d_t\,, \quad t = t_0, t_0+1, \ldots \tag{A.1}$$

where $\alpha_1, \ldots, \alpha_p$ and $\{d_t\}$ are given, is called an *inhomogeneous linear difference equation* of order p ($\alpha_p \neq 0$). If $d_t = 0$, we have the *homogeneous difference equation*:

$$X_{t+p} + \alpha_1 X_{t+p-1} + \cdots + \alpha_p X_t = 0\,, \quad t = t_0, t_0+1, \ldots \tag{A.2}$$

If we make a guess on the *solution* on the form

$$X_t = A\lambda^t\,, \quad t = t_0, t_0+1, \ldots \tag{A.3}$$

and substite into (A.2) we get

$$A\lambda^t\left(\lambda^p + \alpha_1\lambda^{p-1} + \cdots + \alpha_p\right) = 0\,, \quad t = t_0, t_0+1, \ldots$$

Since (in general) $A\lambda^t \neq 0$, it is seen that $\{X_t\}$, defined by (A.3), satisfies (A.2) if

$$\lambda^p + \alpha_1\lambda^{p-1} + \cdots + \alpha_p = 0 \tag{A.4}$$

The equation (A.4) is referred to as the *characteristic equation* for the difference equation (A.2).

The complete solution to the homogeneous difference equation (A.2) is given by

$$X_t = A_1\psi_1 + \cdots + A_p\psi_p \tag{A.5}$$

where the p linear independent functions $\psi_1, \ldots, \psi_p$ are determined by the fact that for each m-double root λ in the characteristic equation (A.4), we have m independent linear functions

$$\lambda^t, t\lambda^t, t^2\lambda^t, \ldots, t^{m-1}\lambda^t\,, \quad t = t_0, t_0+1, \ldots \tag{A.6}$$

The complete solution to the inhomogeneous difference equations (A.1) is found by adding an arbitrary solution to the inhomogeneous equation to the complete solution to the corresponding homogeneous difference equation.

▸ **Remark A.1**

If we have complex roots in the solution to the characteristic equation, then these will appear as complex conjugated, i.e., as $a \pm ib$. This can also be written as

$$re^{\pm i\theta} \quad \text{where} \quad r = \sqrt{a^2+b^2}, \quad \theta = \arg\{a+ib\}$$

The corresponding independent linear functions become

$$\left[\left(re^{i\theta}\right)^t, \left(re^{-i\theta}\right)^t\right] = \left(r^t(\cos(\theta t) + i\sin(\theta t)), r^t(\cos(\theta t) - i\sin(\theta t))\right)$$

where for $r < 1$ we get a damped harmonic. ◂

Example A.1

Find the solution to

$$X_t - 2.9X_{t-1} + 2.8X_{t-2} - 0.9X_{t-3} = 0\,, \quad t = t_0, t_0+1, \ldots \tag{A.7}$$

The characteristic equation is

$$\lambda^3 - 2.9\lambda^2 + 2.8\lambda - 0.9 = 0$$

with roots $\lambda_1 = 0.9$, $\lambda_2 = \lambda_3 = 1$ (one double root). The complete solution to (A.7) becomes

$$\begin{aligned} X_t &= A_1(0.9)^t + A_2 1^t + A_3 t 1^t \\ &= A_1(0.9)^t + A_2 + A_3 t \end{aligned} \tag{A.8}$$

If the initial conditions $X_{t_0}, X_{t_0+1}, X_{t_0+2}$ are given they will uniquely determine the values of A_1, A_2 and A_3.

It is seen that if all the roots of the characteristic equation are inside the unit circle, then the influence from the initial values will vanish and $\{X_t\} \to 0$ for $t \to \infty$. But if just one single root is outside the unit circle, then $\{X_t\} \to \infty$ for $t \to \infty$.

APPENDIX B

Partial autocorrelations

Let $\{X_t\,; t \in \mathbb{Z}\}$ be a stationary stochastic process and let $\{\gamma_k\,; k \in \mathbb{Z}\}$ and $\{\rho_k\,; k \in \mathbb{Z}\}$ denote the autocovariance and autocorrelation functions.

Based on this we introduce the covariance matrix $\mathbf{\Gamma}_k$ and correlation matrix $\boldsymbol{\rho}_k$:[1]

$$\begin{aligned}\mathbf{\Gamma}_k &= \begin{pmatrix} \gamma_0 & \gamma_1 & \cdots & \gamma_{k-1} \\ \gamma_1 & \gamma_0 & \cdots & \gamma_{k-2} \\ \vdots & \vdots & & \vdots \\ \gamma_{k-1} & \gamma_{k-2} & \cdots & \gamma_0 \end{pmatrix} \\ &= \gamma_0 \begin{pmatrix} 1 & \rho_1 & \cdots & \rho_{k-1} \\ \rho_1 & 1 & \cdots & \rho_{k-2} \\ \vdots & \vdots & & \vdots \\ \rho_{k-1} & \rho_{k-2} & \cdots & 1 \end{pmatrix} = \gamma_0 \boldsymbol{\rho}_k \end{aligned} \tag{B.1}$$

Furthermore, we introduce

$$\boldsymbol{\gamma}_k = (\gamma_1, \ldots, \gamma_k)\,, \qquad \widetilde{\boldsymbol{\gamma}}_k = (\gamma_k, \ldots, \gamma_1) \tag{B.2}$$

$$\boldsymbol{\rho}_k = (\rho_1, \ldots, \rho_k)\,, \qquad \widetilde{\boldsymbol{\rho}}_k = (\rho_k, \ldots, \rho_1) \tag{B.3}$$

We *define the partial autocorrelation* $\phi_{k,k}$ as

$$\phi_{k,k} = \text{Cor}\left[X_t, X_{t+k} | X_{t+1}, \ldots, X_{t+k-1}\right] \tag{B.4}$$

i.e., the partial autocorrelation $\phi_{k,k}$ is calculated in a similar way to the partial correlation coefficient (see Johnson and Wichern (2002, p. 406)).

The variance of $(X_t, \ldots, X_{t+k})$ is

$$\begin{aligned}\text{Var}[X_t, \ldots, X_{t+k}] &= \text{Var}[X_t, X_{t+k}, X_{t+1}, \ldots, X_{t+k-1}] \\ &= \left(\begin{array}{cc|c} \gamma_0 & \gamma_k & \boldsymbol{\gamma}_{k-1} \\ \gamma_k & \gamma_0 & \widetilde{\boldsymbol{\gamma}}_{k-1} \\ \hline \boldsymbol{\gamma}_{k-1}^T & \widetilde{\boldsymbol{\gamma}}_{k-1}^T & \mathbf{\Gamma}_{k-1} \end{array}\right)\end{aligned} \tag{B.5}$$

[1]Please note that in order to shorten the notation, we use a slightly different notation for the autocovariance, the autocorrelation, and the partial autocorrelation.

The conditional variance of (X_t, X_{t+k}) given $(X_{t+1}, \ldots, X_{t+k-1})$ becomes

$$\begin{aligned} \operatorname{Var}[X_t, X_{t+k}|X_{t+1}, \ldots, X_{t+k-1}] = & \begin{pmatrix} \gamma_0 & \gamma_k \\ \gamma_k & \gamma_0 \end{pmatrix} \\ & - \begin{pmatrix} \gamma_{k-1} \\ \widetilde{\gamma}_{k-1} \end{pmatrix} \mathbf{\Gamma}_{k-1}^{-1} \begin{pmatrix} \gamma_{k-1}^T & \widetilde{\gamma}_{k-1}^T \end{pmatrix} \end{aligned} \tag{B.6}$$

If we introduce $\boldsymbol{X} = (X_{t+1}, \ldots, X_{t+k-1})$, we get:

$$\begin{aligned} \operatorname{Cov}[X_t, X_{t+k}|X] &= \gamma_k - \widetilde{\gamma}_{k-1}\mathbf{\Gamma}_{k-1}^{-1}\gamma_{k-1}^T \\ \operatorname{Var}(X_t|X) &= \gamma_0 - \gamma_{k-1}\mathbf{\Gamma}_{k-1}^{-1}\gamma_{k-1}^T \\ \operatorname{Var}(X_{t+k}|X) &= \gamma_0 - \widetilde{\gamma}_{k-1}\mathbf{\Gamma}_{k-1}^{-1}\widetilde{\gamma}_{k-1}^T \end{aligned}$$

Since

$$\gamma_{k-1}\mathbf{\Gamma}_{k-1}^{-1}\gamma_{k-1}^T = \widetilde{\gamma}_{k-1}\mathbf{\Gamma}_{k-1}^{-1}\widetilde{\gamma}_{k-1}^T$$

we get

$$\begin{aligned} \phi_{k,k} &= \frac{\gamma_k - \widetilde{\gamma}_{k-1}\mathbf{\Gamma}_{k-1}^{-1}\gamma_{k-1}^T}{\gamma_0 - \widetilde{\gamma}_{k-1}\mathbf{\Gamma}_{k-1}^{-1}\widetilde{\gamma}_{k-1}^T} = \frac{\rho_k - \widetilde{\boldsymbol{\rho}}_{k-1}\boldsymbol{P}_{k-1}^{-1}\boldsymbol{\rho}_{k-1}^T}{1 - \widetilde{\boldsymbol{\rho}}_{k-1}\boldsymbol{P}_{k-1}^{-1}\widetilde{\boldsymbol{\rho}}_{k-1}^T} \\ &= \frac{|\boldsymbol{P}_{k-1}||\rho_k - \widetilde{\boldsymbol{\rho}}_{k-1}\boldsymbol{P}_{k-1}^{-1}\boldsymbol{\rho}_{k-1}^T|}{|\boldsymbol{P}_{k-1}||1 - \widetilde{\boldsymbol{\rho}}_{k-1}\boldsymbol{P}_{k-1}^{-1}\widetilde{\boldsymbol{\rho}}_{k-1}^T|} \\ &= \frac{\begin{vmatrix} \boldsymbol{P}_{k-1} & \boldsymbol{\rho}_{k-1}^T \\ \widetilde{\boldsymbol{\rho}}_{k-1} & \rho_k \end{vmatrix}}{\begin{vmatrix} \boldsymbol{P}_{k-1} & \widetilde{\boldsymbol{\rho}}_{k-1}^T \\ \widetilde{\boldsymbol{\rho}}_{k-1} & 1 \end{vmatrix}} \qquad \text{(see, e.g., Johnson and Wichern 2002, p. 121)} \\ &= \frac{\begin{vmatrix} 1 & \rho_1 & \cdots & \rho_{k-2} & \rho_1 \\ \rho_1 & 1 & \cdots & \rho_{k-3} & \rho_2 \\ \vdots & \vdots & & \vdots & \\ \rho_{k-1} & \rho_{k-2} & \cdots & \rho_1 & \rho_k \end{vmatrix}}{\begin{vmatrix} 1 & \rho_1 & \cdots & \rho_{k-2} & \rho_{k-1} \\ \rho_1 & 1 & \cdots & \rho_{k-3} & \rho_{k-2} \\ \vdots & \vdots & & \vdots & \\ \rho_{k-1} & \rho_{k-2} & \cdots & \rho_1 & 1 \end{vmatrix}} \end{aligned} \tag{B.7}$$

If we have the set of equations

$$\boldsymbol{A}\boldsymbol{x} = \boldsymbol{b}$$

we can determine x_i (the i'th element in $\boldsymbol{x}$) as

$$x_i = \frac{\det \boldsymbol{B}_i}{\det \boldsymbol{A}}$$

where $\boldsymbol{B}_i$ is obtained by substituting the i'th column in $\boldsymbol{A}$ with $\boldsymbol{B}$.

Thus, it is seen from (B.7) that $\phi_{k,k}$ is determined from the equation set:

$$\boldsymbol{P}_k \begin{pmatrix} \phi_{k,1} \\ \vdots \\ \phi_{k,k} \end{pmatrix} = \begin{pmatrix} \rho_1 \\ \vdots \\ \rho_k \end{pmatrix} \tag{B.8}$$

which is seen to be the Yule-Walker equations for an AR(k) process, i.e., $\phi_{k,k}$ can be found as the moment estimate of the last coefficient in an AR(k) model.

Based on (B.8) we can provide a recursive method to determine $\phi_{k,k}$. We introduce

$$\begin{aligned} \boldsymbol{\phi}_k &= (\phi_{k,1}, \ldots, \phi_{k,k}), & \widetilde{\boldsymbol{\phi}}_k &= (\phi_{k,k}, \ldots, \phi_{k,1}) \\ \boldsymbol{\phi}_{k,k-1} &= (\phi_{k,1}, \ldots, \phi_{k,k-1}), & \widetilde{\boldsymbol{\phi}}_{k,k-1} &= (\phi_{k,k-1}, \ldots, \phi_{k,1}) \end{aligned}$$

Writing (B.8) for $k+1$ yields

$$\begin{gathered}
\begin{pmatrix} \boldsymbol{P}_k & \widetilde{\boldsymbol{\rho}}_k^T \\ \boldsymbol{\rho}_k & 1 \end{pmatrix} \begin{pmatrix} \phi'_{k+1,k} \\ \phi_{k+1,k+1} \end{pmatrix} = \begin{pmatrix} \boldsymbol{\rho}_k^T \\ \rho_{k+1} \end{pmatrix} \Leftrightarrow \\
\left.\begin{aligned} \boldsymbol{P}_k \boldsymbol{\phi}_{k+1,k}^T + \widetilde{\boldsymbol{\rho}}_k^T \phi_{k+1,k+1} &= \boldsymbol{\rho}_k^T \\ \widetilde{\boldsymbol{\rho}}_k \boldsymbol{\phi}_{k+1,k}^T + \phi_{k+1,k+1} &= \rho_{k+1} \end{aligned}\right\} \Leftrightarrow \\
\left.\begin{aligned} \boldsymbol{\phi}_{k+1,k}^T &= \boldsymbol{P}_k^{-1}(\boldsymbol{\rho}_k^T - \widetilde{\boldsymbol{\rho}}_k^T \phi_{k+1,k+1}) \\ \widetilde{\boldsymbol{\rho}}_k \boldsymbol{P}_k^{-1}(\boldsymbol{\rho}_k^T - \widetilde{\boldsymbol{\rho}}_k^T \phi_{k+1,k+1}) + \phi_{k+1,k+1} &= \rho_{k+1} \end{aligned}\right\} \Leftrightarrow \\
\left.\begin{aligned} \boldsymbol{\phi}_{k+1,k}^T &= \boldsymbol{\phi}_k^T - \widetilde{\boldsymbol{\phi}}_k^T \phi_{k+1,k+1} \\ \phi_{k+1,k+1} &= \frac{\rho_{k+1} - \widetilde{\boldsymbol{\rho}}_k \boldsymbol{\phi}_k^T}{1 - \widetilde{\boldsymbol{\rho}}_k \boldsymbol{\phi}_k^T} \end{aligned}\right\}
\end{gathered}$$

We get the following recursion formulas for calculation of $\phi_{k,k}$ for $k = 1, 2, \ldots$:

$$\begin{aligned} \phi_{k+1,j} &= \phi_{k,j} - \phi_{k+1,k+1}\phi_{k,k+1-j}\,, \quad j = 1, \ldots, k \\ \phi_{k+1,k+1} &= \frac{\rho_{k+1} - \sum_{j=1}^{k} \phi_{k,j}\rho_{k+1-j}}{1 - \sum_{j=1}^{k} \phi_{k,j}\rho_j} \end{aligned}$$

with the initial value $\phi_{11} = \rho_1$.

APPENDIX C

Some results from trigonometry

The following *identity* holds:

$$\sum_{k=0}^{m} e^{-i\omega k} = \frac{1 - e^{-i\omega(m+1)}}{1 - e^{-i\omega}} \tag{C.1}$$

$$= e^{-i\omega m/2} \frac{e^{i\omega(m+1)/2} - e^{-i\omega(m+1)/2}}{e^{i\omega/2} - e^{-i\omega/2}} \tag{C.2}$$

The first equality is obtained since

$$\left(1 - e^{-i\omega}\right) \left\{ \sum_{k=0}^{m} e^{-i\omega k} \right\} = 1 - e^{-i\omega(m+1)}$$

i.e., the in-between terms vanish. The last equality is seen directly.

Correspondingly we have:

$$\sum_{k=0}^{m} e^{i\omega k} = e^{i\omega m/2} \frac{e^{i\omega(m+1)/2} - e^{-i\omega(m+1)/2}}{e^{i\omega/2} - e^{-i\omega/2}} \tag{C.3}$$

Applying the *Euler relation* yields

$$e^{i\omega} = \cos(\omega) + i \sin(\omega) \tag{C.4}$$

and the *inverse relations*:

$$\cos(\omega) = \frac{1}{2} \left\{ e^{i\omega} + e^{-i\omega} \right\}, \qquad \sin(\omega) = \frac{1}{2i} \left\{ e^{i\omega} - e^{-i\omega} \right\} \tag{C.5}$$

yield

$$\sum_{k=0}^{m} \cos(\omega k) = \cos(\omega m/2) \frac{\sin(\omega(m+1)/2)}{\sin(\omega/2)} \tag{C.6}$$

$$\sum_{k=0}^{m} \sin(\omega k) = \sin(\omega m/2) \frac{\sin(\omega(m+1)/2)}{\sin(\omega/2)} \tag{C.7}$$

For ω equal to integer values times 2π, we have:

$$\sum_{k=0}^{m} \cos(\omega k) = m + 1\,, \quad \text{for } \omega = 2p\pi,\ p \in Z, \tag{C.8}$$

$$\sum_{k=0}^{m} \sin(\omega k) = 0\,, \quad \text{for } \omega = 2p\pi,\ p \in Z. \tag{C.9}$$

From (C.1) and (C.2) as well as the Euler relations, it follows that

$$\begin{aligned} \sum_{k=-m}^{m} \mathrm{e}^{-i\omega k} &= \left[\mathrm{e}^{i\omega m/2} + \mathrm{e}^{-i\omega m/2}\right]\left[\frac{\sin(\omega(m+1)/2)}{\sin(\omega/2)}\right] - 1 \\ &= \frac{\sin\big((m+\frac{1}{2})\omega\big)}{\sin(\omega/2)} = D_m(\omega)2\pi \end{aligned} \tag{C.10}$$

The function $D_m(\omega)$ is called the *Dirichlet kernel of order m*.

The above mentioned relations are commonly applied in spectral analysis together with the well-known trigonometry formulas for addition.

APPENDIX D

List of acronyms

The number following the acronym marks the page where the acronym is first described.

ACF AutoCorrelation Function, *103*

AIC Akaike's Information Criterion, *174*

AR AutoRegressive, *6*

ARI Integrated AutoRegressive, *131*

ARIMA Integrated AutoRegressive-MovingAverage, *9*

ARMA AutoRegressive-Moving-Average, *9*

ARMAX AutoRegressiveMoving Average with eXogenous input, *223*

ARX AutoRegressive with eXogenous input, *223*

BIC Bayesian Information Criteria, *174*

BLUE Best Linear Unbiased Estimator, *37*

CARMA Controlled AutoRegressiveMovingAverage, *223*

CCF Cross Correlation Function, *104*

CLS Conditioned Least Squares Method, *163*

ELS Extended Least Squares, *321*

FIR Finite Impulse Response, *223*

GLM General Linear Model, *33*

IACF Inverse AutoCorrelation Function, *129*

IMA Integrated Moving Average, *131*

LS Least Squares, *34*

MA Moving Average, *117*

MARIMA Multivariate AutoRegressive Integrated Moving Average, *249*

MIMO Multiple-Input, Multiple-Output, *215*

MISO Multiple-Input, Single-Output, *215*

ML Maximum Likelihood, *40*

MLE Maximum Likelihood Estimate, *40*

OE Output Error, *223*

OEM Output Error Method, *223*

OLS Ordinary Least Squares, *35*

PACF Partial AutoCorrelation Function, *124*

RLS Recursive Least Squares, *315*

RML Recursive Maximum Likelihood, *324*

RPEM Recursive Prediction Error Method, *321*

RPLR Recursive Pseudo Linear Regression, *319*

SBC Schwartz's Bayesian Criterion, *174*

SISO Single-Input, Single-Output, *215*

SSE Sum of Squared Errors, *34*

VAR Vector AutoRegressive, *260*

VARMA Vector AutoRegressive-Moving Average, *249*

VARIMA Vector AutoRegressive Integrated Moving Average, *249*

WLS Weighted Least Squares, *35*

APPENDIX E

List of symbols

The number following the symbol description marks the page where the symbol is first mentioned.

$\boldsymbol{X}$	n-variate random variable, 13
$F_{\boldsymbol{X}}(x)$	Joint distribution function of $\boldsymbol{X}$, 13
$\mathrm{P}\{X_1 \le x_1\}$	Probability of $X_1 \le x_1$, 13
$f_{\boldsymbol{X}}(x)$	Joint density function of $\boldsymbol{X}$, 13
$f_{\boldsymbol{S}}(\boldsymbol{x})$	Marginal density function, 14
$F_{\boldsymbol{S}}(\boldsymbol{x})$	Marginal distribution function, 14
$f_{\boldsymbol{X},\boldsymbol{Y}(\boldsymbol{x},\boldsymbol{y})}$	Joint density function of $\boldsymbol{X}$ and $\boldsymbol{Y}$, 14
$f_{\boldsymbol{Y}\mid\boldsymbol{X}=\boldsymbol{x}}$	Conditional density function for $\boldsymbol{Y}$ given $\boldsymbol{X} = \boldsymbol{x}$, 14
$\boldsymbol{\mu}, \mathrm{E}[\boldsymbol{X}]$	Expectation or mean value or first moment of $\boldsymbol{X}$, 15
$\sigma_{x_i}^2, \mathrm{Var}[X_i]$	Variance or second central moment of X_i, 16
$\sigma_{ij}, \mathrm{Cov}[X_i, X_j]$	Covariance between X_i and X_j, 17
$\boldsymbol{\Sigma}_{\boldsymbol{X}}, \mathrm{Var}[\boldsymbol{X}]$	Covariance matrix of $\boldsymbol{X}$, 18
ρ_{ij}	Correlation between X_i and X_j, 18
$\boldsymbol{R}$	Correlation matrix, 18
$\mathrm{E}[\boldsymbol{Y} \mid \boldsymbol{X} = \boldsymbol{x}]$	Conditional expectation of $\boldsymbol{Y}$ given $\boldsymbol{X}$, 20
$\mathrm{Var}[\boldsymbol{Y} \mid \boldsymbol{X} = \boldsymbol{x}]$	Conditional variance of $\boldsymbol{Y}$ given $\boldsymbol{X}$, 21
$N(\mu, \sigma^2)$	Normal distribution with mean μ and variance σ^2, 22
$\chi^2(n)$	Chi-squared distribution with n degrees of freedom, 23
$\boldsymbol{X}^-$	Generalized inverse (g-inverse) of $\boldsymbol{X}$, 24
$\boldsymbol{\theta}$	Parameter vector, 31
$\widehat{\boldsymbol{\theta}}$	Estimator of parameter vector, 34
L	Likelihood function, 40
i.i.d	Independent identically distributed, 48
ε_t	White noise process, 48
λ	Forgetting factor, 50
c	Normalizing constant, 50
$\boldsymbol{L}$	Transition matrix, 53
T_t	Trend, 60
S_t	Seasonal effect, 60
δ_k	Kronecker's delta sequence, impulse function, 60
$y_t, y(t)$	Signal in time domain, 69

$*$	Convolution operator, 70
$h(k), h(u)$	Impulse response function, 70
$\delta(t)$	Dirac delta function, impulse function, 70
$S_k, S(t)$	Step response function, 72
$Y(\omega)$	Signal in frequency domain, 73
$\mathcal{H}(\omega)$	Frequency response function, 73
B, z^{-1}	Backward shift operator, 82
F, z	Forward shift operator, 82
$H(z)$	Transfer function, 83
∇	Difference operator, 87
S	Summation operator, 87
$H(s)$	Transfer function, 92
$\gamma_{XX}, \gamma_{XX}(k)$	Autocovariance function, 99
$\rho(k)$, ACF	Autocorrelation function, 103
$f(\omega)$	Spectrum, power spectrum, 114
ϕ_{kk}, PACF	Partial autocorrelation function, 124
$\rho i(k)$, IACF	Inverse autocorrelation function, 129
∇_s	s-season difference operator, 132
$\widehat{Y}_{t+k\|t}$	k-step predictor, 138
$\boldsymbol{I}(\theta)$	Fisher information matrix, 168
$I(\omega)$	Periodogram, 187
$W(\theta)$	Spectral window, 195
$b(\omega)$	Skewness, 197
$\boldsymbol{\Gamma}_k$	Covariance matrix at lag k, 205
$c_{XY}(\omega)$	Co-spectrum, 206
$q_{XY}(\omega)$	Quadrature spectrum, 206
$\alpha_{XY}(\omega)$	Cross-amplitude spectrum, 206
$\phi_{XY}(\omega)$	Phase spectrum, 206
$w_{XY}(\omega)$	Complex coherency, 206
$G_{XY}(\omega)$	Gain spectrum, 207
$\boldsymbol{K}_t$	Kalman gain, 287
$\widehat{X}_{t\|t}$	Filter estimator, reconstruction of X_t, 290
$\boldsymbol{I}$	Identity matrix, 309

Bibliography

Abraham, B., and J. Ledolter (1983) *Statistical Methods for Forecasting.* New York: Wiley.

Andersen, K., H. Madsen, and L. Hansen (2000) "Modelling the heat dynamics of a building using stochastic differential equations." In: *Energy and Buildings* 31 13–34.

Åström, K. J. (1970) *Introduction to Stochastic Control Theory.* New York: Academic Press.

Åström, K. J., and P. Eykhoff (1971) "System identification – a survey." In: *Automatica* 7 123–162.

Baadsgaard, M., J. N. Nielsen, and H. Madsen (2000) "Estimating Multivariate Exponentail-Affine Term Structure Models from Coupon Bound Prices using Nonlinear Filtering." In: *Econometric Journal* 3 1–20.

Bard, Y. (1974) *Non-linear Parameter Estimation.* London: Academic Press.

Bartlett, M. S. (1946) "On the theoretical specification of sampling properties of an autocorrelated process." In: *Journal Royal Statistical Society B.* 8 27–41.

Bechmann, H., M. K. Nielsen, N. K. Poulsen, and H. M. Madsen (2002) "Grey-box modelling of aeration tank settling." In: *Water Research* 36 1887–1895.

Box, G. E. P., and D. R. Cox (1964) "An Analysis of Transformations." In: *Journal of Royal Statistical Society, Series B* 26 211–246.

Box, G. E. P., and G. M. Jenkins (1970/1976) *Time Series Analysis, Forecasting and Control.* San Francisco: Holden-Day.

Box, G. E. P., and G. C. Tiao (1975) "Intervention Analysis with Applications to Economic and Environmental Problems." In: *Journal of the American Statistical Association* 70 70–79.

Brillinger, D. R. (1981) *Time Series; Data Analysis and Theory.* San Francisco: Holden-Day.

Brockwell, P. J., and R. A. Davis (1987) *Time Series; Theory and Methods.* New York: Springer-Verlag.

Brown, R. G. (1963) *Smoothing, Forecasting and Prediction.* Englewood Cliffs, NJ: Prentice Hall.

Carstensen, N. J. (1990) *Adaptive Monitoring of Air Quality.* Lyngby, Denmark: IMM/DTU.

Chatfield, C. (2003) *The Analysis of Time Series: An Introduction.* 6th ed. New York: Chapman, Hall.

Christensen, R. (2002) *Plane Answers to Complex Questions; The Theory of Linear Models.* New York: Springer.

Conradsen, K. (1984) *En introduktion til Statistik, Bind 1.* 5th ed. Lyngby, Denmark: IMSOR.

Cox, D. R., and H. D. Miller (1968) *The Theory of Stochastic Processes.* London: Chapman, Hall.

Daniells, H. E. (1956) "The approximate distribution of serial correlation coefficients." In: *Biometrika* 43 169–185.

Davis, M. H. A., and R. B. Vinter (1985) *Stochastic Modelling and Control.* London: Chapman, Hall.

Doob, J. L. (1953) *Stochastic Processes.* New York: Wiley.

Dunsmuir, W., and P. M. Robinson (1981) "Estimation of time series models in the presence of missing data." In: *Journal American Statistical Association* 76 560–568.

ENFOR (2007) *Prediction related to energy systems.* 2007 URL: `http://www.enfor.eu`.

Engle, R., and B. S. You (1985) "Forecasting and testing in cointegrated systems." In: *Journal of Econometrics* 35 143–159.

Funch, B. O., and G. K. Hansen (1987) "Varianskomponenter i tidsrække- og interventionsanalyse." MA thesis DTU, Lyngby, Denmark: IMSOR.

Goodwin, C. G., and R. L. Payne (1977) *Dynamic System Identification: Experimental Design and Data Analysis.* New York: Academic Press.

Grimmit, G. R., and D. R. Stirzaker (1992) *Probability and Random Processes.* Oxford: Clarendon Press.

Haahr, A. M., H. Madsen, J. Smedegaard, W. L. P. Bredie, L. H. Stanhnke, and H. H. F. Refsgaard (2003) "Release Measurement by Atmospheric Pressure Chemical Ionization Ion Trap Mass Spectrometry, Construction of Interface and Mathematical Modeling of Release Profiles." In: *Analytical Chemistry* 75 655–662.

Hald, A. (1952) *Statistical Theory with Engineering Applications.* New York: Wiley.

Hannan, E. J. (1970) *Multiple Time Series.* New York: Wiley.

Harvey, A. C. (1996) *Forecasting, Structural Time Series Models and the Kalman Filter.* Cambridge: Cambridge University Press.

——— (1981) *Time Series Models.* New York: Halsted Press.

Harvey, A. C., and R. G. Pierse (1984) "Estimating missing observations in economic time series." In: *Journal of the American Statistical Association* 79 125–131.

Hastie, T., R. Tibshirani, and J. Friedman (2001) *The Elements of Statistical Learning; Data Mining, Inference and Prediction.* New York: Springer.

Holst, J. (1977) *Adaptive Prediction and Recursive Estimation.* Lund, Sweden: Department of Automatic Control, Lund Institute of Technology.

Holst, J., and N. K. Poulsen (1988) *Noter til Stokastisk Adaptiv Regulering.* Lyngby, Denmark: IMM, DTU.

Holt, C. C. (1957) Forecasting trends and seasonals by exponentially weighted moving averaverages 52 ONR Memorandum Pittsburgh: Carnegie Institute of Technology.

Huber, P. T. (1964) *Robust Statistics.* New York: Wiley.

Jazwinski, A. H. (1970) *Stochastic Processes and Filtering Theory.* New York: Academic Press.

Jenkins, G. M. (1979) *Practical Experiences with Modelling and Forecasting Time Series.* Jersey: Gwilym Jenkins, Partners Ltd.

——— (1975) "The interaction between the muskrat and mink cycles in North Canada." In: *8th International Biometric Conference* Constanta, Romania 55–71.

Jenkins, G. M., and A. Alavi (1981) "Some aspects of modelling and forecasting multivariate time series." In: *Journal of Time Series Analysis* 2 1–47.

Jenkins, G.M., and D. G. Watts (1968) *Spectral Analysis and Its Applications.* San Francisco: Holden-Day.

Jensen, H. E., T. Høholdt, and F. Nielsen (1992) *Lineære systemer.* Lyngby, Denmark: Matematisk Institut, DTU.

Johansen, S. (2001) "A Course in Time Series Analysis." In: New York: Wiley Chap. Cointegration in the VAR model.

Johansson, R. (1993) *System Modelling and Identification.* Englewood Cliffs, New Jersey: Prentice Hall.

Johnson, R. A., and D. W. Wichern (2002) *Applied Multivariate Statistical Analysis.* 5th ed. London: Prentice Hall.

Jónsdóttir, H., J. L. Jacobsen, and H. Madsen (2001) "A grey-box model describing the hydraulics in a creek." In: *Environmetrics* 37 347–356.

Kailath, T. (1980) *Linear Systems.* New York: Prentice Hall.

Kalman, R. E. (1960) "A new approach to linear filtering and prediction problems." In: *Trans. ASME Journal of Basic Engineering* 82 35–45.

Kendall, M. G., and A. Stuart (1983) *The Advanced Theory of Statistics.* 4th ed. London: Griffin.

Kristensen, N. R., H. Madsen, and S. B. Jørgensen (2004) "Parameter estimation in stochastic grey-box models." In: *Automatica* 40 225–237.

Ljung, L. (1976) *System Identification; Advances and Case Studies.* New York: Academic Press Chap. On the consistency of prediction error identification methods.

——— (1987) *System Identification: Theory for the User.* New York: Prentice Hall.

Ljung, L., and T. Söderström (1983) *Theory and Practice of Recursive Identification.* Cambridge MA: MIT Press.

Lütkepohl, H. (1991) *Introduction to Multiple Time Series Analysis.* Berlin: Springer.

Madsen, H. (1992) *Projection and Separation in Hilbert Spaces.* Lyngby, Denmark: IMM, DTU.

——— (1985) *Statistically Determined Dynamical Models for Climate Processes.* Lyngby, Denmark: IMSOR.

Madsen, H., and J. Holst (1995) "Estimation of continuous-time models for the heat dynamics of a building." In: *Energy and Buildings* 22 67–79.

——— (1989) "Identification of a time varying transfer function model for the variations of air temperature." In: *Fourth International Meeting on Statistical Climatology* Rotorua 243–247.

Madsen, H., J. Holst, and E. Lindström (2007) *Modelling Non-linear and Non-stationary Time Series.* Lyngby, Denmark: IMM, DTU.

Madsen, H., H. A. Nielsen, and T. S. Nielsen (2005) "A tool for predicting the wind power production of off-shore wind plants." In: *Proceedings of the Copenhagen Offshore Wind Conference & Exhibition* Copenhagen.

Madsen, H., H. Spliid, and P. Thyregod (1985) "Markov models in discrete and continuous time for hourly observations of cloud cover." In: *Journal of Climate and Applied Meteorology* 24 629–639.

Madsen, H., and P. Thyregod (1988) "Modelling the Time Correlation in Hourly Observations of Direct Radiation in Clear Skies." In: *Energy and Buildings* 11 201–211.

Madsen, H., P. Pinson, G. Kariniotaktis, H. A. Nielsen, and T. S. Nielsen (2005) "Standardizing the performance evaluation of short-term wind prediction models." In: *Wind Engineering* 29 475–489.

Madsen, H., J. N. Nielsen, E. Lindström, M. Baadsgaard, and J. Holst (2004) *Statistics in Finance.* Lund, Sweden: University of Lund, KF Sigma.

Makridakis, S., and S. Wheelwright (1978) *Interactive Forecasting.* San Francisco: Holden-Day.

Martin, R. D., and V. Yohai (1985) "Robustness in Time Series and Estimating ARMA Models." In: Vol. 5 Handbook of Statistics Amsterdam: Elsevier Science 119–155.

Maybeck, P. S. (1982) *Stochastic Models, Estimation and Control; Vol 1,2,3.* New York: Academic Press.

McLeod, A. I., and W. K. Li (1983) "Diagnostic checking ARMA time series models using squared-residual autocorrelations." In: *Journal of Time Series Analysis* 4 269–273.

Meinhold, R. J., and N. D. Singpurwalla (1983) "Understanding the Kalman filter." In: *American Statistician* 70 123–127.

Milhøj, A. (1986) *Tidsrækkeanalyse for Økonomer.* Copenhagen: Akademisk Forlag.

Nielsen, H., H. Madsen, and T. S. Nielsen (2006) "Using quantile regression to extend an existing wind power forecasting system with probabilistic forecasts." In: *Wind Energy* 9 95–108.

Nielsen, J. N., M. Vestergaard, and H. Madsen (2000) "Estimation in Continuous-time Stochastic Volatility Models Using Nonlinear Filters." In: *International Journal of Theoretical and Applied Finance* 3 279–308.

Nielsen, T. S., A. Joensen, H. Madsen, L. Landberg, and G. Giebel (1999) "A New Reference for Predicting Wind Power." In: *Wind Energy* 1 29–34.

Øksendal, B. (1995) *Stochastic Differential Equations.* 4th ed. Berlin: Springer.

Olbjer, L., U. Holst, and J. Holst (2005) *Tidsserieanalys.* Lund, Sweden: Centre for Mathematical Sciences, Lund University, KFS AB.

Olsson, U. (2002) *Generalized Linear Models.* Lund, Sweden: Studentlitteratur.

Palsson, O. P., H. Madsen, and H. T. Søgaard (1994) "Generalized Predictive Control for Non-Stationary System." In: *Automatica* 30 1991–1997.

Paltridge, G. W., and C. M. R. Platt (1976) *Radiative Processes in Meteorology and Climatology.* Amsterdam: Elsevier.

Papoulis, A. (1983) *Probability, Random Variables and Stochastic Processes.* New York: McGraw-Hill.

——— (1984) *Signal Analysis.* New York: McGraw-Hill.

Parzen, E. (1963) "On spectral analysis with missing observations and amplitude modulation." In: *Sankhya, Series A* 25 383–392.

——— (1962) *Stochastic Processes.* San Francisco: Holden-Day.

Pawitan, Y. (2001) *In All Likelihood, Statistical Modelling and Inference Using Likelihood.* New York: Oxford University Press.

Porsholt, T. (1989) *Tidsvarierende Økonomiske Modeller.* Lyngby, Denmark: IMM, DTU.

Priestley, M. B. (1981) *Spectral Analysis and Time Series.* London: Academic Press.

Rao, C. R. (1973) *Linear Statistical Inference and Its Applications.* 2nd ed. New York: Wiley.

Shao, Y. E. (1997) "Multiple intervention analysis with application to sales promostion data." In: *Journal of Applied Statistics* 24 181–192.

Shumway, R. H. (1988) *Applied Statistical Time Series Analysis.* Englewood Cliffs, NJ: Prentice Hall.

Sobczyk, K. (1985) *Stochastic Differential Equations for Applications.* Lyngby, Denmark: ABK.

Söderström, T. (1973) An On–line Algorithm for Approximate Maximum Likelihood Identification of Linear Dynamic Systems 7308 Report Lund Institute of Technology, Lund, Sweden: Deptartment of Automatic Control.

Söderström, T., and P. Stoica (1989) *System Identification.* London: Prentice Hall.

Spliid, H. (1983) "A fast estimation method for the vector autoregressive moving average model with exogenous variables." In: *Journal American Statistical Association* 78 843–849.

Subba Rao, T., and M. M. Gabr (1984) *An Introduction to Bisprectral Analysis and Bilinear Time Seris Models.* New York: Springer.

Thyregod, P., and H. Madsen (2006) *General and Generalized Linear Models.* Lyngby, Denmark: IMM, DTU.

——— (1992) "On an Adaptive Acceptance Control Chart for Autocorrelated Processes." In: *Frontiers in Statistical Quality Control* 4 138–154.

Tiao, G. C., and G. E. P. Box (1981) "Modelling multivariate time series with applications." In: *Journal American Statistical Association* 76 802–816.

Vio, R., P. Rebusco, P. Andreani, H. Madsen, and R. V. Overgaard (2006) "Stochastic modeling of kHz quasi-periodic oscillation light curves." In: *Astronomy and Astrophysics* 452 383–386.

Wall, K. D. (1987) "Identification theory for varying coefficient regression models." In: *Journal Time Series Analysis* 8 359–371.

Wei, W. W. S. (2006) *Time Series Analysis.* 2nd ed. Addison-Wesley.

Wiener, N. (1949) *Extrapolation, Interpolation, and Smoothing of Stationary Time-Series.* Cambridge MA: MIT Press.

Winters, P. R. (1960) "Forecasting sales by exponentially weighted moving averages." In: *Management Science* 6 324–342.

Yaglom, A. M. (1962) *An Introduction to the Theory Of Stationary Random Functions.* Englewood Cliffs, NJ: Prentice Hall.

Young, P. C. (1984) *Recursive Estimation and Time Series Analysis.* Berlin: Springer.

Index

T

CHAPMAN & HALL/CRC
Texts in Statistical Science Series

Series Editors
Bradley P. Carlin, *University of Minnesota, USA*
Julian J. Faraway, *University of Bath, UK*
Martin Tanner, *Northwestern University, USA*
Jim Zidek, *University of British Columbia, Canada*

Analysis of Failure and Survival Data
P. J. Smith

The Analysis and Interpretation of Multivariate Data for Social Scientists
D.J. Bartholomew, F. Steele, I. Moustaki, and J. Galbraith

The Analysis of Time Series — An Introduction, Sixth Edition
C. Chatfield

Applied Bayesian Forecasting and Time Series Analysis
A. Pole, M. West and J. Harrison

Applied Nonparametric Statistical Methods, Fourth Edition
P. Sprent and N.C. Smeeton

Applied Statistics — Handbook of GENSTAT Analysis
E.J. Snell and H. Simpson

Applied Statistics — Principles and Examples
D.R. Cox and E.J. Snell

Bayes and Empirical Bayes Methods for Data Analysis, Second Edition
B.P. Carlin and T.A. Louis

Bayesian Data Analysis, Second Edition
A. Gelman, J.B. Carlin, H.S. Stern and D.B. Rubin

Beyond ANOVA — Basics of Applied Statistics
R.G. Miller, Jr.

Computer-Aided Multivariate Analysis, Fourth Edition
A.A. Afifi and V.A. Clark

A Course in Categorical Data Analysis
T. Leonard

A Course in Large Sample Theory
T.S. Ferguson

Data Driven Statistical Methods
P. Sprent

Decision Analysis — A Bayesian Approach
J.Q. Smith

Elementary Applications of Probability Theory, Second Edition
H.C. Tuckwell

Elements of Simulation
B.J.T. Morgan

Epidemiology — Study Design and Data Analysis, Second Edition
M. Woodward

Essential Statistics, Fourth Edition
D.A.G. Rees

Extending the Linear Model with R: Generalized Linear, Mixed Effects and Nonparametric Regression Models
J.J. Faraway

A First Course in Linear Model Theory
N. Ravishanker and D.K. Dey

Generalized Additive Models: An Introduction with R
S. Wood

Interpreting Data — A First Course in Statistics
A.J.B. Anderson

An Introduction to Generalized Linear Models, Second Edition
A.J. Dobson

Introduction to Multivariate Analysis
C. Chatfield and A.J. Collins

Introduction to Optimization Methods and Their Applications in Statistics
B.S. Everitt

Introduction to Randomized Controlled Clinical Trials, Second Edition
J.N.S. Matthews

Introduction to Statistical Methods for Clinical Trials
Thomas D. Cook and David L. DeMets

Large Sample Methods in Statistics
P.K. Sen and J. da Motta Singer

Linear Models with R
J.J. Faraway

Markov Chain Monte Carlo — Stochastic Simulation for Bayesian Inference, Second Edition
D. Gamerman and H.F. Lopes

Mathematical Statistics
K. Knight

Modeling and Analysis of Stochastic Systems
V. Kulkarni

Modelling Binary Data, Second Edition
D. Collett

Modelling Survival Data in Medical Research, Second Edition
D. Collett

Multivariate Analysis of Variance and Repeated Measures — A Practical Approach for Behavioural Scientists
D.J. Hand and C.C. Taylor

Multivariate Statistics — A Practical Approach
B. Flury and H. Riedwyl

Practical Data Analysis for Designed Experiments
B.S. Yandell

Practical Longitudinal Data Analysis
D.J. Hand and M. Crowder

Practical Statistics for Medical Research
D.G. Altman

Probability — Methods and Measurement
A. O'Hagan

Problem Solving — A Statistician's Guide, Second Edition
C. Chatfield

Randomization, Bootstrap and Monte Carlo Methods in Biology, Third Edition
B.F.J. Manly

Readings in Decision Analysis
S. French

Sampling Methodologies with Applications
P.S.R.S. Rao

Statistical Analysis of Reliability Data
M.J. Crowder, A.C. Kimber, T.J. Sweeting, and R.L. Smith

Statistical Methods for Spatial Data Analysis
O. Schabenberger and C.A. Gotway

Statistical Methods for SPC and TQM
D. Bissell

Statistical Methods in Agriculture and Experimental Biology, Second Edition
R. Mead, R.N. Curnow, and A.M. Hasted

Statistical Process Control — Theory and Practice, Third Edition
G.B. Wetherill and D.W. Brown

Statistical Theory, Fourth Edition
B.W. Lindgren

Statistics for Accountants
S. Letchford

Statistics for Epidemiology
N.P. Jewell

Statistics for Technology — A Course in Applied Statistics, Third Edition
C. Chatfield

Statistics in Engineering — A Practical Approach
A.V. Metcalfe

Statistics in Research and Development, Second Edition
R. Caulcutt

Survival Analysis Using S —Analysis of Time-to-Event Data
M. Tableman and J.S. Kim

Time Series Analysis
H. Madsen

The Theory of Linear Models
B. Jørgensen

For Product Safety Concerns and Information please contact our EU representative GPSR@taylorandfrancis.com Taylor & Francis Verlag GmbH, Kaufingerstraße 24, 80331 München, Germany